Fossil Horses

The family Equidae has an extensive fossil record spanning the last 58 million years, and the evolution of the horse has frequently been cited in both textbooks and museums as a classic example of long-term evolution. In recent years, however, there have been many important discoveries of fossil horses, and these, in conjunction with such new methods as cladistics and techniques like precise geochronology, have allowed us to achieve a much greater understanding of the evolution and biology of this important group.

This book synthesizes the large body of data and research from several disciplines relevant to an understanding of fossil horses, including biology, geology, and paleontology. Using horses as the central theme, the author weaves together in the text topics such as modern geochronology, paleobiogeography, climate change, evolution and extinction, functional morphology, and population biology during the Cenozoic era.

Anyone with an interest in vertebrate paleontology, evolutionary biology, or general natural history will find this an absorbing and challenging book.

Fossil Horses

Systematics, Paleobiology, and Evolution of the Family Equidae

BRUCE J. MacFADDEN
Florida Museum of Natural History
University of Florida

CAMBRIDGE
UNIVERSITY PRESS

Published by the Press Syndicate of the University of Cambridge
The Pitt Building, Trumpington Street, Cambridge CB2 1RP
40 West 20th Street, New York, NY 10011-4211, USA
10 Stamford Road, Oakleigh, Melbourne 3166, Australia

First published 1992
First paperback edition 1994

Printed in the United States of America

Library of Congress Cataloging-in-Publication Data

MacFadden, Bruce J.
Fossil horses: systematics, paleobiology, and evolution of the family
equidae / Bruce J. MacFadden.
p.   cm.
Includes bibliographical references and index.
ISBN 0-521-34041-1 (hardback)
1. Horses, Fossil.   2. Paleontology—Cenozoic.   I. Title.
QE882.U6M27   1992   92-15810
569'.72–dc20   CIP

A catalog record for this book is available from the British Library

ISBN 0-521-34041-1 hardback
ISBN 0-521-47708-5 paperback

Contents

Preface

The most difficult part of writing a book, for me at least, was to start it. I am grateful to the University of Florida, and in particular the Chairman of my department, Dr. Norris H. Williams, for giving me the freedom to be on leave during the 1989–90 academic year, at which time I completed the first half of what follows. I had been planning to write this book for several years, and that distance from the normal university commitments provided the impetus to get it under way. I spent that leave at Cornell University, and I thank the Institute for the Study of Continents (INSTOC) for my appointment as Visiting Scientist during 1989–90, and Teresa Jordan for sponsoring my stay there.

There are many people to acknowledge and thank. At Cornell, Wolf Sach and John M. Hermanson of the College of Veterinary Medicine allowed me to sit in (actually stand up) on large-animal anatomy class during the spring 1990 semester. Although that experience gave me a greater understanding of equine anatomy, I must admit that it was impossible for me to keep up with the pace of the course. I greatly value the time spent with Howie Evans discussing book writing and natural history. I thank Chris Chyba and John Chiment, also at Cornell, for discussions on the philosophy of science, as well as my compatriots in the Department of Geological Sciences for various stimulating discussions, most notably Rick Allmendinger, Terry Jordan, Karl Wirth, and Anne Blythe. Art Bloom was an excellent role model in showing me the discipline needed to write a book and keeping me on track during its beginning stages.

I am grateful to my colleagues at the University of Florida, particularly my jogging buddies Dave Webb and Doug Jones. They shared my enthusiasm and listened to me while the book was being written. Some of my best scholarly interactions occurred during that daily one-hour jogging respite. On numerous occasions John Eisenberg discussed with me important aspects of modern mammalogy and ecology, and Neil D. Opdyke provided similar insight on the magnetostratigraphy of land-mammal deposits. I greatly appreciate that sharing of their insight on these subjects. I also benefited from frequent discussions with my former graduate students Richard C. Hulbert, Jr., and J. Daniel Bryant on fossil horses and related topics. I thank Wendy Zomlefer, who was always enthusiastic about preparing beautiful illustrations of fossil horses (particularly all those teeth!), many of which appear here. Deborah Brackenbury of the University of Florida Department of Instructional Resources photographed countless figures from less-than-perfect original publications. Many of the illustrations herein were enhanced by their excellent artistic and photographic skills. Anita Brown helped with the process of obtaining permissions and organizing my library resources. I thank Dianna Carver for preparing the laser-jet "typescript" for the various versions of the manuscript and Linda Chandler for her careful proofreading and editing.

I thank R. C. Hulbert, Jr., C. M. Janis, D. S. Jones, S. D. Webb, and M. O. Woodburne for being brave enough to read earlier versions of the manuscript. I thank Peter-John Leone and Robin Smith at Cambridge University Press for their advice and encouragement. I also extend sincere thanks and appreciation to my family and friends, and in particular my wife, Jeannette, for all of their support and encouragement along the way.

Much of my research, synthesized or presented in original form in this book, was supported by grant funds, including National Science Foundation grants DEB 78-10672 and BSR 85-15003 and American Chemical Society (Petroleum Research Fund) grant 20836-AC8. This is University of Florida Contribution to Paleobiology no. 395.

*   *   *

This book is dedicated to the memory and scientific legacy of Morris F. Skinner (1906–89). While I was a graduate student at Columbia University and the American Museum of Natural History during the early 1970s, Morris introduced me to the wonders of the Frick collection of fossil mammals. Although I recently learned that my first toy (at age four months) had been a stuffed horse named "Geronimo," there followed a considerable hiatus; my interest in fossil horses stems not from equestrian pursuits as a child but rather from a graduate-course term-paper assignment. From the first spark of my interest, Morris shared his enthusiasm with me and nurtured me along as a neophyte "horseologist." I shall never forget "Sunday school" classes devoted to the Frick horses in the tower of the museum. In addition to scholarly tutelage, Morris and his wife, Marie, always provided wonderful respite from the rigors of being a graduate student, frequently engaging in long conversations at afternoon

Morris F. Skinner (1906–89) with the Frick collection of fossil horses. This photograph, taken in the mid-1970s in the tower of the American Museum of Natural History, shows various clades of late Tertiary horses arrayed in a temporal sequence. Skinner spent half a century collecting and curating the Frick horses, which in combination with AMNH specimens provide the foundation for a modern understanding of the systematics, paleobiology, and evolution of the Equidae.

tea in their office atop the museum overlooking Central Park. They always had time to listen, whether or not the subject was horses. After my graduate-student days, Morris continued to encourage me to pursue my fascination with fossil horses. And I am not alone, for his tutelage extended to two generations of students and colleagues as he gave freely of his knowledge and advice to interested scientists. More than any person before or since, Morris knew the value of the Frick collection, having spent more than 50 years in the field excavating the specimens (always with meticulous stratigraphic data) and overseeing their preparation and curation back in New York (see photograph). Although future generations of paleontologists will not have the privilege of knowing Morris Skinner and his love for fossil horses, his spirit and legacy remain for posterity with the Frick collection.

Gainesville, Florida   Bruce J. MacFadden

1

Introduction: Why study fossil horses?

The geological record of the Ancestry of the Horse is one of the classic examples of evolution.   – Matthew (1926:139)

One cannot but hesitate to tell again this oft-told tale, which has been reiterated until it has become a hackneyed commonplace.
                                                                   – Scott (1917:98–9)

By one count it has been estimated that more than forty thousand books have been written on various aspects of horses (Cunningham & Berger 1986), ranging from their husbandry to veterinary applications, as well as their lore in history, and, of great relevance here, their significance for paleontology and evolution. Despite this bewildering array of writings about horses, which range from highly technical to popular, only about a dozen books have been written on fossil Equidae. Nevertheless, because of their widespread biogeographic occurrences over the past 58 million years (myr), horses have provided one of the most frequently cited examples of evolution interpreted from the fossil record. Even in an anthropocentric world, fossil horses are at least on a par with fossil humans and humanlike ancestors as examples of evolutionary patterns and processes. This is certainly not because of lack of interest in the latter group, but in contrast to horses, hominids have a relatively fragmentary fossil record. Before we begin a discussion of fossil horses, however, I would like to review briefly the great role that modern horses have played in human culture.

The lore of horses and humans

Prehistoric associations

Horses have been associated with humans for more than 350 centuries. The oldest known record of this association comes from middle Paleolithic cave paintings in France (Figure 1.1) and

1

Figure 1.1. *Late Pleistocene* Equus *depicted in Paleolithic cave painting from Les Combarelles, France. From Raphael (1945).*

Figure 1.2. *Late Pleistocene cave ceiling mural of Paleolithic fauna, including* Equus, *from Altamira, Spain. From Raphael (1945).*

Spain (Figure 1.2) that probably are between 35,000 and 100,000 years old. These drawings apparently were made by Neanderthal humans, and they depict ancient, wild horses and other late Pleistocene Ice Age mammals (Raphael 1945; Leroi-Gourhan 1965, 1967, 1968, 1982; Ucko & Rosenfeld 1967; Dennell 1986). Although principally concentrated in western Europe, Paleolithic cave art has been found in other areas of the Old World. For example, in 1959, cave paintings depicting mammoths were discovered in the southern Ural Mountains of Russia (Sutcliffe 1985). The horse is the most frequently depicted animal in Paleolithic art (Leroi-Gourhan 1965–82; Sutcliffe 1985), and horses also predominate in many archaeological sites in the Old World (e.g., Olsen 1989). It seems unlikely that all of this art was created with a single purpose in mind. These artists may have attempted to chronicle the presence of indigenous animals as primary sources of food, as fertility symbols, to pay homage to ritualistic beliefs or deities, or perhaps, as in later human history, even for purely aesthetic purposes (Ucko & Rosenfeld 1967; Sutcliffe 1985).

In the New World, horses are believed to have become extinct at the end of the Pleistocene, as part of the "great megafaunal extinction" about 11,000 years ago (Anderson 1984; Grayson 1989), although, in a minority view, Clutton-Brock (1981) believes that they may have persisted into historical times. Therefore, the horses that were so abundant during the late Pleistocene were conspicuously absent from later faunas. Their extinction is also suggested by the fact that no horses are known to have been depicted in pre-Columbian art (Sheets et al. 1984). Horses were reintroduced into the New World by the Spanish explorers during the sixteenth century (Figure 1.3).

In the Old World, however, wild equids, including horses and their relatives (zebras, onagers, and asses), persisted into the Holocene. So far as is known, horses were domesticated during the late Neolithic, probably by the end of the fourth or the beginning of the third millennium B.C. (fairly late relative to other animals). There is not universal agreement, but domesticated horses were probably descended either from stocks of tarpan (which became extinct in the wild in 1851) (Zeuner 1963) in western Europe or from "Przewalski's horse" in the high, arid steppe regions of eastern Asia, where it still is extant. It is also believed that other species of the genus *Equus*, including wild asses and their relatives, were originally domesticated in Africa, and possibly southern Asia.

The wave of domestication of the horse spread rapidly throughout the Old World as humans quickly realized the potential benefits of this animal, particularly as a draft animal for agriculture and transportation. Recent studies of the linguistic beginnings of proto-Indo-European cultures about 6,000 years ago indicate this close association; for example, the words for "horse" (*ek^hos, the phonetic precursor of the Latin *equus*) and "foal" (*p^holo) were already part of their vocabulary (Gamkrelidze & Ivanov 1990). Many Bronze Age petroglyphs (rock engravings), such as those from central Asia, depict humans and horse-drawn chariots (Figure 1.4).

Horses have likewise been closely associated with oriental cultures for millennia. Early evidence for this comes from the spectacular 22-centuries-old Qin "terracotta army" in the People's Republic of China. Discovered in 1974 near Xi'an while

Figure 1.3. *This illustration depicts how horses were transported in ships during the early sixteenth century by the Spanish explorers. The horses were suspended in slings, with their feet off the ground, to prevent broken limbs during rough seas. The horses, however, did not travel well in this mode, and many died before reaching their final destination. From Hudson et al. (1989); reproduced with permission of the University of Florida Press.*

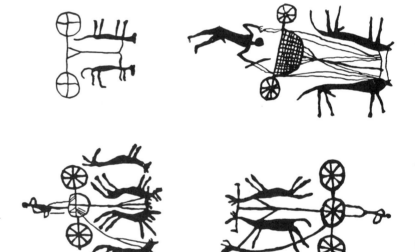

Figure 1.4. *Petroglyphs (rock engravings) of horses and chariots from the Bronze Age of central Asia. From Azzaroli (1985); reproduced with permission of E. J. Brill, Leiden, The Netherlands.*

farmers were digging a water well, only a small portion of this immense site has been excavated; yet it is without a doubt one of the most startling and significant archaeological finds ever discovered. It is estimated to cover some 22 square miles (57 km^2) and to contain life-size terracotta replicas of about 8,000 soldiers and 500 horses entombed with the emperor Qin Shihuangdi, who died in 210 B.C. Among the rows of soldiers, the horses are lined four abreast in front of spaces where wooden chariots, now disintegrated, were placed (Figure 1.5). The details of the terracotta figures are exquisite (Danforth 1982; Holledge 1988); the horses are sculpted in fine detail, and their heavy proportions suggest those of the present-day Przewalski's horse of the Asian steppe. These archaeological remains indicate that during the Qin period there was an exquisite artistic ability to sculpt in terracotta and bronze, the latter representing a relatively recent technology for that time. This spectacular find is but one example of the pervasive use of horses in Chinese culture and art.

Bones and teeth of fossil horses and other extinct mammals have been sold in markets and drugstores in China for centuries, where they have been prized for medicinal purposes (Figure 1.6). As Buffetaut (1987:17) noted, "according to an eighteenth-century medical treatise, dragon bones could be used to cure diseases of the heart, kidneys, intestines and liver. They were prescribed against a variety of illnesses, from constipation to nightmares and epilepsy. They could be taken raw, fried in fat, or cooked in rice wine." The possible pharmacological benefits of fossil bone in twentieth-century Chinese culture are not entirely anecdotal; in addition to any other curative properties they may have, they provide an excellent source of calcium. The alleged medicinal value of mammal bones was not restricted to the Orient; in medieval Europe the unicorn horn was highly prized, both in medicine and as a token of good luck. Even today some German drugstores carry the sign "Einhorn Apotheke," and thus the mythical properties of these substances live on (Sutcliffe 1985).

By the second millennium and thereafter, horses were frequently associated at diverse cultural sites in the western Old World, and they are known to have been widespread in ancient Egyptian, Greek, Etruscan, and Roman civilizations (Figure 1.7). Because of their value in transportation and as beasts of burden, and in sports and war, taboos against eating horses developed early in some cultures (Zeuner 1963; Clutton-Brock 1981), although in others horse meat still remains a delicacy. In a rather amusing passage, Zeuner (1963:381) noted that "the sex life of the ass is spectacular and full of temperament, hence it is only natural that there should be frequent allusions to it, and the beds of Roman married couples were often adorned with asses' heads."

Some of the most spectacular uses of horse images are preserved as "hill figures" in Great Britain. These gigantic icons, some hundreds of meters long, were created in the first century A.D. by cutting away turf to expose the underlying chalk deposits (Baskett 1980) (Figure 1.8). The spectacular Bayeux tapestry from the eleventh century chronicles the Norman invasion of England, and horses are conspicuously depicted therein (Figure 1.9).

Figure 1.5. Portion of the Qin terracotta army from Xi'an, People's Republic of China. The space behind the four horses was where a wooden chariot, now disintegrated, formerly stood. More than 8,000 soldiers and some 500 horses are believed to be represented in this magnificent 22-centuries-old archaeological site. Modified from Danforth (1982).

Figure 1.6. A large open-air "drugstore" in Kweilin, China, in which large quantities of fossil bones and teeth are collected, sorted, and stored. These fossils are later marketed to drugstores around China, as well as throughout the world. The fossils here, which include mastodont and bovid teeth in the foreground, are from late Pleistocene cave deposits from the surrounding region. Photograph courtesy of Prof. Dr. H. D. Kahlke.

Figure 1.7. Bronze model of horse
and archer, Etruscan, ca. 500 B.C.
From Baskett (1980); reproduced with
permission of Little, Brown & Co.

Figure 1.8. The White Horse, near
Uffington, Berkshire, England, first
century A.D. The hill figures of Great
Britain were formed by scraping away
superficial vegetation and exposing the
underlying bedrock, in this case white
chalk. As can be seen from the livestock
in the upper left corner, this figure is
several hundred meters long. The origin and significance of these remain
enigmatic. From Baskett (1980); reproduced with permission of Little, Brown
& Co.

Figure 1.9. "La Tapisserie de Bayeux." This Norman tapestry was fab-
ricated in the eleventh century to
commemorate the conquest of England
by William the Conquerer. This extraordinary work of history and art is
over 70 meters long and is composed of
linen and wool. The part of the tapestry
shown here depicts Harold (William's
brother-in-law) arriving at the gates of
William the Conqueror's castle. Other
portions depict horses on ships crossing
the English Channel (one of the first
times they are known to have been used
in amphibious landings) and in the battle of Hastings. From La Tapisserie
de Bayeux; reproduced with permission of Pedagogia (Caen, France).

Historic phase

There are frequent references to horses, horsemen, and chariots in the Bible. In the Old Testament (Genesis 47:17) Joseph gave the Egyptians bread in exchange for horses. In Chronicles II (9:25), "Solomon had four thousand stalls for horses and chariots, and twelve thousand horsemen." Also since biblical times, the mythical unicorn, which frequently has been considered equine, has captured the imagination of humans (Figure 1.10). Much folklore was developed and much energy spent over the ages in the futile quest to find the unicorn. The legend of the unicorn evolved to such an extent that its purported value to humanity covered a wide range of beliefs: that it could enhance fertility, or ensure chastity, in addition to its medicinal uses. Paleontological expeditions were even initiated, like that of the German Duke of Württemberg in 1700, specifically to find the unicorn (see Chapter 4).

The anatomy of horses has fascinated artists throughout the ages. Baskett (1980:68) noted that "Renaissance artists were partially liberated from the stricter confines of religious painting; they turned to pure interpretation of form using classical art as their example." Thus, Michelangelo's (1475–1564) "Studies of a Horse" and "A Horseman Attacking Footsoldiers," (ca. 1505) are classic studies that blend art and anatomy. Similarly, the series of magnificent eighteenth-century drawings by Stubbs (Figure 1.11) united scientific and artistic portrayals of the horse.

Horses have been used in battle for at least 4,000 years. The cavalry has always embodied the ability of an army to strike quickly and outmaneuver its foe. Virtually all great land battles in historic times have involved, to a greater or lesser extent, mounted horsemen. The twentieth century brought a turning point in the use of the horse in battle. At the beginning of World War I, horses were marshaled in great numbers. For example, during the first two weeks of August 1914 some 120,000 mounts were assembled for the British Expeditionary Force to take to France (Macdonald 1987). The availability of lush pastures for grazing influenced the decision as to when (late summer) the Central Powers would launch their initial campaigns of World War I. With the advent of the mechanized age, featuring tanks and armored vehicles, and the battlefield stalemate of trench warfare that developed during World War I, the military importance of horses began to decline (French 1951). In only a few cases were conditions favorable for the use of cavalry, such as the successful charge at Beersheba (Figure 1.12).

During World War II, horses were relegated primarily to their ancient role of beasts of burden, often hauling supplies or pulling artillery pieces over terrain that was inaccessible to motorized transport. In those instances in which horses were used in battle,

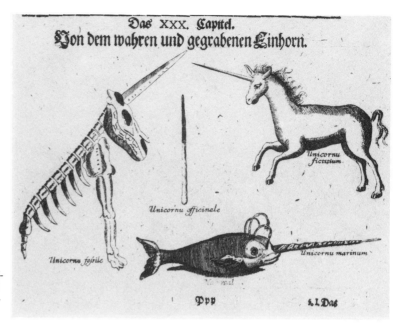

Figure 1.10. *An eighteenth-century engraving showing various types of unicorns, including one composed of fossil specimens (left). From Valentini's Mu-seum* Museum Museorum *(1704).*

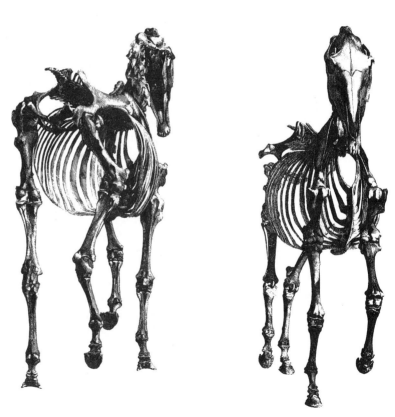

Figure 1.11. *Anatomical illustrations of the horse based on detailed dissections taken from George Stubbs's classic works of the eighteenth century. These particular engravings were done by Stubbs in 1766 on a subscription basis and were sold as part of a set of 18. From* The Anatomical Works of George Stubbs, *by Terence Doherty; © 1975 by Terence Doherty and re-printed by permission of David R. Godine, publisher.*

Figure 1.12. "The Battle of Beersheba 31 October 1917," by George Lambert, painted in 1920, oil on canvas, 123 × 247 cm. This charge was one of the last great, successful uses of cavalry in offensive tactics. "Light horse" were principally mounted infantry, so that horses were used for rapid transport into battle, and then the troops dismounted. However, in 1917 the Australian Light Horse charged and overran fortified and entrenched Turkish positions surrounding the ancient city of Beersheba in Palestine. Cavalry charges against mechanized armored units during World War II met with disastrous results. Reproduced courtesy of the Australian War Memorial.

as, for example, by Russian and Polish forces attacking Germans on the Eastern Front, the results were disastrous; horses were no match for tanks, artillery, and machine guns. Despite the decreasing emphasis on the horse in battle, its legacy of power and good fortune has persisted in military lore and operations. Thus, on several occasions military missions have been named "Operation Pegasus": During World War II, in the early hours of June 6, 1944, British troops transported by glider captured strategically important bridges behind German lines in Normandy (Ambrose 1985). Similarly, "Pegasus" was the name for the attack on North Vietnamese strongpoints around Khe Sanh by units of the U.S. First Cavalry (Airmobile) Division in 1968, this time using troops rapidly deployed by helicopter (Stanton 1987). Many mechanized armored units still are designated as "cavalry."

From this brief introduction it should be clear that an enormous lore has evolved around horses, particularly as they relate to humans. This familiarity has also led to the use of horses to understand evolution. I believe that, in addition to their great influence on human culture, horses play a dominant role in evolutionary theory because of their widespread occurrence as fossils in Cenozoic deposits throughout the Northern Hemisphere.

Rationale, purpose, and scope

As mentioned earlier, there is a bewildering array of books on various aspects of horses. So if horses have already been so thoroughly examined, why study them and write another book about them? Despite the wealth of writings on this general topic, only about a dozen books have been written since the late nineteenth century dealing specifically with fossil horses and evolution. As far as I have been able to determine, the first book written in the Western world that discussed fossil horses was the British zoologist Flower's *The Horse: A Study in Natural History* (1892). The most recent comprehensive book on this subject, *Horses: The Story of the Horse Family in the Modern World and through Sixty Million Years of History*, was published about 40 years ago (Simpson 1951). For various reasons, not least of which is Simpson's reputation as one of the great evolutionary biologists and paleontologists of the twentieth century, his book has been a lasting reference on fossil horses.

A central theme of this book is that paleontology is currently experiencing a renaissance of grand proportions that has resulted from advances in technology and methodology and from new discoveries (see Chapter 2). Virtually all of these advances either were in their infancy (e.g., geochronology) or had not yet occurred (e.g., the widespread availability of computers and methods such as cladistics) when previous books on fossil horses were written.

The purpose of this book is to present fossil horses in a modern context, one that makes use of the recent advances in paleontology and related disciplines. Because horses are familiar and of widespread appeal, this book is intended for the reader with a general, scholarly interest in paleobiology, evolution, and natural history. This book is organized into several informal sections, all of which are tied together by the common thread of fossil horses.

The first part of this book (Chapters 2–4) provides an extended introductory background. In these chapters I discuss the development of scientific theories, the history of science, what is new and exciting in paleontology today, and the importance of fossil collections. In the second part (Chapters 5–7) I present the systematics and distribution of horses in time and space. Although Chapter 5, on systematics, may be too "heavy" for some of the intended audience, it provides a necessary foundation for those chapters that follow. The third part of this book (Chapters 8–10) is intended to examine the many aspects of how fossil horses contribute to a modern understanding of processes and patterns in evolution. The final major section (Chapters 11–13) discusses functional morphology, population ecology, and community ecology. The majority of the book is intended to be a synthesis of previous work. However, some original research results will be presented, particularly in the later chapters.

In this book my emphasis will be on fossil horses from North America. There are two reasons for this: (1) The principal evolutionary history of this group occurred there, and (2) this is the part of their history that I know best. As appropriate, however, I shall also draw upon relevant examples from the Old World and South America. I would be remiss if I did not acknowledge that there is by no means universal agreement among paleontologists about the systematics and paleobiology of the Equidae. Recent debates on these topics are exemplified by MacFadden and Skinner (1982) and MacFadden (1987b), on the one hand, and by Forsten (1982a) and Eisenmann et al. (1987), on the other. Where appropriate, I discuss opposing views on certain issues, but, not surprisingly, this book emphasizes my point of view regarding fossil horses.

2

A renaissance in paleontology

Like all scientific pursuits, paleontology has benefited from numerous advances during the past two centuries. In this chapter we shall first review the growth and development of vertebrate paleontology. This will provide a necessary background for the ensuing discussion of the status of this science in a modern context. I hope that this chapter will convince the reader that because of many recent advances, this is an exciting time to study paleontology. Although this chapter is of broader scope than those that follow, it is intended to provide a summary of the reasons why paleontology is currently undergoing a renaissance.

The growth of vertebrate paleontology

Although humans have been collecting fossils for more than 350 centuries (see Chapter 4), it was not until the Middle Ages and the Renaissance that most natural philosophers began to question the antiquity and interrelationships of fossils. Of relevance to vertebrate paleontology during that time, the bones and teeth of large beasts (mostly late Cenozoic mammoths and mastodons) created considerable interest among scholars and gave rise to the study of "gigantology" (Buffetaut 1987).

Three related fields of scientific inquiry had to be developed before vertebrate paleontology could advance as a science. First, a framework was needed so that organisms could be classified and placed into natural groups. Thus, paleontology benefited greatly from late-eighteenth-century systematic biology, the foundations of which are universally associated with Linnaeus (1707–78). It soon became clear in systematics that there was a need to classify organisms, both living and extinct. Unlike those who classify cars, musical instruments, stamps, and the like, systematic biologists since Linnaeus have also been interested in phylogeny (i.e., the origin and descent within natural groups). In the early nineteenth century, studies of geological successions of extinct organisms preserved in sedimentary sequences provided the first evidence of long-term morphological changes and phylogenetic interrelationships.

Second, paleontologists needed a framework for anatomical and morphological comparisons of living and extinct forms. This framework was provided by the development of comparative vertebrate anatomy, which had its origins in France early in the nineteenth century with the work of Georges Cuvier (1769–1832), one of the great anatomists and "catastrophists" of his day, and was advanced further in the middle part of that century in Great Britain by Sir Richard Owen (1804–92), himself a great anatomist and ardent opponent of Darwin's theory of evolution.

Third, a geological time dimension was needed to better understand the antiquity of the earth. One of the natural philosophers who contributed to the earliest estimates of geological time was Georges Louis Leclerc, comte de Buffon (1707–88), certainly one of the most influential eighteenth-century French scientists. Buffon is generally regarded as believing that the earth was at least 75,000 years old, although in some unpublished manuscripts his estimates ranged up to 3 million years (Buffetaut 1987). That was a radical departure from long-standing ecclesiastical views, the most famous of which was that of Archbishop Ussher of Ireland, who in the 1650s, after careful consideration, stated that the earth had been created in 4004 B.C. (Berry 1987; Gould 1991).

In North America, vertebrate paleontology was studied primarily in natural-history societies and academies in the northeast during the late eighteenth and early nineteenth centuries. Many of the specimens collected at that time came from the eastern part of the continent (see Figure 4.1). Notable amateur paleontologists during that time included Benjamin Franklin, who pondered the significance of elephant fossils found at Big Bone Lick, Kentucky, and Thomas Jefferson, who wrote papers on fossil mammals (Simpson 1942; Lanham 1973; Gregory 1979; Martin 1981). Joseph Leidy (1823–91), a physician and anatomy professor at the University of Pennsylvania in Philadelphia, is generally considered the founder of the scientific study of fossil vertebrates in the United States. Interestingly, Leidy's first paleontological paper (1847) was on Pleistocene *Equus* found near Natchez, Mississippi (Hay 1902), and he later wrote many important papers on fossil horses.

During the middle of the nineteenth century, the study of natural sciences expanded rapidly in Great Britain, France, and Germany, whereas these fields lagged behind in the United States. That difference can be attributed to the presence of long-established programs at the older institutions of higher education in Europe, many of which had chairs of paleontology on their faculties (Hsü 1986), and to the negative influence of the Civil War in the United States. During that time, there were few well-respected scholarly journals devoted to natural sciences in the United States, and with an average life expectancy of six years, many of them faltered. Only Yale's *American Journal of Science* was widely read in Great Britain and Europe. Also, there were few universities in the United States that granted advanced degrees in the natural sciences. Following the German model, Yale was the first to award the Ph.D. degree, in 1861. Prior to that time, an aspiring young scholar who trained in the United States for a career in science usually did so with an M.D. degree (Bruce 1987). The other career track in natural sciences during that time was to obtain a coveted advanced degree from one of the great institutions in England and Europe (during the first half of the nineteenth century, most university professors in the United States were trained in the Old World).

Paleontology and natural history were mostly relegated to hobby status within local scientific societies composed of men of wealth and social status of the leisure class (Bruce 1987). After the Civil War, paleontology began to be recognized in the United States as a formal, scholarly discipline at prominent institutions of higher education. That was exemplified in 1866 by the appointment of O. C. Marsh (1831–99) to the first professorship of paleontology, which was at Yale University. Like many nineteenth-century North American university professors, Marsh received no salary from Yale, and his research was sponsored by his wealthy, philanthropic uncle, George Peabody. (The Yale Peabody Museum of Natural History in New Haven and the Harvard Peabody Museum of Anthropology in Cambridge were named for him.) As part of Peabody's original bequest to Yale, Marsh was not required to teach classes, so that he could more fully devote his time to research (Schuchert & LeVene 1940; Gregory 1979; Miller 1981; Bruce 1987).

Later during the nineteenth century, interest in paleontology (and other natural sciences), both professional and from the lay public, increased and resulted in the founding of many of today's prominent natural-history museums, such as the Smithsonian Institution in Washington, D.C., and the American Museum of Natural History in New York, as well as the Natural History Museum in London. Early on, these museums were eager to obtain spectacular natural-history specimens for their public following. As Bruce (1987:344) noted, "paleontology, straddling the earth and life sciences, captivated Centennial America with its colossal dinosaurs." As a result of this era of expansion, attendance at these larger institutions soared; by 1900 the American Museum of Natural History in New York and the British Museum (Natural History) in London each opened their doors to about 400,000 visitors annually. Also, by 1910 more than 2,000 science museums had been established around the world (Sheets-Pyenson 1987). By the beginning of the twentieth century there was a new public awareness of natural history. Riding that wave of public interest, museums were keen to acquire specimens both for research and for exhibition. Also at that time, the rapidly growing, although not universal, acceptance of Darwinian gradualism (Hull, Tessner, & Diamond 1978) spurred its proponents to seek further support for that theory, and paleontology was recognized to be of paramount importance for understanding the long-term time dimension to evolution. In North America that was the "heroic" era of vertebrate paleontology (Gregory 1979). In the first few decades of the twentieth century, general interest in paleontology subsided somewhat worldwide as society turned its attention from pure science to other preoccupations, including the Great War, then the Great Depression, and then World War II. Those global events in human history slowed the progress of natural sciences on many fronts (e.g., Kragh 1987; Kay 1989).

In the middle part of the twentieth century, paleontology played an essential part in the formulation of the "modern synthesis" (Huxley 1940), which unified evolutionary concepts from systematics, genetics, morphology, and paleontology (e.g., Jepsen, Mayr, & Simpson 1949). During the past few decades, paleontology has been undergoing an exciting renaissance and has contributed greatly to other scientific disciplines. For example, the recent search for the cause of the mass extinctions at the Cretaceous–Tertiary boundary would have been of significantly less interest to nonspecialists and the lay public without the concomitant interest in dinosaurs. In a recent perspective, Buffetaut (1987:199) noted that "some of the long-extinct creatures of the past revealed by vertebrate palaeontology have now become familiar to large numbers of people well outside scientific circles. This is not a negligible achievement for what is sometimes considered as a rather abstruse branch of scientific knowledge."

The past few decades: a modern renaissance

Most of the advances in modern science have resulted from give-and-take between disciplines, and the current renaissance in paleontology could not have occurred without significant advances in related fields. I attribute this to three factors: (1) new technologies, (2) new theories and methodologies, and (3) new discoveries. Some of these will be elaborated in succeeding chapters as they relate to fossil horses, but I would like to discuss briefly some significant factors that have contributed greatly to the current renaissance in paleontology.

New technologies

GEOCHRONOLOGY. Paleontology is a historical science with a long-term time dimension. Accordingly, as the study of geochronology advances, so does our ability to more precisely interpret the fossil record. In the nineteenth century and the first half of the twentieth it was difficult to assess the magnitude of geological time, and therefore studies that were dependent upon that knowledge could not advance rapidly. The greatest advances in geochronology relevant to paleontology occurred after World War II and resulted from the development of radioisotopic and paleomagnetic dating techniques (see Chapter 6). Increased precision of analytical techniques and instrumentation has now progressed to the point that frequently we can discern events separated by 1,000 years in the Pleistocene, and events separated by 0.1–0.5 myr in the early Cenozoic. With refined geochronologies, paleontological events such as mammalian dispersals between continents, which previously were thought to have been geologically instantaneous, are now interpreted to have been time-transgressive. A good example of this with fossil horses is the current calibration of the *"Hipparion"* and *Equus* datum planes (see Figure 6.9). Another important field that has benefited from more precise geochronologies is the study of evolutionary patterns, such as the quantification of morphological and taxonomic changes in fossil horses (Chapter 9). In summary, because the time dimension of evolution is unique to paleontology, the ever-increasing power of geochronological resolution will continue to expand the range of interesting questions that can be asked by paleobiologists.

ISOTOPE GEOCHEMISTRY. Modern paleobiology goes beyond classical systematics and traditional approaches to interpreting the fossil record. One of the newest earth-science disciplines brought into the fold of paleobiology involves the use of low-temperature-stable isotopes (Abelson 1988) to analyze the diets of ancient mammals and the climates in which they lived.

The familiar adage "you are what you eat" also has implications for isotope geochemists, archaeologists, and, if trends continue, vertebrate paleontologists. During the 1980s there were many studies that used isotopes to interpret ancient diets, as reviewed by Van der Merwe (1982) and DeNiro (1987). So far, most of these studies have been limited to analysis of the organic fraction in bone (collagen), principally from Holocene archaeological sites, but recent advances indicate the potential to analyze both the organic (Glimcher et al. 1990) and inorganic, or crystalline, bone minerals (mostly apatite, a hydrated calcium phosphate), as well as the denser dental tissues (i.e., dentine and enamel) (e.g., Thorp & van der Merwe 1987; Ambrose & DeNiro 1989). If the problem of diagenesis can be overcome, then these techniques can be applied to Tertiary mammals to answer many interesting paleobiological questions.

Studies of the isotopic compositions of mammalian skeletal

tissues have been able to determine the relative amounts of C3 and C4 (for all abbreviations, see Appendix I), photosynthetic carbon incorporated into bones. For ancient human populations, for example, such studies can indicate the presence of a pre-maize diet (high C3) versus the advent of a maize-dominated diet (high C4) in the New World. Similarly, carbon ratios in bone can distinguish many types of ungulate mammals that fed on browsing (predominantly C3) versus grazing (predominantly C4) foodstuffs. Simultaneous comparisons of ratios of $^{13}C/^{12}C$ and $^{15}N/^{14}N$ isotopes in bone have been used to interpret the types of proteins eaten by ancient humans (shellfish and vertebrates fractionate these isotopes in different proportions within their bodies). Of relevance here, it would be particularly interesting to know if there were any isotopic shifts, particularly for carbon, associated with the dramatic increase in tooth crown heights in Miocene horses.

In addition to their use for interpreting diet on the basis of bone chemistry, isotopic studies are becoming of great utility in interpreting the ancient climates in which fossil horses and other terrestrial organisms lived (similar techniques have been used in the marine realm for decades, but only recently have they been applied on land). For example, studies of ratios of $^{13}C/^{12}C$ and $^{18}O/^{16}O$ isotopes fractionated in fossil caliches and petrocalcic horizons (soil-forming carbonates) allow interpretations of ancient climate changes (Quade, Cerling, & Bowman 1989a). In Pakistan, Quade, Cerling & Bowman (1989b) and Cerling et al. (1989) found that isotopic shifts were correlated with significant turnover in late Miocene mammalian faunas, which included hipparion horses. Elsewhere, the fractionation of oxygen isotopes ($\delta^{18}O$) has been interpreted to represent significantly colder climates (mean annual temperature estimated to have been less than 5°C) for dinosaur faunas at high (southern) latitudes in Australia (Rich et al. 1988).

THE COMPUTER AGE. In recent years, computers have contributed to the vitality of paleontology as a science. In addition to automating many of the functions of machine-based science described earlier, computers have been immensely beneficial in more traditional paleontology, including taxonomic and evolutionary studies (e.g., Rogers, Fleming, & Estabrook 1967). Statistical analyses of large data sets, which previously were either very time-consuming or impossible to do in precomputer days, are now routine in paleontology and allow the application of many multivariate techniques. Also, phylogenetic computer programs, such as PAUP (Swofford 1985), allow simultaneous analyses of many characters distributed in many more taxa than was previously possible. In recent years, many systematic and revisionary studies of North American fossil horses have employed computer-based programs to analyze characters and hypothesize phylogenetic interrelationships.

OTHER TECHNOLOGIES. Other new technological advances that have recently been applied have led to interesting questions about fossil vertebrates, and as a consequence we have a better understanding of their paleobiology. SEM (scanning electron microscopy) studies of microwear scratches and larger patterns resulting from chewing have resulted in insight into the diets of hominoid fossils (Teaford & Walker 1984), herbivore communities (Solounias, Teaford, & Walker 1988), and horses (Rensberger, Forsten, & Fortelius 1984). Computerized axial tomography (CAT) has been used to investigate the anatomical structures of fossils that otherwise could not be seen or would require extensive and potentially destructive preparation, such as dinosaur eggs (Hirsch et al. 1989), *Archaeopteryx* (Haubitz et al. 1988), and mammal skulls (e.g., Conroy & Vannier 1984; Joeckel 1990). Remote sensing and satellite imagery have been used to map early Tertiary fossil-bearing units in Wyoming (Stuckey et al. 1987; Krishtalka, Stuckey, & Redline 1988). In addition to remote sensing, local seismic acoustic diffraction (a technique analogous to CAT scanning, but obviously on a larger scale) has been used to locate bones of one of the world's largest known dinosaurs, *Seismosaurus* (Gillette 1991). Neither CAT scanning nor remote-sensing techniques have yet been applied directly to studies of fossil horses.

The field of biochemical systematics has expanded rapidly over the past few decades. Up to now, biochemical data have contributed much to our understanding of phylogenetic interrelationships of modern taxa and the hypothesized taxonomic divergence times in the past. Some recent studies of extinct taxa, such as late Pleistocene proboscideans (Shoshoni et al. 1985) and the extinct equid known as the quagga (Higuchi et al. 1984), and the discovery of chloroplast DNA in Miocene magnolia leaves (Golenberg et al. 1990) have provided exciting examples of the potential of biochemical systematics, although the current technology for analyzing soft tissues limits the applicability in fossil specimens.

In summary, given these advances, it is not surprising that in addition to the traditional microscopes and calipers and the now-ubiquitous desktop computers, paleobiologists are incorporating scanning electron microscopes, CAT scanning devices, cryogenic magnetometers, stable-isotope mass spectrometers, and even biochemical systematics laboratories into their research programs. Furthermore, these high-tech instruments are not unique to academic departments at universities; many museums are also going in this direction and thereby integrating the "traditional" natural-history specimens with sophisticated new, exciting ways to study them.

New theories and methodologies

During the past two decades, several new and important theories and methodologies have been developed in paleobiology. Although there is not uniform acceptance for some of these, all have been universally beneficial in stimulating thought about how we interpret the fossil record.

PHYLOGENETIC SYSTEMATICS. The foundation for all organismic biology is taxonomy, more broadly referred to as systematics. Within the past three decades, three schools of taxonomic methodologies have emerged: (1) phenetic, (2) evolutionary, and (3) cladistic. The latter approach, considered a radical, avant-garde method in the 1970s, now is the dominant theoretical framework used by vertebrate paleontologists to construct phylogenies that consist of either all fossil groups or combinations of extant and fossil groups. As mentioned earlier, paleontologists are keenly aware of the importance of a precisely calibrated fossil record for an enhanced understanding of the history and diversity of extinct organisms. It is therefore difficult for some workers to understand the rationale of constructing cladograms, which express the interrelationships of taxa without recourse to the fossil record, to determine character polarities. In order to then construct a phylogeny, or family tree of ancestors and descendants through time, most cladists studying fossil groups then analyze the cladogram with regard to the stratigraphic distribution of taxa. Rather than considering the different geometries produced by cladograms versus stratigraphically calibrated phylogenies as totally independent data sets, they find that these two methods often provide corroboration of each other. Because of the widespread abundance of fossil horses, since the mid-1970s our understanding of the phylogeny of the Equidae has greatly benefited from both new cladistic approaches and a better-calibrated geological time scale.

GRADUAL VERSUS PUNCTUATED EVOLUTION. Much new thought has been generated since the early 1970s about how evolutionary patterns and processes are interpreted from the fossil record. Until that time, Darwinian gradualism was considered the dominant paradigm to explain long-term evolution. However, as Darwin himself noted, the fossil record is less than perfect, and discontinuities in evolutionary sequences were attributed to stratigraphic gaps. On the other hand, the hypothesis of punctuated equilibria (Eldredge & Gould 1972; Gould & Eldredge 1977) asserts that imperfections and gaps are not necessary to explain the somewhat herky-jerky fossil record. Instead, taxa are characterized by relatively long periods of morphological stasis, punctuated by times of rapid evolutionary change to a new state. In this model, times of rapid evolution usually are related to allopatric speciation events resulting from geographic isolation.

In contrast to the Gould and Eldredge model, Gingerich (e.g., 1976, 1977, 1984b) has been a modern advocate for gradual change in the fossil record, a view thus more in line with the prevailing paradigm since the middle of the nineteenth century. He has coined the term "phyletic gradualism" or a "stratophenetic" approach to interpreting the fossil record, which is similar to the views of earlier scientists such as Darwin and Simpson.

Despite some recent statements in the literature about the widespread acceptance of punctuated equilibria (Lewin 1986), many workers, in addition to Gingerich, still advocate gradualism and thus do not think that punctuated equilibria is a valid model to describe and portray evolutionary processes and patterns (Gingerich 1984b; Brown 1987). This debate may also be a moot point, because one hypothesis does not necessarily have to exclude the other (e.g., the idea of "punctuated gradualism") (Malmgren, Berggren, & Lohmann 1984). It is entirely plausible that in some environmental situations, or at different scales within the organic hierarchy, one model could be the dominant mode of evolutionary change, whereas in another situation the other model could be valid. Likewise, because evolution is mosaic (de Beer 1954), within a given lineage it is possible that independently evolving characters can evolve under different modes. As we shall discuss in several chapters that follow, virtually all of these possibilities also exist with fossil horses. The familiar sequence of North American fossil equid genera often is quoted as a prime example of gradual evolution. Yet this, too, may be an artifact of less precise time control or the scale of observation. On the other hand, other examples involving fossil horses, such as body-size evolution, could plausibly be cited in support of long periods of stasis punctuated by times of rapid change.

The philosophy and historiography of science

> *We need not devote any time or space to fresh arguments for the truth of evolution. The demonstration of evolution as a universal law of living nature is the great intellectual achievement of the nineteenth century.*
> – Osborn (1917:viii)

Is the theory of evolution true? Did an asteroid cause the extinction of dinosaurs? Do fossil horses provide proof of long-term evolution? Questions like these are daunting; yet they have important consequences for how science is done and how it is perceived in modern society.

Until recently, the dominant philosophical framework in science was the inductive method, sometimes called the Baconian approach. In the inductive method, observations and data are gathered, patterns emerge, and hypotheses are formulated, the latter often being said to be "verified" or "true." However, philosophers such as Sir Karl Popper have reexamined what science is and how it is done. The essence of the Popperian approach is the hypotheticodeductive method, in which hypotheses are never verified, but must weather the test of repeated attempts at their falsification. In modern parlance, most scientific hypotheses can be corroborated (i.e., they fail to be falsified), but they can never be declared true, except perhaps in mathematics or theoretical physics, where truth and falsity are inherent in the initial assumptions and conditions to be tested. Within the past few decades, paleontologists, particularly those involved in systematic research, have also become interested in the value of the inductive method versus the deductive method as different philosophical approaches to science (e.g., Patterson 1978; Gaffney 1979).

In the purely deductive or hypotheticodeductive approach, hypotheses are constructed and then corroborated by the acquisition of data. In contrast to mathematics and physics, however, this is not what usually happens in the natural sciences. Rarely do descriptive scientists construct hypotheses without some preliminary data. Scientists make observations, from which preliminary, general patterns are suggested; hypotheses are then constructed, and larger data sets are acquired to corroborate or refute the hypotheses. In addition, geology and paleontology, with their distinctly historical component, differ philosophically and methodologically from other kinds of science that lack a time dimension (Kitts 1977; Simpson 1981). Thus, paleontologists are at least as interested in reconstructing the past as they are in studying modern phenomena or predicting the future, which is the case in the natural sciences that lack a historical component.

As a historical pursuit, paleontology faces many of the same philosophical considerations that shape the study of human history. Science, like history, is composed of fact and interpretation. Therefore, it is impossible for paleontologists to identify the actual sequence of events that occurred millions or tens of millions of years ago. For example, although it seems that most dinosaurs became extinct by the end of the Cretaceous era, we are not sure what caused their demise. Accordingly, some scholars, particularly historians, make a distinction between "objective history" (H_1), those events that actually occurred in the past, and historiography (H_2), the writing or chronicling of history (e.g., Kragh 1987).

Kragh (1987:20) noted that

> The object of history (H_2) is thus history (H_1) in the same way as the object of natural science is nature. Just as our (scientific) knowledge of nature is limited to the research results of science that *are* [her emphasis] not nature but a theoretical interpretation of it, so our knowledge of the events of the past is limited to the results of history (H_2) that are not the past, but a theoretical interpretation of it. Radically positivist philosophers have maintained that the existence of an objective nature is a meaningless fiction and that it is impossible to distinguish between nature and our knowledge of it. In the same way, some idealist historians maintained that the distinction between H_1 and H_2 is a fiction that serves no useful purpose; that there is no actual history apart from that which the historian constructs from his sources.

For paleontology, this distinction is important. One of the best examples of this is the current use of "parsimony" in cladistics, where of all the possible combinations, that with the simplest branching pattern is the one normally used in phylogeny reconstruction. The analogy here is that parsimony is a scientific basis for reconstructing the past (H_2), and in all probability it does not approximate the actual ("true") sequence of evolutionary events (H_1) that occurred in the past. Therefore, for fossil horses the often-cited phylogeny presented in textbooks (e.g., Figures 3.13 and 5.14) is our best hypothesis based on the most up-to-date data, but not every feature of it is totally correct or true.

Kragh (1987:22) also stated that "science (S_1) can be regarded as a collection of empirical and formal statements about nature, the theories and data that, at a given moment in time, comprise accepted scientific knowledge. According to this view, science will typically be a finished product, as it appears in textbooks and articles." However, this notion of final knowledge tends to deter the growth of scientific thought.

Some colleagues lament that calling natural science "hypothetical" rather than "true" somehow lessens the value of the endeavor. Some geology professors emphasize the "law of superposition," or Cope's law (see Chapter 10), although these are not strictly "laws" in the sense of the philosophy of science (i.e., where they are true in all cases, without exception) (e.g., Rieppel 1988). Many students at universities want to be taught the truth about the extinction of dinosaurs or the actual descent of our human ancestors. Likewise, the lay public reads about advances in paleontology as if they are simple truths (e.g., an asteroid killed the dinosaurs; humans are descended from the apes). Granted, undoubtedly there was a true sequence of historical events that resulted in the extinction of the dinosaurs, and another sequence led to the advent of humans. Natural science, like history, is composed of fact and interpretation, but actual events may be difficult or impossible to identify. As McKenna (1987:82) noted, "as in murder investigations, reconstruction of phylogenetic history after the event has its difficulties. There may be several versions of what supposedly has happened. Nevertheless, only one sequence of events actually occurred."

New fossil discoveries

A cardinal joy of paleontology is the thrill of discovery – being the first to see a new fossil.   – Voorhies (1981:70)

The science of paleontology continues to be enhanced by the influx of important and exciting specimens collected from the field and by their curation in museums. All specimens housed in permanent collections are important; yet paleontology advances most significantly by the discovery of truly extraordinary specimens. Although the importance of collections will be discussed in Chapter 4, it is relevant to mention briefly that this is an important aspect that contributes greatly to the current renaissance in paleontology.

As is also the case for many other taxa that have left a fossil record (e.g., hominids and dinosaurs), our understanding of the Equidae has been advanced greatly by new discoveries over the years. In addition to large new collections from previously poorly represented regions (e.g., Asia), additional specimens from classic localities have shifted the study of fossil horses from the previous predominant use of teeth to combined analyses that also include cranial and postcranial characters. These new looks into equid systematics have resulted in many new phylogenetic, evolutionary, and paleobiological hypotheses concerning this important group. Without these, much of the research presented here could not have been done.

In summary, the fundamental basis of paleontological research is provided by well-documented and well-curated museum collections. This international resource contained in the major natural-history museums of the world is at least an order of magnitude larger and certainly better documented than it was a century ago. As other facets of paleontology advance, so too must collections grow.

The new paleobiology

I conclude this chapter with a brief discussion of the meanings of two words. Although frequently they are used interchangeably, for our purposes here a distinction will be made between "paleontology" and "paleobiology" to demonstrate where the study of fossils is going. One might consider the first word, "paleontology," literally "the study of the old," to include the more traditional pursuits of identifying, classifying, and interrelating extinct organisms. Despite the analogy to stamp collecting made by some critics (Lewin 1982), paleontology as defined here is of great importance as the foundation of the study of extinct organisms. These tasks and the accumulated knowledge resulting from them are important pursuits and are the necessary prerequisites for other studies. On the other hand, paleobiology, literally "the study of ancient life," is a more dynamic term whose meaning includes all of paleontology, as defined earlier, plus the goal of bringing fossils "to life" to better understand their evolution, adaptations, community interactions, and distribution in space and time. All of the advances in methodologies and technologies and the new discoveries described earlier contribute to an exciting renaissance for the study of fossils in a dynamic paleobiological context that would have been impossible several decades ago.

3

Orthogenesis and scientific thought: old notions die hard

The most widely cited example of orthogenesis, in any sense of the word, is the evolution of the horse. — Simpson (1944:157)

Orthogenesis, or a sequence of "straight-line evolution" with no side branches from ancestors to descendants, was embodied in the thinking of many nineteenth-century paleontologists. As Simpson noted, over the years fossil horses have been cited as a prime example of orthogenesis because this group is so abundantly represented in the geological succession on several continents (most notably North America, but also Europe, as will be discussed later) and also because these sequences can be connected up with the familiar extant form, *Equus*. Some paleontologists might say that the orthogenetic paradigm is dead now. On the contrary, I would argue that unfortunately it has become more ingrained than it should be, both in modern scientific thought and in everyday society in general. In this chapter I review the development of orthogenesis, cite many examples of its use over the years, and discuss why it can no longer be considered a valid theory to explain both the patterns and processes of evolution. I also discuss how paleontological research is conveyed by the scientists who do the original work and how their results are interpreted by other scientists and the lay public. To jump the gun a bit, we find that once a notion becomes part of accepted scientific knowledge, it is very difficult to modify or reject it as further advances are made. This intellectual inertia is pervasive, and it ranges from most scientists not directly involved in the original research to the lay public.

The foundations of orthogenesis

As with most conceptual foundations of natural history in Western civilization, we must go back to the great Greek philosophers for the origins of orthogenesis. There can be no question that Aristotle (384–322 B.C.) had the most profound impact on subsequent thought in Western civilization. Many have called Aristotle the first natural historian, and his thinking pervades every aspect of this science. Of relevance to orthogenesis, some of Aristotle's ideas embodied unidirectional evolution from less complex (and more lowly) organisms, such as polyps, to the most complex (and therefore most advanced), namely, humans. Aristotle was a firm believer in an underlying design, order, and perfection in nature and the universe. Furthermore, he believed that nature always strives toward these goals and that organic evolution is therefore guided by an internal perfecting tendency (Osborn 1908).

Early systematists of the Linnean era frequently followed Aristotelian views. That resulted in the concept of the *scala naturae*, an ordered, hierarchical sequence of life. It should be noted that these early taxonomists arranged only living forms in a structural progression. Buffetaut (1987) states that Geoffroy Saint-Hilaire (1772–1844) was the first to propose an ancestral–descendant sequence of fossil forms arranged in time. In 1825, Saint-Hilaire wrote that "it is not repugnant to reason, that is to physiological principles, that the crocodiles of the present epoch could be descended through an uninterrupted succession from the antediluvian species" (Buffetaut 1987:69).

By the middle of the nineteenth century, several schemes depicting orthogenetic sequences were based on Tertiary mammals from the Old World. Although those occurrences of fossiliferous outcrops are more sporadic than outcrops in the rich sequences of western North America, sufficient collections had been made in Europe by that time to demonstrate the widespread distribution of fossil horses. As early as 1851, Sir Richard Owen (Figure 3.1) recognized an orthogenetic series beginning with *Palaeotherium* and ending with *Equus* in which descendants became progressively better runners through reduction of their metapodials toward the monodactyl condition (Desmond 1982). Gaudry (1867) described a sequence of European fossil horses from the Eocene to Pliocene depicting a straight-line, ancestral–descendant sequence from *Pachynolophus* to *Palaeotherium* to *Anchitherium* to *Hipparion* to *Equus* (Figure 3.2). Also using the European sequence, the Russian Kowalevsky (Figure 3.1) published a classic and often-quoted paper (1873) on fossil horses. Although its primary goal was a thorough description of *Anchitherium*, that work also served as a forum for discussing the interrelationships of other Equidae. Like Gaudry, Kowalevsky (1842–83) usually is considered an advocate of orthogenesis in fossil horses; however, careful reading of his work indicates that he was more perceptive. He realized that both temporal and morphological sequences were represented from *Palaeotherium* to *Anchitherium* to *Hipparion* to *Equus*, but he did not believe that each older genus necessarily gave rise directly to its successor.[1] Regrettably, after publishing several classic works in paleontology, Kowalevsky fell upon hard times, was unable to

[1]"L'apparition des quatre genres de la série: Paléothérium medium, Anchithérium, Hipparion et Cheval dans des couches qui se succèdent dans le temps, tend encore plus à confirmer les résultats auxquels on arrive par l'étude de leurs ossements. Cependant, rien n'est plus loin de moi que l'idée qu'un animal que nous nommons Palaeoth. *medium* a donné directement naissance à un Anchithérium, celui-ci à un Hipparion (peut-être Meryhippus [*sic*] Ld.) et ainsi de suite." (Kowalevsky 1873:3)

Figure 3.1. *Nineteenth-century European contributors to an understanding of fossil horses. Left: Sir Richard Owen (1804–92), although an ardent antagonist of Darwin's theory, wrote many important papers on the taxonomy and morphology of extinct mammals, including fossil horses (such as the specimens of Pleistocene* Equus *collected from Argentina by Darwin). From Illustrated* London News, *1872. Right: V. Kowalevsky (1842–83); in 1873 he published a classic monograph on fossil horses from the Old World. From Franzen (1984).*

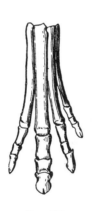

Figure 3.2. *Gaudry's "pseudophylogeny" of fossil horses based on specimens from Eocene–Pliocene deposits in Europe. Although arranged in an orthogenetic, and therefore presumed ancestral–descendant progression, it is now known that this sequence represents several successive dispersal events, rather than* in situ *evolution. From Gaudry (1896).*

Fig. 182.	Fig. 183.	Fig. 184.	Fig. 185.	Fig. 186.
Pachynolophus (*Orohippus*) *agilis.*	*Palæotherium crassum.*	*Anchitherium aurelianense.*	*Hipparion gracile.*	*Equus Stenonis.*
Éocène moyen.	Éocène supérieur.	Miocène moyen.	Miocène supérieur.	Pliocène.

secure an academic position, and ultimately died by chloroforming himself (Desmond 1982). It was that phylogeny of Old World fossil horses, now known to represent a discontinuous series of unrelated immigrant genera, that influenced other workers of the day, such as Huxley. That sequence also was an integral part of the foundation of orthogenesis interpreted from the fossil record.

Marsh and fossil horses

In 1866, O. C. Marsh became the first professor of paleontology in North America, at Yale University. Soon thereafter, he started to collect fossils in western North America for the new Peabody

Figure 3.3. O. C. Marsh (1831–99; back row, center) and the Yale expedition of 1872. Long before the days of government-funded paleontological digs, Professor Marsh enlisted Yale undergraduates for summer expeditions to the western territories to collect fossil vertebrates. Reproduced with permission of the Peabody Museum of Natural History and Yale University.

Figure 3.4. Thomas Henry Huxley (1825–95), probably taken in the early 1860s. From Desmond (1982); reproduced courtesy of the Wellcome Institute Library, London.

Museum at the university. He undoubtedly was stimulated by the fascinating discoveries made by the Warren-Hayden surveys of Nebraska, the Dakotas, and adjacent territories, which began in 1853 and ended in 1878 (Lanham 1973). During his first expedition to these rich Tertiary deposits in 1868, he was very impressed by the remains of a Pliocene horse skeleton from western Nebraska (Rainger 1981). Subsequent Yale expeditions, with student assistants (Figure 3.3) and often with cavalry escorts, were led by Marsh during the early 1870s (Lull 1931; Schuchert & Le Vene 1940). As a result of those endeavors, by the mid–1870s a considerable collection of Tertiary fossil mammals from the western United States was housed in the Yale Peabody Museum, and because of their widespread abundance in that region, fossil horses were well represented in that collection.

In 1876, Thomas Henry Huxley (1825–95; Figure 3.4), considered by many to be Darwin's bulldog, an ardent advocate of evolution, and principal target for nonevolutionists of the day, such as Sir Richard Owen (Desmond 1982), made a trip to the United States to visit museums, to lecture, and to attend the founding ceremonies for The Johns Hopkins University, where he presented a keynote address (Gould 1987). Upon his arrival he visited Marsh in New Haven for three weeks and thus had an opportunity to view the collection of fossil horses in the Yale Peabody Museum. Prior to that time Huxley had been wrestling with the question of the descent of the horse based on the presumed evolutionary sequence uncovered in the Old World (Gaudry 1867; Kowalevsky 1873). However, after studying the Yale collection he became convinced that the principal evolutionary radiation of horses had occurred in the New World (with the fossil record from the Old World representing several distinct dispersal events from North America). Huxley was so taken with this new evidence for horse evolution that he changed his viewpoint in a famous lecture delivered soon thereafter in New York. Marsh (1895:181) later recounted that after seeing the Yale horse collection, Huxley believed that these specimens "demonstrated the evolution of the horse beyond question, and for the first time indicated the direct line of descent of an existing animal."

An interesting exchange occurred between Huxley and Marsh as they were studying at the Yale Peabody Museum (Schuchert & Le Vene 1940:236):

> One day, as he and Marsh sat discussing possible ancestral horses, then unknown, Huxley fell to sketching on a sheet of brown paper, and presently he said: "That is my idea of 'Eohippus.'" Then he added, "But he needs a rider," and with a few more pencil strokes he supplied the lack. He and Marsh joked about the sketch, and finally Marsh said: "The rider also

Figure 3.5. *"Eohippus and Eohomo" sketch done in 1876 at the Yale Peabody Museum during a meeting between Thomas Henry Huxley and O. C. Marsh. Reproduced with permission of the Peabody Museum of Natural History and Yale University.*

must have a name. What shall we call him?'' ''Call him 'Eohomo,' '' said Huxley, and Marsh wrote the names underneath the sketch.

Whether that is the exact sequence of events, no one will ever know. However, this now-famous sketch is preserved for posterity at the Yale Peabody Museum (Figure 3.5). Later that year, Marsh (1876) formally proposed the name *Eohippus* for the earliest known genus of North American Equidae, although it is now universally considered a junior synonym for *Hyracotherium* Owen 1840.

In 1879, Marsh published an article entitled ''Polydactyle Horses, Recent and Extinct,'' where he discussed the evidence for gradual evolution and presented an illustration of the morphological and presumed phylogenetic progression of North American fossil horses based on the collection at the Yale Peabody Museum (Figure 3.6). This figure depicts the sequences of changes in metapodials, radii and ulnae, tibiae and fibulae, and upper and lower molars in representative equid genera (Marsh also realized that others were known, e.g., *Epihippus*, but they were not included). An important point that will be discussed later is that this figure is not drawn to scale. For example, the upper molars of *Orohippus* and *Equus* are graphically depicted as the same, yet in reality there is a fourfold difference in sizes

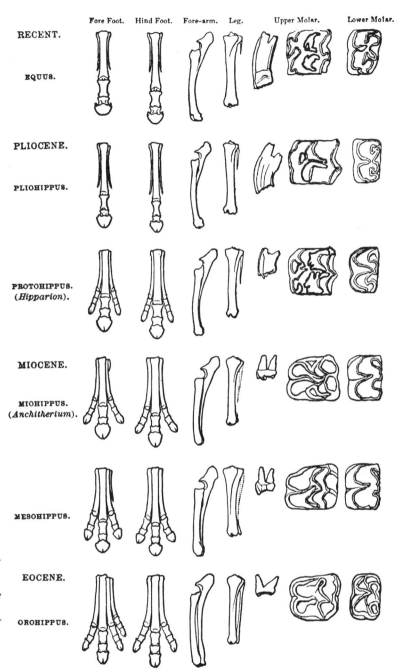

| | Fore Foot. | Hind Foot. | Fore-arm. | Leg. | Upper Molar. | Lower Molar. |

RECENT.

EQUUS.

PLIOCENE.

PLIOHIPPUS.

PROTOHIPPUS.
(*Hipparion*).

MIOCENE.

MIOHIPPUS.
(*Anchitherium*).

MESOHIPPUS.

EOCENE.

OROHIPPUS.

Figure 3.6. *O. C. Marsh's horse phylogeny (1879) based on collections then available at the Yale Peabody Museum. Although published in 1879, this figure was originally drafted for Huxley in 1876 for his lecture in New York. Oro-hippus was used instead of* Eohippus *for the earliest horse because the latter had not been described when this figure was drafted.*

(with lengths from about 10 to 40 mm, respectively) (MacFadden 1988b). As will become clear in the discussion that follows, Gould (1987:18) is correct that this is "one of the most famous illustrations in the history of paleontology."

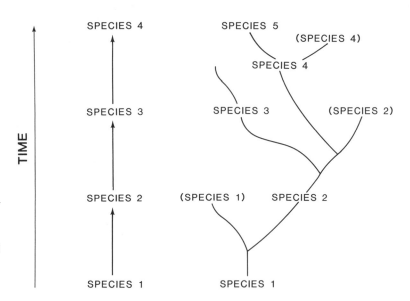

A. ORTHOGENESIS B. BRANCHING PHYLOGENY

Figure 3.7. Models of ancestry and descent. In orthogenesis (A), ancestor evolves into descendant, resulting in nonoverlapping straight-line evolution via anagenesis. It is now believed, however, that orthogenesis is not the dominant mode; rather, lineages and clades evolve by branching speciation (B) via cladogenesis, in which there is temporal overlap of ancestors and descendants. Numbers in parentheses represent continuations of ancestral species.

Orthogenetic concepts and causality

The concept of orthogenesis (Figure 3.7) was first put forth by Haacke (1893) and others in Europe (Simpson 1944), and it was embodied, without a name, in the thought of many nineteenth-century paleontologists (e.g., Lamarck and Saint-Hilaire). In the United States, the prominent paleontologist Henry Fairfield Osborn (1857–1935; Figure 3.8) championed orthogenetic thinking, although he used his own terminology. Simpson (1944, 1953a), who presented one of the most coherent discussions of orthogenesis, considered Osborn's "aristogenesis" (1934) to be conceptually similar, even though Osborn's article seems more concerned with the mechanism of phyletic evolution rather than with straight-line descent.

Any discussion of orthogenesis should clearly separate two prominent aspects of this concept: (1) the graphic depiction of evolution from ancestor to descendant and (2) the underlying causal explanation for this pattern. Strictly speaking, graphic depiction of an orthogenetic sequence of taxa arrayed in time is not incorrect. Indeed, the modern literature offers many examples of interpretations of unilateral ancestral–descendant or anagenetic speciation that represent valid hypotheses. The more difficult question, however, is what such an orthogenetic pattern may imply about underlying mechanisms and causality in evolution. Unfortunately, this distinction is rarely made, and these two concepts of orthogenesis are frequently considered to be one.

Figure 3.8. *Some of the twentieth-century paleontologists who have contributed to an understanding of fossil horses.* Top left. *Henry Fairfield Osborn (1857–1935) wrote classic monographs on craniometry (1912b) and systematics (1918) of fossil Equidae. From Osborn (1930).* Top right. *William Diller Matthew (1871–1930) top right. Courtesy of Margaret Colbert.* Bottom left. *R. A. "Stirt" Stirton (1901–66). Courtesy of Joseph T. Gregory. Matthew and Stirton wrote many of the seminal papers on fossil horse systematics and evolution that provide the foundation for our current understanding of the group.* Bottom right. *George Gaylord Simpson (1902–84). From the Simpson Paleontological Library. Although Simpson did not actively research horses as did Matthew and Stirton, he used the Equidae as prime examples supporting his ideas of evolution and in 1951 published the important popular book entitled* Horses.

Graphic orthogenesis

No one can fault Marsh and other paleontologists of the late nineteenth and early twentieth centuries for believing that orthogenesis was a valid concept to describe the long-term evolution of fossil taxa. As relevant specimens and collections were accumulated, it is understandable that a general view of evolution and the fossil record would embrace the elegant explanation of orthogenesis. In this regard I disagree with Stanley's view of the status of natural science during the late nineteenth century (1981:75): "Thus, paleontology joined the conceptually weak and speculative areas of biology that late in the nineteenth century engendered disillusionment with evolutionary research."

With the late-nineteenth-century interest in Darwin's ideas, it is understandable that paleontologists believed that orthogenesis also meant gradual evolution. The problem with the acceptance of orthogenesis in modern scientific thought stems from how some natural historians in the early twentieth century viewed the causal processes of evolution.

Causal orthogenesis

> *Nature produces those things which, being continually moved by a certain principle in themselves, arrive at a certain end.*
> – Aristotle (Osborn 1917:9)

Since the first publication of *On the Origin of Species* in 1859, Darwinian evolution has had a profound impact on stimulating scientific thought. A fundamental problem with Darwinian theory, as originally presented, was a lack of a plausible mechanism for inheritance through generations, which of course was later explained by Mendelian genetics. Many of the nineteenth-century biologists who attempted to explain this enigma were influenced by religion, by their prior philosophical beliefs, and by scientists' anthropocentric view of nature. Marsh's horses fitted in quite well with such preconceived notions about how evolution should proceed.

Lamarck followed Aristotle in believing that organisms were driven by a vital force (*élan vital*) to make them more nearly perfect, more competitive, and hence better suited to survive (Figure 3.9). For example, it has been traditional to believe that the human species is the ultimate evolutionary product of progressively refined forms, starting with lowly unicellular organisms and ultimately culminating with *Homo sapiens* sitting atop the *scala naturae*.

It is easily understood why, in the minds of many nineteenth-century pro-Darwinists, Marsh's fossil horses seemed to provide definitive evidence of long-term evolution as some predestined, unidirectional trend toward organismic progress and improvement. Bigger meant better (the only trend left out of Marsh's famous illustration, Figure 3.6), and therefore Cope's law (see Chapter 10) was a corollary to orthogenesis in the view of many scientists. Reduction of toes and limbs (the ulnae and fibulae) and elongation of the metapodials provided for more efficient running and higher-crowned teeth allowed for more efficient

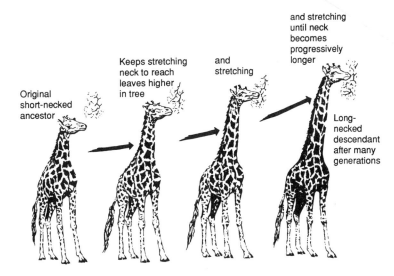

Figure 3.9. *Lamarck's giraffes. Lamarck (1744–1829) and his followers believed that life is driven by an inner need (élan vital) to become better, so that successive generations of giraffes evolved longer necks to reach higher vegetation in the trees. Modified from Savage (1977).*

processing of abrasive foodstuffs, both of which were advantageous to species living in the more open grassland habitats. That way of thinking seemed consistent with the preconceived notion of evolutionary destiny, and it was supported by an elegant temporal sequence of fossils. Elegant or not, orthogenesis was at best a gross oversimplification of the actual phylogenetic pattern of the Equidae. Despite the fact that randomness was anathema to many prominent nineteenth-century natural historians (Bowler 1989), we now know that evolution results from many random genetic changes, some of which become incorporated into subsequent generations. Evolution is not driven by an ordered, inner need to improve the lot of a species, as was believed by Lamarck and his followers. Many scientists have also traditionally equated evolution with progress but even that has more recently been called into question (e.g., Bowler 1976; Nitecki 1988a) (see Chapter 9). In summary, few scientists today would consider evolution to be an orderly, directed process resulting in perfectly adapted species.

Other views of fossil horses: a complexly branching tree

I find my conclusions so different from those of Doctor Matthew, for example, that I could not write a joint work with him. All of his conceptions of evolutions are quite different from mine and probably from yours also.
– Henry Fairfield Osborn writing to William Berryman Scott in 1921 (Rainger 1986:453)

Many nineteenth-century paleontologists embraced orthogenesis as a valid explanation of long-term evolution. However, soon

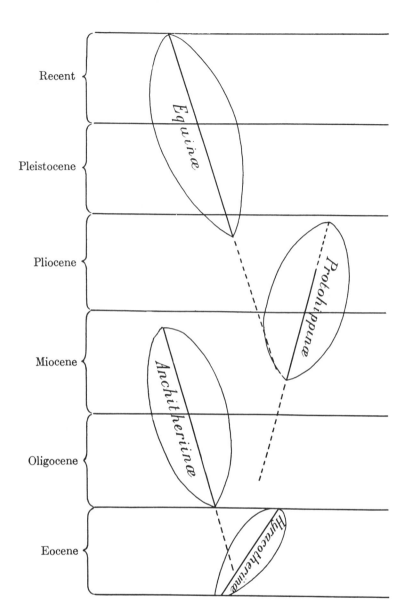

Figure 3.10. *Gidley's illustration (1907) of the phylogeny of the four subfamilies of the Equidae. In contrast to previous orthogenetic, nonoverlapping illustrations (e.g., Figure 3.6), this is one of the first figures to depict fossil horses in a branching scheme in which there is temporal overlap (between the Protohippinae and Anchitheriinae and Protohippinae and Equinae).*

after Marsh's espousal of straight-line horse evolution, as additional fossil collections were being accumulated, other paleontologists were formulating a far more complex view of this group. By the turn of the century this more complex view of horse phylogeny was making its way into the original research literature. Rainger (1986), for example, sees William Diller Matthew (Figure 3.8) as one of the young turks during the early twentieth century who, because of his keen awareness of stratigraphy, developed a view of evolution different from the paradigm of the old guard (e.g., Marsh and Osborn).

With regard to fossil horses, Gidley (1907) was one of the specialists to graphically depict the branching of various contemporaneous lineages (Figure 3.10). Thus, the Anchitheriinae partially overlap the Protohippinae, and the Protohippinae partially overlap the Equinae. In a classic paper on fossil horses, Matthew (1926) sent mixed messages to his readers. On one hand, his evolutionary ladder indicated a direct line of successive grades, but on the other he also realized that horse phylogeny was a complex branching pattern of contemporaneous lineages (e.g., on p. 160 he noted the coexistence of *Pliohippus, Protohippus,* and *Hipparion* during the Miocene). This phylogenetic complexity is embodied in subsequent phylogenies (Figure 3.11) of the North American Equidae (e.g., Matthew 1930). After Matthew moved from New York to Berkeley in 1927, he greatly influenced his graduate student Ruben A. Stirton (Figure 3.8), one of the most active researchers on fossil Equidae of North America during this century. In one of his classic papers, Stirton (1940) depicted the complex branching phylogeny of horses (Figure 3.12) in which, during their heyday, eight contemporary genera occurred during the "Lower Pliocene" (now late Miocene). Stirton's phylogeny is virtually unique in presenting the interrelationships of known equid species as well as genera. The studies of workers such as Gidley, Matthew, and Stirton were synthesized and simplified in the classic horse phylogeny of Simpson (Figure 3.8). Like Marsh's orthogenesis, Simpson's horse phylogeny (e.g., Simpson 1951) has been used repeatedly in textbooks as a prime example of long-term evolution (Figure 3.13). However, unlike Marsh's view, Simpson's general pattern depicting horse evolution as a complex, branching phylogeny is still valid today, although many new details have been added since the first half of this century.

The spread of orthogenesis: old notions die hard

Although most palaeontologists now reject such views, they sometimes still find an echo in popular palaeontological literature.
— Buffetaut (1987:157)

Marsh's illustration of the orthogenetic sequence for horses (Figure 3.6) has been reproduced (sometimes exactly as it was published in 1879, sometimes in modified form) countless times in books dealing with paleontology, evolution, and natural history, and even today it is seen in college-level textbooks in geology and zoology (Figure 3.14). I agree with Gould (1987:20), who wonders "what the textbook tradition of endless and thoughtless copying has done to retard the spread of original ideas." Probably more than any other single figure depicting a fossil group, Marsh's illustration of the "Genealogy of the Horse" has for over a century greatly influenced scientific thought related to

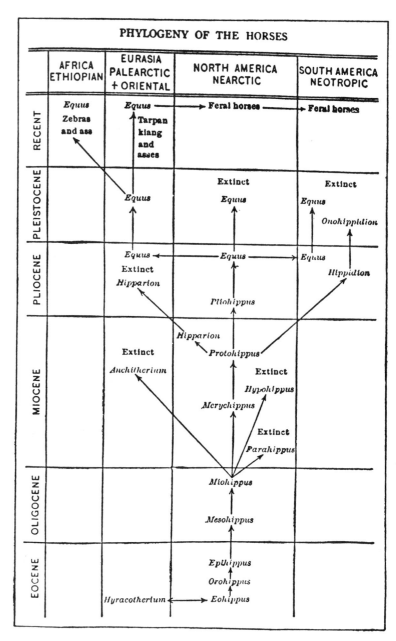

Figure 3.11. *Matthew's generic phylogeny (1930) of the Equidae. The essentials of this figure were later modified by Simpson (e.g., 1951, 1953a) and subsequently adapted (e.g., in textbooks and museum exhibits) numerous times in synthetic discussions of horse phylogeny.*

evolution. This stems from the fact that fossil horses are abundant, and they provide one of the best examples of a group that has a well-preserved fossil record as well as extant forms. Furthermore, this figure, which cannot be faulted for its interpretation of what was known over a century ago, has done much to delay the progress of science and obscure new evolutionary ideas that have been developed since then. In particular, orthogenesis, fossil horses, and even gradual evolution have been intimately associated since those ideas were presented in the

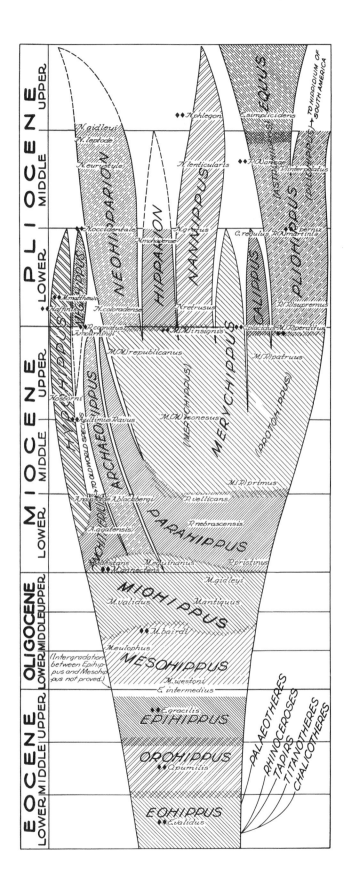

Figure 3.12. Stirton's phylogeny (1940) of North American Equidae taken from his classic paper. Although he shows fuzzy boundaries between genera, this illustration is virtually unique in showing species-level ancestors and descendants.

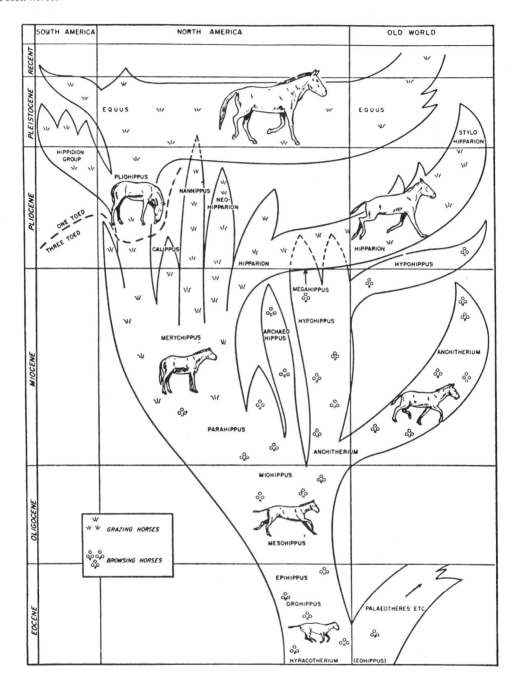

Figure 3.13. *Simpson's phylogeny (1951) of the Equidae, published in his book* Horses. *In the caption for another, very similar version of this illustration, Simpson (1953a:261) states that it is a "greatly simplified representation of the phylogeny of the Equidae.*

Only generic branches are indicated, and even at this level many students would recognize more branches. Each generic area should be thought of as made up of from several to thousands of different strands, the real lineages." It is unfortunate that this caveat has not

also been carried along with the graphic in the numerous instances in which this figure has been copied in textbooks. From Simpson (1951); reproduced with permission of Oxford University Press.

Figure 3.14. Right: *Patterns of horse evolution depicted in the textbook* In-troduction to Evolution, *by Fred A. Racle; © 1979, p. 71; reproduced with permission of Prentice-Hall, Englewood Cliffs, New Jersey.*

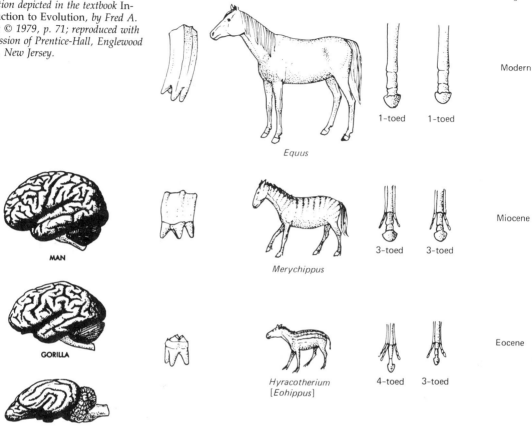

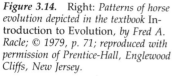

Figure 3.15. *Vertebrate brain evolution, arrayed in an orthogenetic scheme. From* Scientific American.

late nineteenth century. This is despite the fact that many pa-leontologists since that time, particularly those working on fossil horses, have realized that the story is much more complex.

In addition to horses, the orthogenetic dogma has spread to other groups, whether it be vertebrate brain evolution (Figure 3.15), titanotheres (Figure 3.16), or the anthropocentric pro-gression of human evolution. The orthogenetic template also has influenced millions of lay people, many of whom visit natural-history museums with turn-of-the-century exhibits that convey 100-year-old ideas. The late nineteenth century was an important formative period in the development of thinking about paleontology and evolution. As is still the case today, the curators at natural-history museums were responsible for the scientific content of their institutions' exhibits. Thus, at the Mu-sée d'Histoire Naturelle in Paris, for example, the exhibits that still can be seen today were influenced by nineteenth-century curators such as the paleontologist Gaudry. Buffetaut (1987:142) noted that ''Gaudry's museological conception was a direct re-

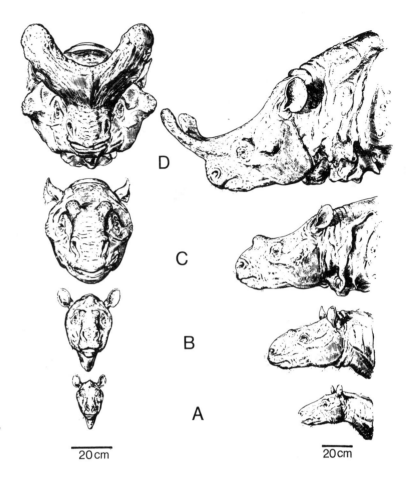

Figure 3.16. *Orthogenesis in North American titanotheres.* (A) Eotitanops borealis, *early Eocene;* (B) Manteoceras manteoceras, *middle Eocene;* (C) Protitanotherium emarginatum, *late Eocene;* (D) Brontotherium platyceras, *early Oligocene. From Stanley (1974b); reproduced with permission of the Society for the Study of Evolution.*

20 cm   20 cm

flection of his philosophical ideas on evolution: the vast, undivided palaeontology gallery was to give the visitor a vivid impression of the progress of life through geological time."

Despite the fact that paleontologists know that it is incorrect, the linkage among horses, orthogenesis, and progress still remains a pervasive, dominant theme in many exhibits in natural-history museums today (Figure 3.17). In addition, the combination of orthogenesis and progress is also found in many of the simplified phylogenies and evolutionary histories depicted for other taxonomic groups, and that further contributes to a pervasive pattern of orthogenesis. Some museums have begun major renovations in their exhibits so that they will no longer emphasize orthogenesis, but many decades will pass before this pattern fades from the public's view.

Given the manner in which information on science is conveyed to the public, the lingering strong association between evolution and orthogenesis as interpreted from the fossil record is not

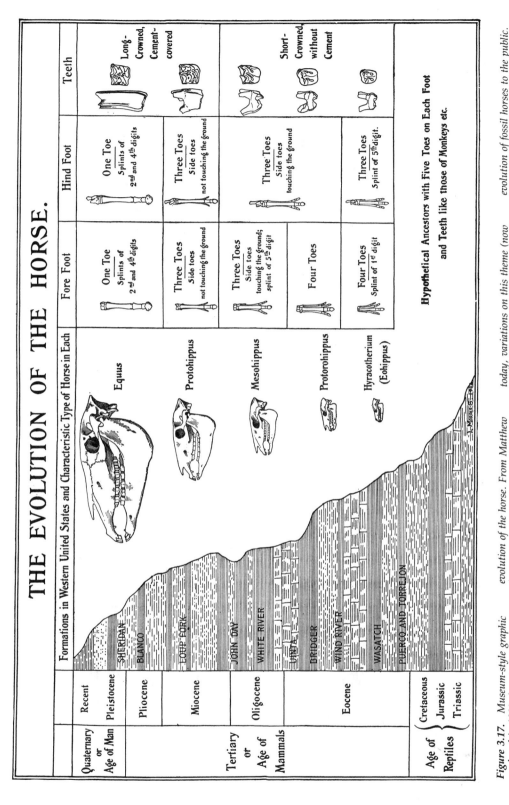

Figure 3.17. Museum-style graphic produced in 1902 that illustrates the evolution of the horse. From Matthew (1903, 1926). In some museums even today, variations on this theme (now almost a century old) still convey the evolution of fossil horses to the public.

Figure 3.18. *Orthogenesis embodied in a syndicated science column from* The Gainesville Sun, *1985. Reproduced with permission of Philip Seff.*

Figure 3.19. *Orthogenesis in the public sector. Cartoonist Larry Johnson prepared this illustration, which appeared in* The Boston Globe *just prior to a Patriots–Raiders football game. Reproduced with permission of* The Boston Globe. *Also see Gould (1989b).*

surprising. This is exemplified by a newspaper column on horse evolution written by a geology Ph.D. (Figure 3.18). As somewhat amusingly described by Gould (1989b), orthogenesis has also impacted other aspects of our society, including commercial advertisements and professional sports columns (Figure 3.19).

If we know that orthogenesis is a concept fraught with problems, then why is it still so pervasive in science writing? I believe that the root of the problem is the quest for simplicity. Scientists are continually confronted with the problem of conveying complex data and interpretations in an intelligible way to their colleagues, to other specialists, and often to the public. In many cases the best way to do this seems to be to distill the essence of the research to a simplified form. Although the original researchers have always been keenly aware of the complexity of horse phylogeny (e.g., Simpson 1951) since the early part of this century, much has been lost in subsequent reproductions in textbooks and other scientific syntheses (see caption for Figure 3.13). Therefore, even today orthogenesis goes hand in hand with simplification because together they provide such an elegant interpretation of the almost impossibly complex evolution of the Equidae.

In summary, a potential consequence of simplifying research findings so that they can be more easily communicated to others is that it often retards the progress of science. It is appropriate to end this discussion with a quotation from the prominent British philosopher Alfred North Whitehead (1957:163): "The aim of science is to seek the simplest explanations of complex facts. We are apt to fall into the error of thinking that the facts are simple, because simplicity is the goal of our quest. The guiding motto in the life of every natural philosopher should be, Seek simplicity and distrust it."

4

Collections, museums, and exceptional discoveries

The antiquity and development of collecting fossil vertebrates

The urge to collect interesting objects, other than to provide food or shelter, is one of the oldest forms of human diversion. So far as we know, humans have collected fossils for at least 350 centuries. The oldest evidence of this avocation comes from Middle Paleolithic (ca. 100,000–35,000 years ago) deposits in a cave at Arcy-sur-Cure in France (Dennell 1986). That modest beginning, consisting of a gastropod and coral, probably was made by Neanderthals. Fossil shark teeth are fairly common at numerous late Paleolithic sites in southern France, where they are found considerable distances (up to 150 km) from the nearest fossil localities, indicating a deliberate human activity of collecting. Some of these are perforated, suggesting that they were used as necklaces or pendants and possibly for ritualistic purposes.

In predynastic and dynastic Egypt, fossil shark teeth were used as amulets (to ward off evil spirits) and sometimes were set in metal. On the Greek island of Kos, in the Aegean, in the 1920s Barnum Brown, the colorful paleontologist from the American Museum of Natural History, found a fossil elephant tooth (of Pleistocene *Elephas*) at the ruins of Asklepieion, a famous medical school where Hippocrates is supposed to have studied (Brown 1926; Buffetaut 1987). In Roman times, the emperor Augustus kept a collection of huge fossil bones, purported to be the remains of giants, that had been found on the island of Capri. In the New World, one of the earliest fossils known to have been collected was of the early Tertiary mammal *Phenacodus*, collected by Native Americans, probably between 700 and 900 A.D., from the San Juan Basin in what is now New Mexico (Simpson 1942).

During the Middle Ages, the Renaissance, and up until the seventeenth century in Europe, large fossilized mammal bones, mostly of late Cenozoic proboscideans, were generally thought to be the remains of monsters, dragons, slain warriors, or the mythical unicorn (see Figure 1.10). The cultural fascination (and

49

reverence) for these in the Western world led to the study (few would consider it a science in the modern sense) of "gigantology." Collections of giant bones often were kept in churches, castles, and public buildings. By Renaissance times, many among the royalty and nobility kept natural-history "cabinets" (some of which were maintained by the forerunners of today's natural-history curators), a practice that also spread to the universities in Europe (Buffetaut 1987).

According to Othenio Abel, an Austrian vertebrate paleontologist, the oldest known paleontological reconstruction was based on the skull of an extinct wooly rhinoceros (*Coelodonta antiquitatus*). That specimen was the basis for a model of a winged dragon displayed in the town square of Klagenfurt, southern Austria, that was reconstructed by the sculptor Ulrich Vogelsang beginning in 1590 (and completed after his death in 1636). The fossil skull had been collected some 250 years earlier (probably about 1335) from some nearby Pleistocene terrace deposits. The earliest known vertebrate fossil reconstructed in life pose is the famous Brú specimen of *Megatherium* (collected from what later became Argentina), which was described by Cuvier in 1796 and mounted in the Royal Collection of Natural History in Madrid, Spain (Simpson 1984; Buffetaut 1987). Thus, J. B. Brú, who worked in that collection, can be considered to have initiated the tradition of vertebrate preparation. In North America, the earliest skeletal reconstruction on public display was Peale's American mastodon (Figure 4.1), *Mammut americanum*, collected in 1799–1801 from a peat bog in upstate New York and mounted a few years later in Philadelphia (Simpson 1942; Buffetaut 1987).

The earliest recorded systematic paleontological excavation occurred in 1700 near Stuttgart, Germany, and lasted for six months. It was ordered by the duke of Württemberg in his search for the mythical unicorn. Those excavations recovered numerous fossil mammoth specimens, including more than 60 tusks (Sutcliffe 1985; Buffetaut 1987). Undoubtedly the largest single excavation over a brief period was at Tendaguru in German East Africa (now Tanzania). Between 1909 and 1912 German paleontologists, employing as many as 500 local laborers, conducted extensive excavations there for late Jurassic terrestrial vertebrates, including dinosaurs. Like the number of man-hours involved, the quantity of fossils collected was extraordinary. During the first three field seasons a total of 140 tons of bones, all of which were carried by porters from the inland site to the coast (a several-day trip), was shipped back to Berlin (Buffetaut 1987).

Certainly one of the most bizarre circumstances under which fossil vertebrates were collected occurred during World World I. In 1915 French troops were sent to occupy Macedonia, in

Figure 4.1. *Excavation of the Peale mastodon near Newburgh, New York, in 1799–1801, from a painting by Rembrandt Peale. In order to retrieve the bones, water is being removed from the pit by the buckets of a human-powered treadmill, and mud is removed by shoveling onto a series of platforms (Sutcliffe 1985). The painting belongs to Mrs. Bertha White; photograph courtesy Department of Library Services, negative no. 310511, AMNH.*

northern Greece. One of the officers of the 156th Division, Camille Arambourg, was a vertebrate paleontologist in civilian life. While digging trench lines, French soldiers discovered fossil bones at several localities some 30 km to the northwest of Saloniki. Arambourg obtained permission from the commanding officer not only to excavate the fossils but also to remove them from the front. Between December 1915 and April 1916, during lulls in military operations, Miocene fossil mammals were excavated under Arambourg's direction. They were stored in a depot in Saloniki until after the war, when they were shipped to Algeria for preparation and preliminary study. In 1919 the specimens were shipped to the Natural History Museum in Paris, where they found a permanent home. Those collections

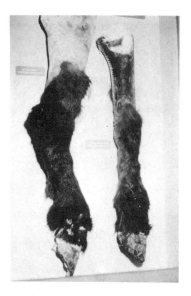

Figure 4.2. *Limb bones, with soft tissue preserved, of Pleistocene* Equus *"mummies" from the permafrost of Selerikan, USSR. Specimens are on exhibit in St. Petersburg. Photograph courtesy of A. J. Sutcliffe.*

formed the basis for the rich Neogene mammalian fauna (including the horse *Hipparion*) of Saloniki, which was studied further by Arambourg and colleagues after the war (Arambourg & Piveteau 1929; Buffetaut 1987).

Outside of the Western world, fossil collecting has been practiced in China for countless centuries. Among such pursuits, none is more famous than the search for "dragon teeth" (*long chi*) and "dragon bones" (*long gu*), which are the remains of late Cenozoic mammals, including those of the classic *Hipparion* fauna. Ground up as powder, these bones have been thought to cure a multitude of ailments (see Figure 1.6). These dragon bones have long been collected in the countryside by the locals and then sent to market, often by itinerant wholesalers. Often they ended up in drugstores in the large cities, where they mystified Western paleontologists until the beginning of the twentieth century, when their origins came to be understood. The prices of these remedies have been determined by their quality. Not unlike mammalian paleontologists, although for different reasons, the Chinese value teeth more highly, and teeth are therefore more expensive than bones. The resulting commerce has been considerable. In 1885 some 20 tons of fossil bones passed through various Chinese ports. Of relevance here, at the turn of the twentieth century a large collection of Chinese Cenozoic mammals obtained from the "drugstore commerce" was shipped to Munich. It contained nearly 1,000 teeth of the Miocene three-toed horse *Hipparion* (Buffetaut 1987). As might be expected, all of those early specimens lacked locality data, thus rendering them of dubious scientific value. Although the medicinal use of fossil bones and teeth lives on in Chinese culture, the purchase of Neogene mammals by Western paleontologists has declined over the twentieth century since the first expeditions to fossil sites in that country were sanctioned. Those expeditions were led by foreign nationals (e.g., Sefve 1927, on fossil horses), as well as local priests and missionaries, the most notable of whom was Pierre Teilhard de Chardin.

Some of the most interesting and extraordinary preservation of Pleistocene extinct mammals has occurred in the permafrost regions at high latitudes, most notably in Siberia. Although the mammoth and mastodon "mummies" from Siberia are best known, specimens of *Equus* with soft tissue have also been preserved (Sutcliffe 1985) (Figure 4.2). Tar seeps, of which La Brea is the best known, have led to exceptional preservation of bones and frequently have attracted the attention of humans, either because of their economic importance or because of their record of extinct life.

In summary, for millennia fossil vertebrates have attracted the attention of humans of diverse cultural backgrounds, and these objects have been used for a variety of purposes, including jew-

elry and decorative ornamentation, rituals, mythology, thera-
peutic agents, and scientific (*sensu lato*) inquiry about nature. As
the latter inquiry expanded during the seventeenth and eighteenth centuries, many cabinets and personal collections of
natural-history specimens grew into the forerunners of present-
day natural-history museums (Figure 4.3).

The importance of museum collections

*Where Marsh and Cope and Leidy depended upon a few specimens,
mostly fragmentary, we have today great series of complete skulls and
skeletons.* – Matthew (1926:139)

*The Museum, in its largest sense, is a diffuse, sprawling, complex and
ineffable thing, whose real boundaries extend to the farthest corners of
the earth.* – Preston (1986:xvi)

Systematic collections in natural-history museums provide the
fundamental basis and repository for paleontological data, and
that is also the case for fossil horses. We saw in Chapter 3 how
in the 1870s at the Yale Peabody Museum, Marsh's early collections of fossil horses from the western territories, which certainly
are small by today's standards, greatly influenced the evolutionary thought of that time.

Since the latter half of the nineteenth century, natural-history
museums have been founded in many parts of the world, and
systematic collections for research and exhibition have grown.
This growth continues today as many collections in vertebrate
paleontology are rapidly expanding. As part of this continuing
expansion, many collections, particularly those specializing in
Cenozoic mammals, have increased their holdings of fossil
horses. As Sheets-Pyenson (1987:289) noted in her article on
nineteenth-century museums, "impressive façades, appealing
interior designs, and renowned staff all helped to make a positive
first impression on museum visitors. Yet the true measure of
the importance of a museum was the overall excellence of its
collections."

Considering all institutional collections (museums, govern-
ments, and universities) in North America, there are about half
a million fossil horse specimens in the scientific domain. Of
these, only about one-quarter to one-third are fully prepared
and catalogued. The geographic and temporal coverage of fossil
horses on this continent is unrivaled and extends to many localities in southern, central, and western North America (Figure
4.4). Depending upon the circumstances of fossilization and collection, fossil horse specimens can range from isolated teeth and
postcrania to partial dentitions and partially associated postcrania, as well as virtually complete skeletons. As with most
fossil mammals, the durable teeth are the most frequently pre-

served parts of the skeleton, and, as seen earlier, teeth have provided the basis for many schemes to describe the phylogeny and evolution of horses. As a result of the advent of computers, great scientific importance is now placed on large statistical samples as an accompaniment to the more traditional desire to obtain morphologically complete specimens. Knowledge of the exact provenience of fossils in space and time has become increasingly important scientifically, and many modern collections are significantly better documented than were the collections made a century ago. All of these advances have yielded an enhanced understanding of the systematics, paleobiology, and evolution of the Equidae.

In the remaining part of this chapter I have selected two topics for discussion: (1) the importance of museum collections and (2)

Figure 4.3. *"The Artist in His Museum," a self-portrait painted in 1822 by Charles Wilson Peale. The Peale Museum, founded in the 1780s by the painter and naturalist Peale, was the first natural-history museum open to the public in North America. General admission was 25 cents; special exhibits (e.g., a mammoth skeleton) were 50 cents (Sellers 1980). Reproduced with permission of the Pennsylvania Academy of the Fine Arts, Philadelphia (gift of Mrs. Sarah Harrison; The Joseph Harrison, Jr., Collection).*

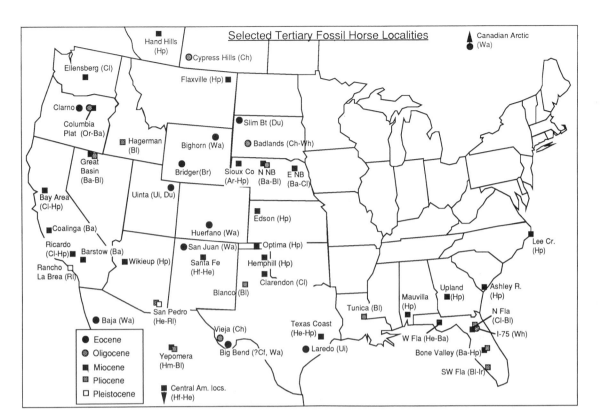

Figure 4.4. *Selected localities of Tertiary horses in North America, with emphasis on the United States. This shows many of the important or classic localities described in the literature and mentioned in this book. Some other smaller or less well known localities (e.g., Baja, Mexico, or Hand Hills, Canada) are included to show paleobiogeographic extremes. The abbreviations next to the localities represent the prin-* *cipal land-mammal ages found there: Ar, Arikareean; Ba, Barstovian; Bl, Blancan; Br, Bridgerian; Cf, Clarkforkian; Ch, Chadronian; Cl, Clarendonian; Du, Duchesnean; He, Hemingfordian; Hp, Hemphillian; Ir, Irvingtonian; Or, Orellan; Rl, Rancholabrean; Ui, Uintan; Wa, Wasatchian; Wh, Whitneyan. Except for Rancho La Brea, which is described in this chapter, and San Pedro Valley, Arizona, other Pleisto-* *cene localities are excluded from this map because they are essentially found everywhere. Many of the localities named here represent an extensive series of sites; for example, "N NB" (north-central Nebraska) includes the prominent Burge and Minnechaduza quarries and other contemporaneous localities, and "Columbia Plat" (Columbia River Plateau, Oregon) includes the John Day and Mascall beds.*

extraordinary discoveries of fossil horses. Because it would be impossible to discuss all important facets of these topics, my examples are selected principally for their remarkable circumstances of fossilization and scientific potential. Related to museum collections, I shall first focus on the extraordinary Frick collection of fossil horses in New York and then briefly mention other important collections elsewhere in North America and around the world. Related to the second topic, I shall discuss the importance of half a dozen truly extraordinary discoveries of fossil horses and the research that has resulted from them.

The Frick horse collection

The Department of Vertebrate Paleontology at the American Museum of Natural History (AMNH) in New York is the largest of its kind in the world (Preston 1986). It is housed in a modern 10-story building, the Childs Frick Wing, within the museum complex devoted to research and exhibits. Seven of these floors, each approximately 5,000 square feet in size, are devoted to specimen storage (Nicholson et al. 1975). The entire second floor is devoted to the fossil horse collection, thereby making it larger than most other entire collections of fossil vertebrates at many institutions. A conservative estimate of the size of this horse collection is that it contains about 75,000 to 100,000 catalogable specimens. Hundreds of scholarly papers on the collection have been published. Thus, the Frick horses represent the most comprehensive and best-documented collection of fossil Equidae gathered in a single institution in the world. They provide the foundation for our understanding of the systematics, evolution, and paleobiology of the Equidae, particularly as represented in North America. As such, it is appropriate that we examine the development of this collection.

Almost from the beginning of the AMNH, which was founded in 1869, paleontological field expeditions were led by curators and staff. The Warren-Hayden surveys, sponsored by the U.S. government during the 1850s and 1860s, brought back indications of vast paleontological riches in the western territories, and like most natural-history institutions of that time, the AMNH concentrated much of its effort in those regions of North America. The principal emphasis during that time was on the earlier Tertiary sequence, and the collection rapidly grew to international preeminence, although some of the staff paleontologists, notably Gidley, Osborn, and Matthew, also had an interest in the later Tertiary. The acquisition of the E. D. Cope collection in 1895 further strengthened the AMNH holdings, particularly because it contained numerous important type specimens.

In the first two decades of the twentieth century, the wealthy industrialist and philanthropist Childs Frick became interested in the collection and study of fossil mammals. In 1919 he was appointed a trustee of the AMNH, thereby formalizing a connection with that institution that lasted until his death in 1965. He financially supported many field expeditions led by the AMNH to collect fossil mammals, including those of Barnum Brown to the Siwalik foothills of India (now Pakistan) and Samos (Greece), some of Roy Chapman Andrews's expeditions to Mongolia and western North America, and Matthew and Gidley's 1924 fieldwork in the classic Blanco Beds of Texas and the San Pedro Valley of Arizona. To a far lesser degree, Frick also supported field expeditions led by other institutions.

Figure 4.5. *Personnel of the Frick Laboratory, 1947. Back row (left to right): John Ottens, Ove Kaisen, Frank Miller, John Kolwiski, Joe Rooney, ?Harry Scanlon. Front row: Dick Kovelli, Ted Galusha, Morris Skinner, Floyd Blair, Nina Blair, Beryl Taylor, Charlie Falkenbach, Hazel de Berard. Courtesy of Marie Skinner.*

In the 1920s, Frick's consuming interest in fossil mammals continued, and he began to hire paleontologists of his own. They worked independently from the Department of Vertebrate Paleontology in separate quarters within the AMNH (Galusha 1975). Frick recognized that the Cenozoic sequence in North America was both very rich and of fundamental importance to an understanding of fossil mammal systematics and evolution. Given the prior emphasis at the AMNH on the earlier part of the Cenozoic, Frick planned fieldwork to exploit the later Tertiary deposits of North America, which were relatively underrepresented in the collections.

For a half-century the Frick Laboratory was actively engaged in extensive field excavations throughout North America and preparation, curation, and study of the fossil mammals back in New York. In its heyday, the Frick Laboratory consisted of a large staff (Figure 4.5) that at times grew to 30 employees, both in the field and in New York. Although concentrating principally on North America, the Frick collectors also conducted international expeditions, most notably to China in the 1930s.

Prior to the mid-1960s the collections amassed by the Frick

Laboratory had not completely entered the scientific domain (i.e., access was restricted), and some paleontologists of the day were concerned about the stratigraphic documentation that did or did not accompany the collections. It is now clear, however, that the Frick collectors, most notably Ted Galusha and Morris Skinner, took exceedingly precise field notes, and specimens were carefully tied to their individual geographic and stratigraphic locations (Figure 4.6). Field excavations were careful and oftentimes extensive and thorough. The Frick collectors developed their own field techniques, including the use of dynamite and a model-T slip-grading blade to remove overburden from quarries. Although Frick took an active research interest in many fossil groups (e.g., his horned-ruminant monograph, Frick 1937), members on his staff were assigned curatorial responsibility for various groups. Of relevance here, Morris F. Skinner developed a primary interest in fossil horses, and for six decades (since the 1930s) he put much effort into that part of the collection (see Preface and Figure 4.7).

Following Frick's death in 1965, the entire Frick collection was donated to the AMNH along with an endowment to support research on it. With these funds and others from the AMNH, the Childs Frick Wing, a new building devoted entirely to vertebrate paleontology, was completed in 1972. This new facility integrated all the fossil holdings from the Frick and AMNH collections and allowed unrestricted access for research. With regard to the horses, the two collections complemented one another nicely, with the strength of the former being in the earlier Cenozoic, and that of the latter in the later Cenozoic.

The Frick collection has been estimated to encompass a mass of some 600 tons and to contain a quarter of a million specimens (the latter figure seems low). It is impossible to place a dollar amount on the value of the Frick collection. There is no question that it is of immense scientific importance on an international scope. Galusha (1975:15) concludes his history of the Frick Laboratory with the following:

> Private support of research in vertebrate paleontology on a scale provided by Childs Frick will probably never again appear, for he was a scientist who came on the world scene at the right time, with the proper regards for the scientific requirements of his chosen field, and with the personal financial resources to do what needed doing. His legacy will be perceived only dimly by the scientists of the future, but each time paleontologists publish the results of their research on the Frick Collection, their contribution will stand as a monument to Frick's profound regard for natural history and, in particular, to his love for vertebrate paleontology.

Figure 4.6. *Specimens collected by the Frick field personnel are characterized by excellent provenience data. Most specimens had the detailed field locality written directly on the specimen, and these were cross-referenced in detailed stratigraphic sections and field catalogues. This specimen, a right mandibular ramus of* Calippus placidus, *was collected in 1938 from the early Clarendonian of Nebraska under the direction of Morris Skinner and shipped from Ainsworth ("Ains.") in crate #475. "Head of Pole Cr.," "Timm Ranch," and "N.W. Cherry Co. Nebr." all signify locality data, whereas "slightly above Cap Rock" indicates its position above a prominent regionally extensive stratigraphic unit. Later this specimen became F:AM 117005. Photograph by Deborah Brackenbury, Office of Instructional Resources, University of Florida.*

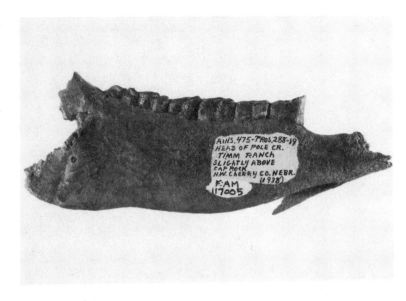

Figure 4.7. *Morris F. Skinner pointing to the position of the unborn, fossilized fetus in the mare specimen (F:AM 60353) of* Protohippus supremus *from the Burge quarry, late Miocene of Nebraska. Photograph taken in 1968 and courtesy of Marie Skinner.*

Other collections in North America

Most vertebrate paleontology collections in North America contain fossil horses, whether they be some of the ubiquitous White River Badlands material, some Pleistocene teeth and metapodials, or more extensive collections of other research potential. In addition to the Frick horses, there are several other collections that are noteworthy because they contain historically important (type) specimens and/or unique samples of regional significance (also see Simpson 1951, Appendix A, "Where to See Fossil Horses").

As mentioned earlier, the Yale Peabody Museum (YPM) is of great historical importance because of the collections assembled by Marsh during the latter part of the nineteenth century and the many types represented there (MacFadden 1977). In addition to an extensive series from the Eocene of the Rocky Mountains, the YPM also has important specimens from the Oligocene–Miocene of the John Day Beds of Oregon and an important but relatively unstudied collection of middle Pleistocene horses from Rock Creek, Texas (Troxell 1915). The Academy of Natural Sciences of Philadelphia, though having a relatively small collection of fossil mammals, is important because many of Leidy's early type specimens of horses are housed there (Gillette & Colbert 1976).

The Natural History Museum at the Smithsonian Institution has a large collection of fossil mammals. Of relevance to the study of fossil horses, its holdings are of particular importance because some early types are deposited there and also because of its collections from the White River Badlands and from the late Miocene at Long Island, Kansas, as well as a spectacular quarry sample of *Equus* from the Pliocene of Idaho (Gazin 1936).

The Florida Museum of Natural History, although a newcomer on the scene, has rapidly amassed an important collection of fossil horses. Particularly large samples have been collected from the early Miocene Thomas Farm and the late Miocene Love Bone Bed, both in northern Florida, as well as the late Miocene–Pliocene Bone Valley deposits of the central Florida phosphate-mining district and many important Pleistocene localities throughout the state. In addition to offering the marine/nonmarine temporal tie-ins, the unique Florida samples also demonstrate southern endemism that seems to have existed along the Gulf Coast savanna corridor (Webb 1974). The Texas Memorial Museum also has good samples from the Miocene Gulf Coast deposits (Quinn 1955), a significant collection from the Miocene panhandle of Texas, and large collections of Pleistocene *Equus*.

The Nebraska State Museum has exceedingly large holdings of fossil horses, and those from the late Cenozoic are second only to the Frick collection in potential importance. The emphasis in this collection is, not surprisingly, on the Great Plains, particularly Nebraska, and like the Frick, it is of great importance for an understanding of the major adaptive radiation of Miocene horses. The Nebraska collection is also noteworthy because it contains many important quarry samples featuring extraordinary preservation, in terms of both quantity of specimens and quality of morphological representation, as discussed later.

In the western United States, two important collections of fossil horses are located in California. The Natural History Museum of Los Angeles County (LACM) has a large collection, with particularly important holdings from the Miocene of the Mojave Desert and the Pliocene of Mexico (previously part of the California Institute of Technology collections, later sold to the LACM) (Lance 1950), and an extensive sample of *Equus* from the La Brea Tar Pits. The Museum of Paleontology at Berkeley (UCMP) is of great importance both because of the paleontological representation in the collections and because of the research emphasis on fossil horses in that collection by two of this century's leading scholars in the field: William Diller Matthew and Ruben Arthur Stirton, both professors there in the Department of Paleontology. In addition to overlapping with some of the holdings mentioned earlier from the LACM, the UCMP has important samples of fossil horses from the panhandle of Texas (Matthew & Stirton 1930), the northern Great Plains (e.g., Gregory 1942; Webb 1969a), and several localities in northern California, particularly Black Hawk Ranch (Richey 1948). The University of Oregon Condon Museum contains some important early types and other specimens from the John Day Beds in eastern Oregon described by Marsh (Gray 1978).

There are many other smaller research collections that contain important samples of fossil horses. These include the Museum of Comparative Zoology (Harvard University), the Carnegie Museum and Field Museum, the University of Michigan and University of Wyoming collections, and, for regional samples from the Great Plains sequence, the collections of the University of Kansas, the University of Oklahoma, and the Panhandle Plains Museum (Figure 4.8).

Important collections of fossil horses worldwide

Although the North American stratigraphic sequences and museum collections of fossil horses from this continent are unparalleled elsewhere in the world, there are certain other collections that should be mentioned in this context, of which the most important are as follows: The British Museum of Natural History, London, and the Natural History Museum, Paris, have classic collections, most notably of Eocene horses and palaeotheres, Miocene *Hipparion*, and Pliocene–Pleistocene *Equus*. Some smaller collections in Germany (Karlsruhe and Darmstadt) have excellent samples of more than a dozen articulated skeletons of primitive *"Hipparion" primigenium* from the middle Miocene quarry site at Höwenegg (Figure 4.9). The natural-history

Figure 4.8. *C. Stuart and Margaret Johnston at Gidley's* Equus scotti *quarry in the Texas panhandle. During the 1930s the Johnstons were very active in excavating fossil mammal sites from the rich late Cenozoic sequence of the Texas panhandle, often with student and Works Progress Administration assistants. C. Stuart Johnston was an inspiring teacher who published several important papers on fossil horses. After his death in 1939, his widow continued some of the fossil excavations. In addition to the better-known paleontologists working at the longer-established universities and museums, unsung heroes and heroines like the Johnstons have done much to advance the growth of research collections and consequently an understanding of horse evolution. Photograph courtesy of the Panhandle-Plains Historical Society, Canyon, Texas.*

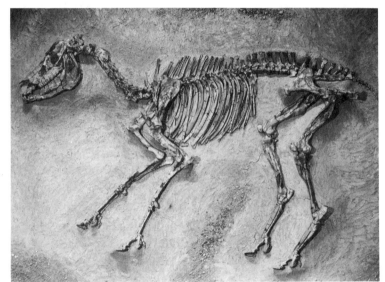

Figure 4.9. *Skeleton of middle Miocene "Hipparion" primigenium from the Höwenegg quarry, Germany, on exhibit at the Karlsruhe Museum. Photograph by Mr. H. Heckel and provided courtesy of Drs. Rietschel and Mittman.*

museums in Leiden and Utrecht have important collections of Neogene horses from Africa and western Europe. The Paleontological Institute in Uppsala, Sweden, has equid collections from the Pliocene–Pleistocene of South America and the Neogene of China, as described in classic monographs (Sefve 1912, 1927). Neogene horses from Russia and adjacent regions are contained in several paleontological institutes and museums, most notably in Moscow, and have been described in classic monographs (e.g., Gromova 1952). Some of the first reported collections containing Neogene horses from the Siwaliks are housed in the Geological Survey of India in Calcutta. Although Neogene mammals are kept in many smaller provincial museums within the People's Republic of China, by far the largest and best-studied (e.g., Qiu, Huang, & Guo 1987) collection of fossil horses from that country is housed in the Institute of Vertebrate Paleontology and Paleoanthropology in Beijing. The South African Museum in Cape Town has large collections of Neogene horses (e.g., *"Hipparion"*) (Boné & Singer 1965) from southern Africa. In South America, Pliocene–Pleistocene horses from that continent are housed principally in Buenos Aires and La Plata. As in the case of the North American collections, this brief survey is intended to focus on the larger, principal collections containing fossil horses that are known in the literature. It is also recognized that there are scores of other museums worldwide with smaller, less well known collections of fossil horses with important research potential.

Fossil horses, exceptional discoveries, and *Lagerstätten:* six case studies

Quarry samples adequate for the study of . . . population characteristics in fossils will always be rare. The effort involved in collecting these and preserving all necessary information is great. However, when carefully collected and studied, large quarry samples promise to contribute disproportionate insight into the paleobiology of species.
— Gingerich (1981:453)

All fossil remains are of scientific value, whether they be foraminifera, a pollen grain, a shell, or bones and teeth, and whether they be fragmentary or in an excellent state of preservation. To a classical paleontologist, some specimens may be of greater importance for morphology or biostratigraphy; yet others may contribute to important statistical samples or be of importance for isotopic studies or some analytical technique not yet developed. Most paleontologists, however, if asked to name some truly extraordinary discoveries, would mention localities and finds such as the Vendian Ediacara fauna of Australia (Glaessner 1984), the Cambrian Burgess Shale of Canada (Whittington 1985; Gould 1989b), the Pennsylvanian Mazon Creek of Illinois (Nitecki 1979), the Jurassic Holzmaden (Hauff & Hauff 1981) and Solnhofen lithographic limestone of Germany (e.g., Viohl 1985), the Cretaceous dinosaur "mummies" from western North America (Osborn 1912a; Lull & Wright 1942), the Eocene Grube Messel in western Germany, and the late Pleistocene La Brea Tar Pits of California (e.g., Harris & Jefferson 1985). Many of these are characterized by extraordinary preservation of both hard and soft tissues, so that, for example, the impressions of *Archaeopteryx* feathers and pterodactyl hair from Solnhofen are preserved in exquisite detail, and the stomach contents of Eocene mammals from Messel are fossilized. Localities such as these, featuring both extraordinary preservation and abundance, have been termed *Konservat-Lagerstät* or *Lagerstätten* (Seilacher 1970; Seilacher, Reif, & Westphal 1985; Allison 1988), the latter being "mother lode" in a liberal translation (Gould 1989b). The scientific value of these localities as unique opportunities to better understand the paleobiology of past life is extraordinary. Gould (1989b:61) says that "*Lagerstätten* are rare, but their contribution to our knowledge of life's history is disproportionate to their frequency by orders of magnitude."

In this section we shall examine six truly extraordinary fossil horse localities from diverse ages and areas. We shall depart from a generally North American emphasis to include a description of a horselike palaeothere from Messel, Germany. Of these, perhaps only Messel and the Ashfall Fossil Beds in Nebraska qualify as *Lagerstätten;* yet the others represent exceedingly rich and important quarry samples that allow unique glimpses into the paleobiology of extinct horses. Some of the fossils collected from these sites will also be discussed as they are relevant to the topics of later chapters, but this account introduces their importance within the context of exceptional discoveries.

There are, in the simplest view, two different kinds of fossil quarry samples, namely, those due to catastrophic or attritional accumulations (see Figure 12.8). Each of these has its unique characteristics, and consequently different paleobiological questions can be asked about them. A catastrophic assemblage presumably is a sample of the herd or population of species caught in the midst of some hazardous event in earth history, such as a flood, avalanche, mudslide, volcanic eruption, or perhaps even the impact of an asteroid. Therefore, these samples are believed to have accumulated in an instant of geological time, usually within hours to days. From these kinds of assemblages, patterns of age structure and resulting population dynamics can be inferred for the species preserved. In contrast, an attritional assemblage frequently results from predator accumulation and the consequent mortality of prey species, and therefore it tends to preserve individuals of certain ages, rather than representing all ages. In the prevailing view, this kind of assemblage usually will favor the relatively young and old individuals, which presumably would have been selected more frequently by predators. In most cases these also are thought to have accumulated over short periods of geological time, although in some cases (e.g., Pleistocene cave bears) the temporal resolution is sufficient to indicate that they were accumulated over many tens of thousands of years. Thus, with attritional samples, other questions, themselves no less interesting, can be asked about such paleobiological phenomena as predator–prey dynamics and natural selection within populations. In the six case studies that follow there are examples of both kinds of accumulations.

Castillo Pocket, early Eocene of Colorado

The rich and widespread sequences of early Eocene (Wasatchian) intermontane basin deposits in the Rocky Mountain region of the western United States have provided extensive collections of the earliest horse, *Hyracotherium* (Figure 4.4). Most of these localities, however, have been surface discoveries, with specimens usually consisting of isolated teeth, jaw and skull fragments containing molar series, or dissociated postcranial remains. In 1952 a quarry was discovered in the Wasatchian Huerfano Basin of Colorado that contained extraordinary and unique samples of two species of *Hyracotherium*. Castillo Pocket, as it is now called, was discovered during the 1952–4 field seasons by George Simpson, George Whittaker (who single-handedly excavated a one-ton, bone-bearing quarry block, encased it in plaster, and drove it back to New York City), and assistants (Figure 4.10). Although characteristic Wasatchian mammals were collected from this site, the large majority of specimens are of *Hyracotherium*, represented by a minimum of 32 individuals. This quarry sample therefore allows a unique window into the paleobiology of the oldest horse. After its collection it remained largely unprepared and undescribed, except for a preliminary description by Kitts (1956), until an elegant study by Gingerich (1981), on which the remainder of this section is based.

The sample seems to be a catastrophic assemblage consisting of two species of *Hyracotherium*. The less common and smaller species, *H. vasacciense*, is represented by eight individuals, and unfortunately that is too small a sample for an in-depth study of population dynamics. The more common and larger species, *H. tapirinum*, is represented by the remaining 24 individuals, a sample large enough to allow interesting questions about the morphological variation, population dynamics, and inferred social structure of the earliest equid genus.

Metric analyses have shown that *H. tapirinum* was highly dimorphic, with the males being about 15% larger in cranial dimensions and 40% larger in canine dimensions. Assuming no preservational biases, the sex ratio seems to be about 1 male to about 1.5 to 2 females. By modern analogy, *H. tapirinum* may have had a polygynous breeding system, namely, one male with a harem of a few or several females. The large male canine suggests that males were territorial, were engaged in intraspecific competition for mates, and used this tooth for both display

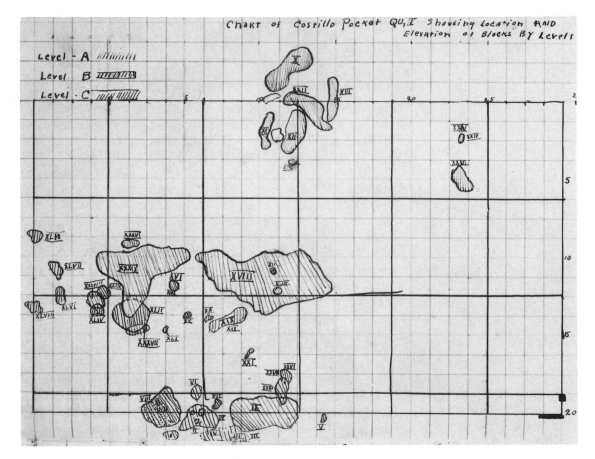

Figure 4.10. *Sketch map showing blocks and levels of* Hyracotherium *specimens collected during 1952–4 from Castillo Pocket, early Eocene (Wasatch- ian) of the Huerfano Basin, Colorado. (Whittaker, unpublished field notes in archives, file number 6:5:3, at the De- partment of Vertebrate Paleontology, AMNH). Reproduced courtesy of the Department of Vertebrate Paleontology, AMNH.*

and agonistic behavior (i.e., fighting with other males). Asso- ciated paleobotanical evidence suggests a more open environ- ment than the traditional notion of widespread dense forests during the early Tertiary. Therefore, judging from these data and from modern mammalian analogues, *H. tapirinum* may have occupied open-park woodland habitats, undergone seasonal mi- grations, and fed on herbaceous dicots. These inferences are in contrast to the traditional view that primitive horses were forest- dwelling browsers and fed predominantly on the leafy vegeta- tion of woody plant species.

The importance of the Castillo Pocket sample of *Hyracotherium* cannot be overemphasized. It is unique among all Eocene equid sites in North America, and it gives us an extraordinary oppor- tunity to interpret the population biology of an extinct species at the base of the Equidae (see Chapter 12).

Grube Messel, middle Eocene of Germany

The Grube Messel, located some 30 km southeast of Frankfurt am Main, contains one of the most spectacular Tertiary fossil vertebrate localities in the world. This site, which for many years was scheduled to become a garbage dump, has been worked intensively since the mid-1970s by the Forschunginstitut Senkenberg and the Hessisches Landesmuseum Darmstadt, under the direction of, respectively, Jens L. Franzen and Wighart von Koenigswald (Franzen 1976, 1977, 1984, 1985).

The sediments from Grube Messel consist predominantly of middle Eocene oil shales, representing an ancient shallow lake only a few square kilometers in area and probably about 10 m deep. Based on oxygen isotope data, the climate during that time is interpreted to have been subtropical, with a mean annual temperature of 20°C. Although there seems to have been a connection to a free-flowing river (as interpreted from the presence of gars and crocodilians), the lake seems to have been relatively stagnant, and those conditions undoubtedly contributed to the excellent preservation of the vertebrate carcasses, as well as insects (Franzen 1985).

The most common vertebrates encountered are birds and bats, with each represented by more than 150 individuals. For most of the fossils discovered there, the quality of preservation is extraordinary, with the outlines of anatomical details enhanced by fossilized bacteria. Thus, there are complete skeletons with even the impressions and outlines of soft tissues, including eyes in anurans, bird feathers, hair, and, in some taxa, stomach contents. The bats are of particular importance because, like birds, their delicate bones are so rarely fossilized. Details of the ear region indicate that by the middle Eocene, microchiropterans had the ability to echolocate (Novacek 1985). The stomach contents of the commonest species, *Palaeochiropteryx tupaiodon*, indicate that it fed exclusively on butterflies (Franzen 1985).

The discoveries of the Messel palaeotheres are equally spectacular (Figure 4.11). Palaeotheres (family Palaeotheriidae) are an extinct group of Eocene–Oligocene perissodactyls that share a common ancestry with the family Equidae, and the two taxa are sometimes combined into a superfamily Equoidea (see Chapter 5). Virtually complete individuals of *Propalaeotherium parvulum* have been found with the impressions of fossilized stomach contents, including leaves, grape pits, and phosphatized bacteria (Koenigswald & Schaarschmidt 1983; Schmitz-Münker & Franzen 1988), thus suggesting a mixed-feeding diet, as was suggested by Gingerich (1981) for *Hyracotherium*. This discovery is of great interest because it tests some century-old hypotheses about the dietary adaptations of the earliest horses and their close relatives. Kowalevsky (1873) believed that palaeotheres were omnivorous and fed on a variety of browse, other herbs, and possibly fruits. On the other hand, based on the low-crowned teeth of primitive horses, paleontologists in North America (e.g., Matthew 1926) believed that they were predominantly browsers, with more restricted diets. The palaeothere specimens from Messel thereby corroborate the hypothesis of Kowalevsky (1873).

A particularly interesting skeleton of *Propalaeotherium* discovered by Franzen (1985:183) "included some unexpected bones when it was excavated by our team in 1980. My first idea was that the very fragmentary and tiny supernumerary bones were those of an embryo . . . a series of isolated cusps of lower deciduous molars allowed the identification of the skeleton as that of a pregnant female."

In summary, the Messel site has truly extraordinary preservation and is worthy of status as *Lagerstätten*. This site also represents a critical time in the evolution of early perissodactyls, from which we can gain greater understanding of the paleobiology of palaeotheres and probably also their closest relatives, the Equidae.

Thomson quarry, early Miocene of Nebraska

During the early part of this century, the AMNH established an active field program in western Nebraska, particularly in the central and northern parts of the panhandle within Sioux County (Figure 4.4). Many fossil mammals were collected there around the turn of the twentieth century, including specimens from the spectacular Agate Springs quarry. In addition to recovering faunal samples, a specific objective of that fieldwork was to obtain more specimens of *Hesperopithecus* (literally "western ape"), which was collected from Olcott Hill and originally described, based on a single very worn tooth, as a hominid fossil (Osborn 1922). Subsequent work revealed that *Hesperopithecus* represented the remains of a Miocene peccary, *Prosthennops*. Regardless of that folly (Gould 1989a), the discoveries from this region have greatly enhanced our understanding of Miocene and Pliocene mammals from the middle of North America, as described in classic papers (e.g., Matthew 1924). During the early 1920s, an important fossil site was discovered and excavated at Stonehouse Draw in Sioux County. It has since been termed Thomson quarry, named after an AMNH collector, Albert Thomson, and rightfully deserves a place of distinction in discussions of the fossil record of horses in North America.

Thomson quarry was originally worked in 1922–3 by the AMNH (Figure 4.12); it was reopened in the 1930s and 1940s by crews from the Frick Laboratory. Because of its seemingly inexhaustible supply of fossils, collectors have referred to Thomson

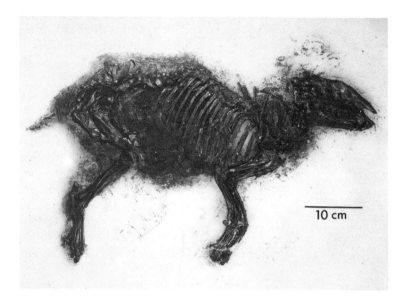

Figure 4.11. *Fossilized skeleton and carbonized impression of soft anatomy of* Propalaeotherium parvulum *from the middle Eocene Messel locality in Germany. The preservation is so extraordinary that in some specimens the fossilized stomach contents contain grape pits, indicting part of the diet of this palaeothere. This specimen, SMF-ME 1285, was collected in 1982 and is housed in the Senkenberg Museum, Frankfurt. Photograph courtesy of Senkenberg Museum.*

Figure 4.12. *Early quarry excavation at Stonehouse Draw, Sioux County, Nebraska. This was later expanded by the Frick field crews, and the productive horizon that was excavated, which included Thomson quarry, extended some 200 m laterally. Photograph from the Department of Vertebrate Paleontology archives, AMNH.*

quarry as "Old Faithful" (Skinner, Skinner, & Gooris 1977). The quarry excavations were extensive, sometimes extending laterally for 200 m. Although at various times many specimens were sent to other museums and even public schools (Van Valen 1964), most of the collection currently is in the AMNH. The assemblage is dominated by a single species, *Merychippus primus*, a three-toed horse that phylogenetically lies near the base of the principal adaptive radiation of Miocene grazing equids in North America (MacFadden & Hulbert 1988; Hulbert & MacFadden

1991). The collection in the AMNH contains about 125 skulls, and judging on the basis of teeth it is estimated to represent a minimum of 150 individuals. The state of preservation of the fossils is generally very good to excellent, and although the postcranial elements are disarticulated, there was little postdepositional sorting, and therefore most of the bones normally preserved as fossils are represented in the collection. As has been recognized in several studies (e.g., Matthew 1924; Kurtén 1953), the *Merychippus primus* sample from Thomson quarry is important for at least three reasons: (1) It is statistically one of the largest samples in existence for a Tertiary horse species that preserves dentitions, skulls, and postcranial elements. (2) It represents a species at a critical time in horse evolution. (3) Some fascinating, pioneering paleobiological work has been done on this sample, as will be described later.

In the early 1960s, Leigh Van Valen, then a Columbia graduate student, was interested in general evolutionary questions about selection and comparisons of that phenomenon in such disparate populations as the extant fruit fly *Drosophila* (the mainstay of genetic research) and fossil horses. To investigate the latter, he chose the *Merychippus primus* sample from Thomson quarry (his other sample was of *Pseudhipparion gratum*, from Burge quarry, another superb site in Nebraska, farther to the east). In addition to describing his investigation of the strength of selection and associated phenomena in *M. primus*, Van Valen's classic articles (1963, 1964, 1965) give us a fascinating interpretation of the general paleobiology of this fossil horse. Van Valen interpreted the Thomson quarry accumulation to be a representative sample of the once-living herd that probably approximated a population in the modern sense (MacFadden 1989) and therefore was not biased toward certain age classes as would be the case in predator–prey accumulations. He found discrete age classes (based on degree of tooth wear), implying annual cohorts and seasonal (rather than continuous) breeding cycles. The youngest recognizable cohorts were the newborns (which are not preserved in most fossil sites because of their delicate, unfused bones), estimated to be one month old, and the oldest, just over 10 years old. Van Valen asserted that if ontogenetic and geographic variations in dental characters could be accounted for, then the remaining variation would be related to the intensity of selection. [That idea had previously been proposed by the prominent British population biologist Haldane (e.g., 1954) and had been tested, prior to much of Haldane's work, by Kurtén (1953) in a pioneering study of fossil mammal populations from the Old World.] Thus, the amounts of variation in dental characters among the cohorts of a given population (as in *M. primus*) could provide a measure of the relative intensities of selection (also see Marcus 1964). As might be expected from general theory, different age classes experienced different selection intensities, with higher intensities particularly in the young cohorts. From these results, mortality was interpreted to be 30–40% for juveniles, dropping to 10% in middle age, and finally increasing slightly during old age. The maximum potential longevity for a given individual under optimal conditions was about 11 years, compared with 20–25 years for *Equus*. Based on those parameters, and with an estimated onset of reproductive maturity at about two years, Van Valen speculated that any given female might have had three progeny that would have survived past the first year of life, sufficient to maintain a stable population structure.

Van Valen concluded that there were discernible differences in selection within the cohorts of *M. primus*. However, relative to other taxa, he stated that (Van Valen 1964:106) "weak natural selection is adequate to account for the most rapid known evolutionary change in the Equidae." *Merychippus primus* is near the base of the adaptive radiation of Miocene grazing horses. Despite explosive speciation at that time, only weak natural selection occurred, and that resulted in relatively small amounts of observed changes in dental morphology (MacFadden & Hulbert 1988).

Ashfall Fossil Beds, late Miocene of Nebraska

As mentioned earlier, the Cenozoic sediments of the Great Plains of North America have yielded extraordinarily rich sequences of terrestrial mammals, including, during the late Miocene, a diversity of taxa rivaling that of the present-day African savannas (Webb 1977). In the late 1970s, Michael Voorhies, curator of vertebrate paleontology at the Nebraska State Museum, found a truly extraordinary quarry in the 10-myr-old Ash Hollow Formation in northeastern Nebraska ("E NB," Figure 4.4). This locality, originally known as Poison Ivy quarry, now Ashfall Fossil Beds State Historical Park, is probably the only pre-Pleistocene terrestrial-mammal-bearing locality currently known in North America that can truly be called a *Lagerstätten*. Complete bird skeletons are preserved, along with feather impressions, tendons, and cartilage. Among the mammals, the delicate hyoid and internal ear bones are preserved. Foodstuffs preserved in the teeth and the presumed digestive tract of the dominant herbivore, the rhinoceros *Teleoceras major*, indicate a diet predominantly of grasses (Voorhies & Thomasson 1979; Voorhies 1981). The rapid burial seems to indicate a catastrophic death assemblage that sampled a cross section of the resident species populations and coexisting communities. To date, over 40 tons of plaster jackets have been collected, including 200 skeletons of virtually complete individuals (Figure 4.13).

Figure 4.13. *Excavation of articulated skeletons of Miocene rhinoceros from Poison Ivy quarry, now Ashfall Fossil Beds State Historical Park, northeastern Nebraska. Reproduced with permission of Annie Griffiths Belt.*

The deposit seems to represent a small pond where mammals, birds, and reptiles were drawn to water. Above the pond and associated fluvial deposits is a 2-m-thick bed of relatively pure volcanic ash. Although the ashfall was not instantaneous by modern standards, it accumulated over a short geological interval and in so doing probably was responsible for the death of the vertebrate fauna. Voorhies (1981), however, speculates that the ash was not a hot gaseous cloud causing instantaneous death (as was experienced by the inhabitants of Pompeii); rather, the siliceous dust probably induced mortality over a period of weeks or months. Nevertheless, the relatively rapid (by most paleontological standards) burial, followed by virtually no postdepositional disturbance, resulted in the extraordinary state of preservation.

With regard to the mammals, the traditional notion has been that horses were dominant during that time in North America. Judging from the sample at the Ashfall site, however, at least for what is now the northern Great Plains, the dominant ungulate (in biomass) during the late Miocene was *Teleoceras*. The population dynamics suggest a behavioral ecology rather different from that for modern rhinoceroses in Africa and Asia. In contrast to the situation of modern rhinos, the sex ratio of about one male to six females preserved at Ashfall suggests polygynous mating, and the presence of discrete tooth-wear classes indicates synchronized breeding cycles. The presence of fossil grass anthoecia (fertile lemmas and paleas) in the hyoid and stomach regions of *Teleoceras* confirms previous hypotheses that the high-crowned cheek teeth of *Teleoceras* indicate a predominantly grazing diet. The nearby presence of cold-intolerant land tortoises and alligators suggests an ancient climate for the midcontinent that was more equable than today (Voorhies 1981).

Most of the research to date on the fauna from the Ashfall site has concentrated on the large sample of *Teleoceras*. In addition, the potential for insight regarding the horses in that fauna is extraordinary. Several equid species are represented there, including *Pseudhipparion gratum, Cormohipparion occidentale, Protohippus supremus,* and *Astrohippus* sp. (Voorhies & Thomasson 1979; M. Voorhies personal communication 1991). Two more, probably belonging to *Neohipparion* and *Dinohippus*, may also be represented there. Given the quality of preservation, morphologically it represents one of the most important assemblages of Tertiary horses in existence anywhere and from any time period. The Ashfall site is rapidly becoming a classic locality in vertebrate paleontology and will receive increasing scientific notoriety as studies of the various groups are presented in the future.

Hagerman Horse quarries, Pliocene of Idaho

In 1928, Elmer Cook, a resident of Hagerman, Idaho, and an amateur fossil collector, discovered the remains of late Tertiary fossil horses in some hilly slopes near that town. Those specimens were brought to the attention of the U.S. National Museum. Starting in 1929 and continuing into the mid-1930s, paleontologists from that institution conducted excavations at the now-famous Hagerman Horse quarries (Figures 4.4 and 4.14)

4.14) under the direction of J. W. Gidley and after his death, C. Lewis Gazin.

Those excavations yielded more than 130 skulls, numerous associated jaws, and about 8–10 skeletons, all representing a single species, *Equus simplicidens* (Winans 1985, 1989), called *E. shoshonensis* by some other authors. As had been suspected by some other paleontologists, Skinner (1972) stated that the Hagerman *Equus* may be closely related to the extant African zebra, *Equus (Dolichohippus) grevyi*, thus suggesting further complexity in the biogeographic history of this genus. The quality of preservation is excellent, with even the delicate hyoid bones remaining, although the discovery is noteworthy more for the completeness and abundance of the specimens than for exquisite fossilization as seen at Messel or Ashfall Fossil Beds.

The associated sediments and vertebrate fauna suggest a mode of accumulation around a small pond or watering hole. It is unclear whether this was a catastrophic, rapid assemblage or was a sample formed over a longer period of geological time (Gazin 1936). Paleomagnetic stratigraphy and the associated fossil mammals indicate a Blancan age for this locality, about 3.5 Ma (Neville et al. 1979; Lundelius et al. 1987).

The significance of the Hagerman *Equus* sample is extraordinary. It is the earliest well-represented sample of the extant genus and therefore facilitates an understanding of the beginnings of the modern adaptive radiation. Based on these specimens, *Equus simplicidens* is estimated to have been about the size of a modern Arabian horse (Gazin 1936) and to have had a mean body mass of about 425 kg (MacFadden 1987a). In many ways that species was morphologically similar to modern *Equus*. Studies of dental and some cranial characters of that ancient species show that the amounts of variation (with coefficients of variation, or *V*s, less than 10) seen in *Equus simplicidens* from Hagerman were similar in magnitude to the same characters in extant samples of *Equus* (MacFadden 1989). Recent studies of the functional morphology of the upper forelimb (Hermanson & MacFadden 1992) suggest that within the Equidae the biomechanical staying mechanism, which conserves energy during standing, was first well developed in Pliocene *Equus* such as *E. simplicidens*. In summary, the collections from the Hagerman Horse quarries are important because they provide a rare opportunity to study an early representative of the extant genus *Equus*.

La Brea Tar Pits, late Pleistocene of California

The La Brea Tar Pits, located within the city of Los Angeles (Figure 4.4), unquestionably constitute one of the largest and most abundant fossil localities ever excavated. Almost a century of collecting there has yielded about 565 species of plants and animals preserved in the tar, including well over a million fossil vertebrate specimens.

As early as 1769 the Spanish explorer Gaspar Portola was aware of the extensive tar seeps. During the nineteenth century some of the tar was recovered and sold to make road asphalt. In the latter part of the nineteenth century the wealthy businessman and philanthropist G. Alan Hancock recognized the scientific importance of Rancho La Brea, and in 1913 he gave the county of Los Angeles 23 acres encompassing that locality, which was later to become Hancock Park. Extensive excavations began soon thereafter, and under the leadership of J. C. Merriam in 1913 alone more than three-quarters of a million bones were excavated (Figure 4.15). Since that time, field crews from numerous institutions have worked La Brea, including the University of California (both Los Angeles and Berkeley), the University of Southern California, the Southern California Academy of Sciences, and the Natural History Museum of Los Angeles County. Many prominent twentieth-century paleontologists have studied specimens from La Brea, and extensive, detailed studies by paleontologists such as J.C. Merriam and Chester Stock (e.g., Stock 1925; Merriam & Stock 1932) stand out as notable contributions to our understanding of late Pleistocene fossil mammals. The biochronology of fossil mammals from this locality has given rise to the term "Rancholabrean" for the youngest North American land-mammal age (Savage 1951; Lundelius et al. 1987).

Although the skeletons usually have been found disarticulated, the quality of preservation of the vertebrate remains from La Brea is otherwise extraordinary. The sheer volume of specimens has resulted in statistical samples of extinct Pleistocene mammals second to none, probably approached only by some of the European cave faunas. Radiocarbon (^{14}C) age determinations indicate that all of the fossils are of late Wisconsinian or Holocene age, spanning an interval from 40,000 to 4,000 years ago.

Unlike most mammal faunas, 90% of the specimens represent carnivores, of which the four most commonly found (in order of abundance) are the dire wolf (*Canis dirus*), saber-toothed cat (*Smilodon fatalis*), coyote (*Canis latrans*), and American lion (*Panthera atrox*). The currently accepted explanation for this overabundance of carnivores (which normally are about 10% in natural mammalian communities) is that these predators were attracted to prey species that had become mired in the tar seeps.

The La Brea horse fossils feature an overwhelming abundance of a single species, *Equus laurentius* [following Winans (1985, 1989); *Equus occidentalis* of earlier authors], although Scott (1990) recently reported the presence of a second species, *E. "conversidens"* [probably equals *E. alaskae, sensu* Winans (1985, 1989)].

Figure 4.14. *Horses being used to collect fossil* Equus *at the Hagerman excavations, Pliocene (Blancan) of Idaho (Gazin 1936, plate 23.2). Note* in situ *bones in the right foreground. Photograph courtesy of Department of Paleobiology, NMNH, Smithsonian Institution, photograph no. 33668D.*

Figure 4.15. *Specimens of Pleistocene mammals from the Rancho La Brea excavations of 1913–15 being sorted in the basement of the Natural History Museum of Los Angeles County. Photograph courtesy of the George C. Page Museum of La Brea Discoveries, Natural History Museum of Los Angeles County.*

E. laurentius averaged a height of 14.5 hands at the shoulder (about 1.5 m) and was generally about the size of a modern Arabian, although of slightly stouter proportions (Figure 4.16). Like the other mammals from La Brea, the fossil horses represent one of the largest statistical samples for any extinct species of the Equidae (Harris & Jefferson 1985).

The scientific potential of the La Brea Tar Pits continues to be extraordinary. This site has yielded enormous samples of a rel-

Figure 4.16. *Restoration of* Equus laurentius *(= E. occidentalis of earlier authors), the late Pleistocene horse from Rancho La Brea. From Harris and Jefferson (1985); reproduced with permission of the George C. Page Museum of La Brea Discoveries, Natural History Museum of Los Angeles County.*

atively complete community (including plants, invertebrates, and vertebrates). This gives paleontologists much greater potential to understand late Pleistocene paleobiology than otherwise would have been possible with other, less well preserved sites.

Epilogue: conservation of threatened sites

In addition to the extraordinary samples of fossil horses cited here, there are many more important sites far too numerous to mention, but their significance will become clear in subsequent chapters as they contribute to an understanding of the systematics, evolution, and paleobiology of the Equidae.

Many fossil sites, regardless of their current relative importance, are threatened by the spread of human activities. Two of those described earlier have recently been slated to be reclaimed in the interest of human "progress." During the 1980s, Messel was scheduled to become a garbage dump for the town of Darmstadt, and the U.S. Army Corps of Engineers had planned to build an irrigation canal through Hagerman. Fortunately, the status of both of these important sites is currently secure for the posterity of science as a result of concerted efforts to convince government officials of the need for their conservation. Similarly, the Ashfall site has recently been established as a state park in Nebraska. The La Brea Tar Pits provide an excellent example of how a unique, important fossil site should be conserved. Located in the middle of urban Los Angeles, it easily could have been completely covered over by parking lots, new buildings, or subway tunnels. Rather, the beautiful George C.

Page Museum, located in Hancock Park, was opened to the public in 1977. This exceptional museum has modern exhibits devoted to fossils from this site, and excavations continue at the nearby tar pits. As such, it is a paleontological success story that will conserve the site and educate the public about its immense scientific importance. It is to be hoped that in the future more threatened fossil sites will be conserved as their research potential becomes known and public awareness of local paleontological resources is increased.

5

Systematics and phylogeny: Ungulata, Perissodactyla, and Equidae

Because of their great popularity with humans and their widespread abundance, it is not surprising that horses have been of interest to natural historians since before the beginning of recorded history. Aristotle understood that there were several different kinds of horses. He stated (Gregory 1910:10) that "there is a genus of animals that have manes, as the horse, the ass, the *oreus*, the *ginnus*, the *innus*, and the animal which in Syria is called *heminus* (mule)." *Equus* was included in the original classification of Linnaeus in 1735, which contained only 33 genera of mammals (Gregory 1910).

Cuvier (Figure 5.1) named the first fossil horselike (equoid)[1] genus *Palaeotherium* in 1804. The family Equidae was proposed by Gray in 1821, in which he included the genus *Equus*, but no fossil forms; he was the first to standardize family endings with -*idae* (Gregory 1910). In 1840 Owen described the first fossil horse genus (*sensu stricto*): *Hyracotherium*, from the early Eocene London Clay. In 1847 Dr. Hiram Prout, a St. Louis physician, described the remains of a "Palaeotherial bone" that had been collected by a friend from a location east of the Black Hills, in a region corresponding to the now classic Oligocene Badlands (*mauvaises terres* of the early French traders) of present-day South Dakota (Prout 1847). A few years later, Leidy (1851) assigned the scientific name *Palaeotherium proutii* to this specimen, although it later was realized to be a titanothere, not an equoid (Lanham 1973). In 1854 Leidy (Figure 5.1) described the first fossil horse genus from North America as *Hippodon*, a name that has not been widely used in the literature, and in 1857 he described *Merychippus*, a mainstay of fossil horse phylogeny since then. The specimens used for those descriptions probably came from middle Miocene deposits in eastern South Dakota (Skinner

[1]Horses (family Equidae) and their closest relatives, the extinct palaeotheres (family Palaeotheriidae), usually are united into the superfamily Equoidea, and hence the adjective "equoid."

Figure 5.1. *Left: Baron Georges Cuvier (1769–1832) is generally considered to have been the founder of the science of comparative anatomy. From Cuvier (1836). He also described important fossils such as Palaeotherium (Figure 5.2) from the Paris Basin and surrounding regions. Right: Joseph Leidy (1823–91) is generally considered to have been the founder of the science of vertebrate paleontology in North America. Leidy published many important papers on fossil horses. Modified from a rare photograph in Colbert (1968); reproduced with permission of E. P. Dutton & Co.*

& Taylor 1967). As that work progressed, fossil horse systematics became increasingly more complex, because two principal groups of paleontologists, separated by the Atlantic Ocean, were simultaneously working on different collections to formulate their ideas. It is clear from their writings that even during the middle of the nineteenth century each school early on knew what the other was doing, yet most workers seem to have believed that they were working on separate taxa because of the vast biogeographic separation; few accepted the possibility of Cenozoic intercontinental dispersal of mammals. A notable exception to that was Leidy's implicit belief that *Palaeotherium* was represented in the New World (although he did not mention the significance of that in his 1851 description), and Kowalevsky (1873) realized that *Anchitherium* occurred in both the Old World and the New World. However, that state of isolation was to change. For example, Huxley (see Chapter 3), so influential in his day, did much to advance transatlantic scientific awareness and communication after his historic trip to the United States in 1876. Soon thereafter, most paleontologists from both schools accepted the idea that the principal adaptive radiation of post–early Eocene fossil horses occurred in North America, with parallel evolution of the Palaeotheriidae in Europe. Thereafter in the Old World the presence of fossil horses indicates immigration from the New World.

The purpose of this chapter is to present the systematics of fossil horses and place this group in its context with other perissodactyls and ungulates. After a discussion of the history of development of systematic thought, I shall examine the phylogenetic position of the pivotal Eocene genus *Hyracotherium* and then present cladistic analysis of the Equidae. Because this chapter is already long and detailed, I have chosen not to include an introduction to cladistics here because I assume that most readers are conversant in its methodology or that they can refer to one of the many detailed works on this subject published elsewhere (e.g., Wiley 1981; Schoch 1986).

History of development of systematic thought: Ungulata and Perissodactyla

The orders of mammals have probably been fairly well determined in the majority of instances. It is not likely, for example, that the Perissodactyla will ever be split up into several orders. . . . The superorder, on the contrary, as here used, stands only for an hypothesis of common origin. . . . The history of classification warns us against taking superordinal groupings too seriously.

– Gregory (1910:462)

Because of the importance of the horse in human culture, knowledge of the evolution of thought with regard to *Equus* and its relatives is fundamental to an understanding of the development of the order Perissodactyla, as well as the interrelationships with other ungulate mammals (e.g., Schoch 1989). In 1766 Linnaeus[2] proposed the Ungulata (Table 5.1) as one branch of a tripartite division of eutherian mammals (the other two being Unguiculata and Mutica). Within his concept of Ungulata he did not recognize the current concept of Perissodactyla as a natural group (that was to come a half-century later). He allied horses with the hippopotamus and tapirs (as a species of the latter) within a taxon Belluae, which was defined by him with "front teeth several, obtuse. Gait heavy. Sustenance by pulling up vegetation" (Gregory 1910:33). Thus, Linnaeus classified these animals based on what they did as well as on morphological characters. The rhinoceroses were included in the widely separate Glires (normally thought to include Rodentia and Lagomorpha).

In 1792 Vicq d'Azyr was the first to use the term "Solipèdes," a gallicized derivation of the prior Solidungula, as in Blumenbach's classification of 1779, meaning monodactyl. In 1816 de Blainville classified the soliped horses (i.e., *Equus*) and pachyderms within the "Impairs," literally, those forms with non-paired digits. Other mammals within Impairs included rhinoceroses, tapirs, and hyraxes (Gregory 1910; Simpson 1945). The position of the latter is interesting because some taxonomists have come full circle on hyraxes. Thus, after decades of pondering their affinities [Gregory (1910:90) believed them to be "wholly extraneous" to perissodactyls], some workers now believe hyraxes to be close extant relatives of perissodactyls, as discussed later.

Cuvier's classification of 1817, *Le Règne Animal*, was widely used among taxonomists, and in some circles it was as popular as Linnaeus's *Systema Naturae*, although Gregory (1910:80) believed it to be "conservative in form but transitional and con-

[2]Although the 10th edition of *Systema Naturae* (Linnaeus 1758) usually is used as the starting point for modern systematics, Linnaeus had published numerous versions between 1735 and 1766, and those had evolved from a 12-page brochure to a 2,400-page tome (Gregory 1910).

Table 5.1. *Portions of selected classifications of mammals relevant to an understanding of the taxonomy of Perissodactyla*

1. Linnaeus (*Systema Naturae*, 1735–66)
Ungulata (1766)
  Pecora; including *Camelus, Moschus, Cervus, Capra, Ovis, Bos*
  Belluae; including *Equus, Hippopotamus* (including tapirs)

2. Blumenbach (1779)
Solidungula; *Equus*

3. Vicq D'Azyr (1792)
Solipèdes; *Equus*

4. de Blainville (1816)
Ongulograd[es]
  Normaux, doigts
    Impairs
      Pachydermes
      Solipèdes
    Pairs
      Non Ruminans ou Brutes
      Ruminans
  Anomaux, pour nager
    Les Lamantins

5. Cuvier (1817)
Pachydermes
  Proboscidiens
  Pachydermes ordinaires; including various fossil and extant artiodactyls of twentieth-century usage, as well as rhinoceroses, hyrax, tapirs, and †*Palaeotherium*
  Solipèdes (Chevaux)

6. Owen 1868
Ungulata
  Order Artiodactyla Owen 1848
  Order Perissodactyla Owen 1848; including †*Coryphodon,* †*Pliolophus,* †*Hyracotherium,* †*Lophiodon,* †*Palaeotherium,* †*Paloplotherium,* †*Macrauchenia,* †*Elasmotherium, Rhinoceros,* etc., *Hyrax,* †*Anchitherium,* †*Hipparion, Tapirus,* †*Toxodon,* †*Nesodon*
  Order Proboscidea Illinger 1811

7. Simpson 1945
Cohort Ferungulata, new
  Superorder Ferae Linnaeus 1758; including "creodonts," carnivores, and "pinnipeds"
  Superorder Protungulata Weber 1904; including condylarths, litopterns, notoungulates, astrapotheres
  Superorder Paenungulata, new; in addition to those that follow, also including pantodonts, uintatheres, pyrotheres, embrithopods
    Order Proboscidea Illinger 1811
    †Order Embrithopoda Andrews 1906
    Order Hyracoidea Huxley 1869
    Order Sirenia Illinger 1811
  Superorder Mesaxonia Marsh 1884
    Order Perissodactyla Owen 1848
      Suborder Hippomorpha Wood 1937; including horses, palaeotheres, titanotheres, and chalicotheres
      Suborder Ceratomorpha Wood 1937; including tapirs and rhinoceroses

Table 5.1 (*cont.*)

Superorder Paraxonia Marsh 1884
    Order Artiodactyla Owen 1848
8. McKenna (1975b)[a]
Grandorder Ungulata Linnaeus 1766, new rank
    Mirorder Eparctocyona, new; several orders, including Artiodactyla
    Mirorder Cete Linnaeus 1758, new rank; including creodonts and
    cetaceans
    †Mirorder Meridiungulata, new; including numerous South
    American ungulates
    Mirorder Phenacodonta, new
        †Order Condylarthra Cope 1881, emended
        Order Perissodactyla Owen 1848
            Suborder Hippomorpha Wood 1937
            †Suborder Ancylopoda Cope 1889
            Suborder Ceratomorpha Wood 1937
        Order Hyracoidea Huxley 1869
    Mirorder Tethytheria, new
        Order Proboscidea Illinger 1811
        Order Sirenia Illinger 1811
        †Order Desmostylia Reinhart 1953
9. Prothero & Schoch (1989b)
Superorder Ungulata Linnaeus 1766
    Order Tethytheria McKenna 1975
    †Order Embrithopoda Andrews 1906
    Order Perissodactyla Owen 1848
        Suborder Hyracoidea Huxley 1869
        Suborder Mesaxonia Marsh 1884
            Infraorder Hippomorpha Wood 1937
                Superfamily Pachynolophoidea Pavlow 1888
                Superfamily Equoidea; other taxa also, but including
                "*Hyracotherium,*" in part (no family designation)
                    Family Equidae Gray 1821; also including *Hyracotherium*, in
                    part (?*Protorohippus*), and 29 other genera
                    †Family Palaeotheriidae Bonaparte 1850
            Infraorder Moropomorpha Schoch 1984; to include tapirs,
            rhinocerotoids, and chalicotheres
            Infraorder Titanotheriomorpha Hooker 1989, new rank;
            including titanotheres

Sources: From Gregory (1910), Simpson (1945), McKenna (1975b), and Prothero and Schoch (1989b).
Note: A dagger (†) denotes an extinct taxon.
[a] The many dichotomous branches of McKenna's classification produced an extensive series of nested hierarchies. Therefore, new higher-rank names were proposed (in addition to the common Linnean class, order, and derivatives thereof), including superlegion, legion, sublegion, supercohort, cohort, subcohort, magnorder, grandorder, and mirorder. Some critics (e.g., Van Valen 1978) cite the proliferation of these new rank names as a weakness of McKenna's classification and a general problem with the cladistic methodology. However, more recent cladists have dispensed with the need to name every new rank within a classification (e.g., Wiley 1981; Schoch 1986).

fused in principle." Cuvier's scheme seems to have been the first attempt to ally fossil and extant taxa together within Pachydermes, although within this group *Equus* was separated off in Solipèdes, distinct from "Pachydermes ordinaires," which included, in part, rhinoceroses, tapirs, and early Tertiary *Palaeotherium* (Figure 5.2).

Despite de Blainville's insight (his distinction between "Impairs" and "Pairs"), Owen (1848, 1868) is usually credited with both the *nomen* and concept of Perissodactyla and Artiodactyla as, respectively, odd- and even-toed ungulates. Like previous workers, Owen included both fossil and extant taxa in the Perissodactyla. In addition to those that would make sense today, however, he included the Eocene pantodont *Coryphodon*, the late Cenozoic South American litoptern *Macrauchenia*, and notoungulates *Toxodon* and *Nesodon* in his concept of the Perissodactyla.

Marsh (1884) proposed two new superordinal ranks, Mesaxonia and Paraxonia, to describe the position of the axis of symmetry in, respectively, Perissodactyla and Artiodactyla (Figure 5.3). In the mesaxonic condition the principal axis of symmetry runs through the central digit, MP III, whereas in the paraxonic condition the axis of symmetry runs between MP III and IV. However, in most classifications [Prothero & Schoch (1989b) being an exception] these terms do not further clarify hierarchical relationships; they are essentially synonymous with Perissodactyla and Artiodactyla. After Owen's classification in 1868, these two taxa were in most cases stabilized as orders or suborders of ungulate mammals, although their interrelationships continue to be debated.

On the basis of some fossil evidence available at the turn of the twentieth century it was suggested that the Artiodactyla should be separated from other ungulate mammals and allied with arctocyonids, primitive carnivorous mammals (Gregory 1910). Rather than expand the prior concept of the Ungulata, Simpson (1945) proposed the *nomen* Ferungulata to include ungulates and carnivores (Table 5.1). His reason was because (1945:174) "the words 'Ungulata' and 'ungulates' have always carried the clear significance of mammals normally hoofed and the taxonomic idea of a group not only excluding but also diametrically contrasting with the carnivores. It would be confusing and ridiculous to call the cohort as here constituted 'Ungulata' and to speak, for instance, of dogs or of lions as 'ungulates.' " Simpson was very concerned about any taxon whose constituents were separated by too much "morphological distance," that is, a larger range of morphological adaptations than one might consider "normal" for a group. Subsequent systematic schemes eschewed this limitation in favor of forming natural clades based on common descent (e.g., McKenna 1975b; Prothero, Manning, & Fischer 1988). In recent classifications, the Perissodactyla have

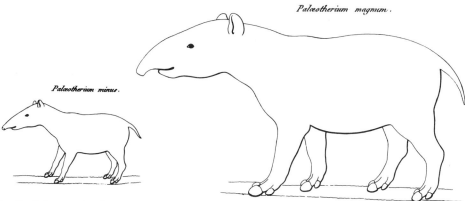

Palæotherium magnum.

Palæotherium minus.

Figure 5.2. *Cuvier's reconstructions (1836) of* Palaeotherium magnum *and* P. minus *from the Eocene Montmartre gypsum. These are believed to have been the first scientific reconstructions of fossil mammals (Buffetaut 1987).*

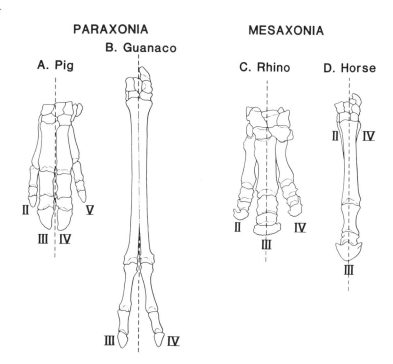

Figure 5.3. *Paraxonic versus mesaxonic character states represented in lower forelimb of primitive (A) and derived (B) artiodactyls and primitive (C) and derived (D) perissodactyls. The axis of limb symmetry runs between MC III and IV in artiodactyls and through MC III in perissodactyls*

been included within a revised concept of ungulate mammals [e.g., the "grandorder Ungulata" (McKenna 1975b) or "superorder Ungulata" (Novacek 1986; Prothero & Schoch 1989b)] along with a taxonomic myriad of other forms encompassing what would, in prior classifications, have been considered a large range of morphological distance. However, with the recent emphasis on phylogenetic classifications in which genealogical descent is reflected, these objections find less support among systematists.

Monophyly and close relatives of the Perissodactyla

Since its initial description by Owen in 1848, Perissodactyla has gained widespread acceptance among systematists as a mono-phyletic clade. One of the few serious challenges to this notion was proposed by Butler (1952). On the basis of deciduous pre-molar morphology and homologies he suggested that palaeo-theres had an origin separate from other perissodactyls, and therefore, as constituted, Perissodactyla was polyphyletic. That idea, however, has not been widely accepted in subsequent dis-cussions of perissodactyl monophyly.

In contrast to Butler's view, there are several shared-derived character states (synapomorphies) that corroborate a hypothesis of perissodactyl monophyly. So far as is known (obviously for soft anatomical structures this means only living representa-tives), these include the following: (1) the saddle-shaped (con-cave) distal navicular facet on the astragalus (Figure 5.4); interestingly, hyracoids also seem close to this condition (Ra-dinsky 1966; MacFadden 1976); (2) the mesaxonic limb symmetry (Figure 5.3) coupled with the extent of reduction of lateral me-tapodials (MacFadden 1976); (3) the expanded cecum (Figure 5.5) in which hindgut fermentation occurs in all perissodactyls (Mitchell 1905; Janis 1976); and (4) the morphology of the lower cheek teeth (Figure 5.6).

Traditionally it was thought that the closest fossil outgroup of the perissodactyls was found within advanced taxa of phen-acodontid condylarths (Table 5.1), as in McKenna's classification (1975b). Although it is a paraphyletic and somewhat heteroge-neous assemblage, the central concept of this family is the genus *Phenacodus,* known from early Eocene deposits in western North America (Radinsky 1966; West 1976; Thewissen 1990). Most of the described species of this genus, however, are already too specialized to be the closest perissodactyl outgroup. However, McKenna (1960:101) mentions a single specimen, questionably allocated by him to *Phenacodus* sp., from the Wasatchian Four Mile Local Fauna of Colorado, and he notes that "this tantalizing animal may well represent the most perissodactyl-like phena-codont known." In his recent revision of phenacodontids, Thew-issen (1990) states that their relationship with perissodactyls remains an interesting but unresolved question.

At least two derived characters are shared between advanced phenacodontids and perissodactyls: (1) mesaxonic limb sym-metry in the latter, and a trend toward such symmetry in the former, and (2) reduction of the cusps of anterior cheek teeth, namely, the paraconid and paralophid complex (MacFadden 1976). Granted, these characters show a more advanced state in perissodactyls, but the condition in phenacodontids is never-theless revealing of what would be expected from a primitive, closest outgroup. On the other hand, Prothero and Schoch

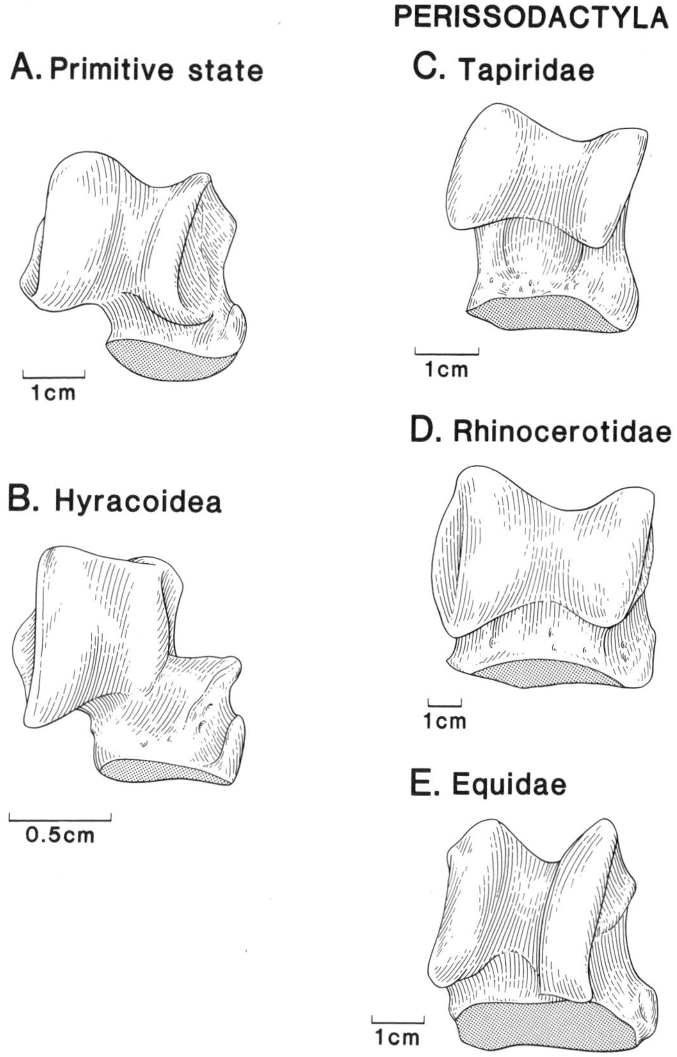

PERISSODACTYLA

A. Primitive state

C. Tapiridae

B. Hyracoidea

D. Rhinocerotidae

E. Equidae

Figure 5.4. *Derived state of perissodactyl astragalus. In the primitive condition in mammals (A) the navicular facet (shaded) is convex. In the derived condition in Perissodactyla (C–E) the navicular facet is concave. Some workers have suggested a close relationship between the Perissodactyla and Hyracoidea, and the hyracoid astragalus (B) shows a similar condition of the navicular facet. Taxa and specimens: A, Ursus americanus, UF-M 18939; B, Dendrohyrax dorsalis, UF-M 20500 (reversed); C, Tapirus sp., UF-M 8130; D, Diceros bicornis, UF-M 24369; E, Equus caballus, UF-M 6684.*

(1989a) do not accept the phenacodontids and perissodactyls as sister taxa (Table 5.1). They suggest the novel hypothesis that among extinct taxa the Embrithopoda (Novacek 1986), a low-diversity group of extinct early Tertiary horned tethytheres, *sensu* McKenna (1975b), are most closely related to perissodactyls (assuming that hyracoids are perissodactyls).

Some critics of the cladistic method might scoff at the notion that closest relatives, or sister taxa, are contemporaneous; for example, phenacodonts and *Hyracotherium* are both found in early Eocene deposits, or if Prothero and Schoch (1989a,b) are

Figure 5.5. *Gastrointestinal (GI) tracts of perissodactyls: (A) the dog (Canis familiaris), an example of the primitive condition for the mammalian GI tract; (B) the Equidae (Equus burchelli); (C) Rhinocerotidae (Diceros bicornis). Although an outpouching, or cecum, occurs at various sites within the lower GI tract in several clades of mammals, in the Perissodactyla (including Equidae, Rhinocerotidae, and Tapiridae, the latter not pictured here) the expanded cecum, which is the principal site of digestion (Janis 1976), is located at the junction between the small and large intestines. Modified from Stevens (1988); reproduced with permission of Cambridge University Press.*

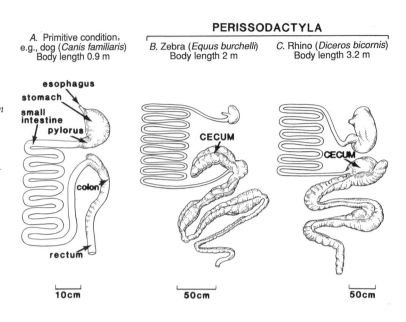

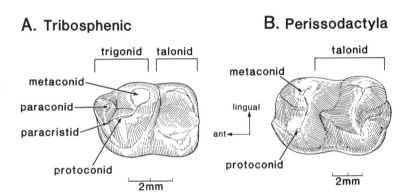

Figure 5.6. *Morphology of perissodactyl lower cheek tooth. (A) In primitive tribosphenic mammals, such as the Paleocene insectivore* Pentacodon *(AMNH 17038, composite left m2 and m3, drawn from cast), the tooth had an anterior trigonid and posterior talonid. The trigonid consists of three principal cusps, the paraconid, protoconid, and metaconid; a small crest, the paracristid, runs between the paraconid and protoconid. (B) In perissodactyls, such as* Hyracotherium *(left m2, UF 513, from Huerfano, Colorado), the primitive tribosphenic trigonid is modified, with reduction of the paracristid and loss of the paraconid.*

correct, embrithopods occurred later than the first perissodactyls. However, that reasoning is constrained by older thinking in which the goal was the search for ancestors lined up in a progressive temporal sequence. This question is less important in a cladistic framework, where these two taxa plausibly could have shared a common ancestor not yet recognized and of greater antiquity.

There is somewhat less agreement as to the closest extant outgroup of the Perissodactyla. Results from protein sequencing (Miyamoto & Goodman 1986) and "traditional" anatomical and paleontological characters (Novacek 1982, 1986; Wible 1987; Novacek, Wyss, & McKenna 1988) corroborate more conservative views of a close relationship between Perissodactyla and Artiodactyla, with Cetacea as the next closest outgroup (in some schemes these are depicted as an unresolved trichotomy) (Figure 5.7A). On the other hand, several characters suggest that Hyracoidea is the closest outgroup of the Perissodactyla and Artiodactyla (Figure 5.7B), including (1) the similarity of the saddle-shaped navicular facet on the astragalus (Radinsky 1966; MacFadden 1976), (2) the derived condition of the internal carotid artery (Wible 1986), and (3) the locomotory system, the presence of the eustachian sac, and the maxillary morphology (Prothero et al. 1988; Fischer 1989; Prothero & Schoch 1989b) (Table 5.1). McKenna's classification (1975b) summarizes the general arguments in this section concerning the interrelationships of the principal fossil and extant outgroups discussed here (Table 5.1), wherein Phenacodonta contains three clades: the †Condylarthra, Perissodactyla, and Hyracoidea.

Perissodactyl clades: the cast of characters

In de Blainville's classification of 1816 the three traditionally recognized higher taxa of extant perissodactyls (horses, tapirs, and rhinoceroses) were united within the Impairs (Simpson 1945). At about the same time, the close affinities of fossil perissodactyl taxa were first formally recognized when Cuvier included †*Palaeotherium* in his 1817 classification. After 1821, when the families within Perissodactyla began to be recognized (Gray 1821), several other taxa were established for extinct forms, of which the following are most notable. In 1872 Gill proposed the family Chalicotheriidae for the enigmatic and usually rare Tertiary clawed chalicotheres (which are essentially synonymous with the order Ancylopoda Cope 1889, literally, "clawed foot"). In 1873 Marsh proposed the family Brontotheriidae for the brontotheres, or "titanotheres," which are very common in late Eocene and early Oligocene deposits in western North America (although rare, they are also known from the Old World). These *nomina* have been used at different levels within various classifications, but in general these higher-level categories have remained conceptually similar since about the turn of the twentieth century. In Simpson's classification (1945), the order Perissodactyla is subdivided into two principal suborders, named by Wood (1937) the Hippomorpha ("horse-shaped," including horses, titanotheres, and chalicotheres) and the Ceratomorpha ("horn-shaped," including tapirs and rhinoceroses). More recent classifications, however, reflect more complexity; McKenna (1975b) placed the enigmatic chalicotheres in their own clade, the Ancylopoda. Prothero and Schoch (1989b), while retaining some of the older *nomina* (e.g., Hippomorpha, although with considerably emended definitions), significantly revamped the classification and traditional phylogenetic groupings within the Perissodactyla (Table 5.1).

The phylogenetic position of *Hyracotherium*

It is quite possible that certain species of Eohippus, *when more intensively studied, will appear to be more directly in the line of ancestry of the horse, others of the tapir or rhinoceros or of some of the extinct phyla of Perissodactyls.*
— Matthew (1926:153)

In 1838 a Mr. Colchester was digging for brick clay near the town of Woodbridge in Suffolk, England, when he discovered a small fossil tooth in some associated sands. Further sifting resulted in the discovery of an additional specimen, a jaw fragment containing a tooth. Those fossils (Figure 5.8) were brought to the attention of Sir Richard Owen, and his first comparisons suggested that they belonged to the "quadrumanous tribe," namely, monkeys (Simpson 1940a). Along with subsequent specimens collected from the London Clay during the next few years, Owen (1840a) proposed the genus *Hyracotherium*, with its first species being *H. leporinum*. Owen refrained from indicating the exact relationships of *Hyracotherium*, although the generic etymology implies similarity to hyraxes, which is not surprising given the other discussions in this chapter. Subsequent workers have almost universally regarded *Hyracotherium* as a primitive member of the Equidae. The story has already been told in Chapter 3 for North America, where Marsh named *Eohippus*, the "dawn-horse," in 1876 based on specimens that probably were collected by David Baldwin from New Mexico [Simpson (1940a) suggested that the lack of detailed locality information was to deter Marsh's rivals from raiding his sites]. As with *Hyracotherium*, several other species of *Eohippus* were named from early Eocene deposits of the western United States (Kitts 1956). Not surprisingly, Marsh continued to use *Eohippus* as a formal scientific generic *nomen* distinct from *Hyracotherium*. However, at the same time, other North American paleontologists, including Cope (Figure 5.9), Granger, and Matthew (1903), referred to specimens collected from the western United States as *Hyracotherium*, thus indicating an awareness of Old World affinities.

Subsequent studies of relevant collections of *Hyracotherium* and *Eohippus* led most paleontologists to conclude that the two genera were the same (Forster-Cooper 1932; Simpson 1952a), and therefore *Eohippus* was relegated to the status of a junior synonym (e.g., Kitts 1956). Nevertheless, the latter has become widely ingrained in the fabric of both professional and popular literature in paleontology, and "eohippus" continues in common usage as a vernacular name and is even included in some English dictionaries (e.g., Neilson 1952).

It may seem that too much time is being devoted to *Hyracotherium*, or eohippus. On the contrary, being at or near (depending upon phylogenetic interpretations) the base of both the perissodactyl and equid radiations, it is central to an understanding

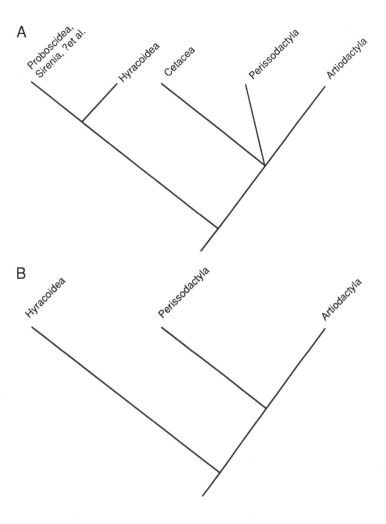

Figure 5.7. *Differing views of interrelationships of Perissodactyla and other extant orders of mammals. (A) On the basis of biochemical and certain morphological characters, the Perissodactyla, Artiodactyla, and Cetacea seem closely related, although some workers show them as an unresolved trichotomy (e.g, Novacek 1982, 1986; Miyamoto & Goodman 1986; Wible 1987; Novacek, Wyss, & McKenna 1988). (B) Another possibility, based on several morphological characters, is that hyracoids are the closest outgroup of Perissodactyla and Artiodactyla (Radinsky 1966; MacFadden 1976; Wible 1986; Fischer 1989).*

Figure 5.8. *Original specimen of Hyracotherium from the Eocene of Suffolk, England, described by Sir Richard Owen in 1840. From Simpson (1940a) and reproduced with original caption from that article.*

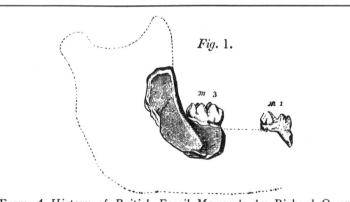

(From *A History of British Fossil Mammals,* by Richard Owen)

WHEN A HORSE was mistaken for a monkey: the first teeth of the dawn-horse which a century ago were thought to prove that monkeys once inhabited England

of the higher-level interrelationships of the Perissodactyla and the definition of the family Equidae. A recent detailed cladistic analysis (Hooker 1989) of critical new and previously studied specimens resulted in a radical rethinking of the concept of the genus *Hyracotherium* and the base of the perissodactyl radiation (Figure 5.10). Using the PAUP algorithm (Swofford 1985), Hooker analyzed 67 character states distributed in 40 relevant Eocene perissodactyl taxa. His results are far-reaching in their significance and exceedingly complex, and they are still being discussed by subsequent workers (e.g., Prothero & Schoch 1989a). Regardless of the exact details of all the new hypotheses that have been or will be generated from his work, the following conclusions from Hooker's studies are important here: (1) In contrast to previous dogma, there are several fossil perissodactyl taxa (such as *Cymbalophus*) that are dentally more primitive than *Hyracotherium*. (2) *Hyracotherium* is a horizontal, paraphyletic "wastebasket" taxon, with different species currently allocated to this genus giving rise to different perissodactyl clades (not unlike what is described in the foregoing quotation from Matthew).

To avoid confusion about nomenclature and meanings, the following distinction is made in this chapter: (1) "*Hyracotherium*" is used in the broad sense to indicate the more inclusive, polyphyletic concept of this genus, and (2) *Hyracotherium* is used in a more restricted sense to include that part of the genus with equid affinities. Most of the subsequent discussion of the North American radiation will be concerned with *Hyracotherium*.[3]

It is frequently stated in the older, widely cited synthetic literature (e.g., Simpson 1951, 1953a; Romer 1966; Colbert 1980) that *Hyracotherium* is the most primitive known perissodactyl, and therefore it was a common ancestor from which all more advanced taxa within this order descended (Figure 5.11A). From the point of view of classification, this is perplexing, because how could *Hyracotherium* be the ancestral perissodactyl and still be a horse? Simpson (1953a:345) exemplifies this paradox:

> In effect, there was no family Equidae when eohippus lived. The family and all of its distinctive characters developed gradually as time went on. Eohippus is referred to the Equidae because we happen to have more nearly complete lines back to it from later members of this family than from other families. There is no particular time at which the Equidae became a family rather than a genus or a species; the whole process is gradual and we assign the categorical rank after the result is before us.

I have examined this question for primitive perissodactyls (MacFadden 1976). Indeed, *Hyracotherium* is dentally very prim-

[3]There is as yet no consensus on a suitable replacement name for the restricted concept of North American *Hyracotherium*. Several older generic names are available, and Prothero and Schoch (1989b) place *Protorohippus* Wortman 1896 in that position (Table 5.1). However, this is by no means fixed, and further work (e.g., Froehlich 1989) must be done to resolve this issue.

Figure 5.9. *First reconstruction of skeleton of* Hyracotherium *"venticolum" [=* H. vasacciense *of Kitts (1956)], from Cope's "bible" (1884). This composite specimen was based on fossils collected from the Eocene of Wyoming.*

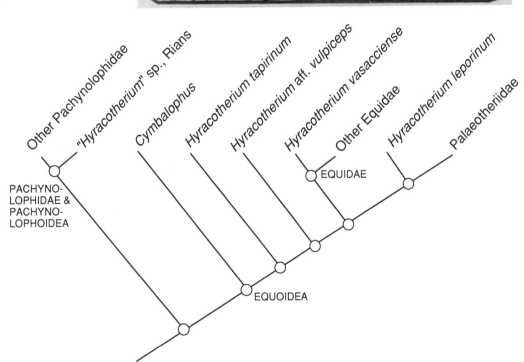

Figure 5.10. *Cladogram of the interrelationships of early Eocene perissodactyls with pachynolophoid and equoid affinities, modified from Hooker (1989), which includes the synapomorphies. Note, in particular, the paraphyly of the genus* Hyracotherium.

itive; in particular it lacks the relatively well developed crests (lophs and lophids) that connect various cusps (depending upon the clade) in other perissodactyls. Not surprisingly, that condition has been given considerable weight by mammalian paleontologists. In a cladistic framework, one solution would have been to separate out *Hyracotherium* based on its primitive dentition as the plesion for all other perissodactyls. However, there are two basicranial characters that elucidate its advanced position among perissodactyls: (1) In the primitive condition for mammals, the foramen ovale, located in the ventral basicranium and housing the mandibular branch of the trigeminal (V) cranial nerve, is separate from the medial lacerate foramen (Figure 5.12A). In the derived state (Figure 5.12B) the foramen ovale is absent; it can be thought of either as lost or as having become confluent with the medial lacerate foramen. (2) In the primitive condition, the optic foramen in the posterior portion of the orbital region, which houses the optic (II) cranial nerve, lies roughly midway between the ethmoid foramen and the anterior opening of the alisphenoid canal (Figure 5.12C). In the derived condition, the optic foramen is located more posteroventrally, where it lies close to the anterior opening of the alisphenoid canal (Figure 5.12D). The distribution of these two character states indicates that the primitive condition is found in fossil and extant specimens of "nonhorse" perissodactyls (a taxon *faute de mieux*). In contrast, the derived state for both of these characters is found in *Hyracotherium* and all other fossil and extant horse specimens that I have examined. Thus, *Hyracotherium* shares derived character states with other horses that justify its inclusion within the Equidae (Figure 5.11B). This result has created another problem: If *Hyracotherium* is no longer the most primitive perissodactyl (because it already demonstrates equid specializations), then what taxon is relegated to this status? Some recent work has shed light on this question. First, McKenna et al. (1989) have described the new genus *Radinskya* [aptly named to honor the many contributions of Leonard B. Radinsky (1937–85) to perissodactyl systematics and paleobiology], a primitive perissodactyl-like mammal from the late Paleocene of China. Second, Hooker's work (e.g., 1984, 1989) has radically revised our ideas of the interrelationships of primitive perissodactyl taxa. Hooker (1984) described the genus *Cymbalophus*, from the late Paleocene–early Eocene of England, which he considers to be very close to the most primitive perissodactyl. Furthermore, he has reorganized the concept of *Hyracotherium*, parts of which have been excluded from the Equidae. He also believes that two other Old World genera, *Pachynolophus* and *Anchilophus*, previously considered palaeotheres, *sensu* Simpson (1945), are also at the base of the Perissodactyla (Figure 5.10, "Other Pachynolophidae"). In summary, there are several late Paleocene and early Eocene perissodactyls that on the basis of their dentitions seem to be more primitive than the equid *Hyracotherium* (Figure 5.10). However, the exact interrelationships of these pivotal taxa will require further investigation.

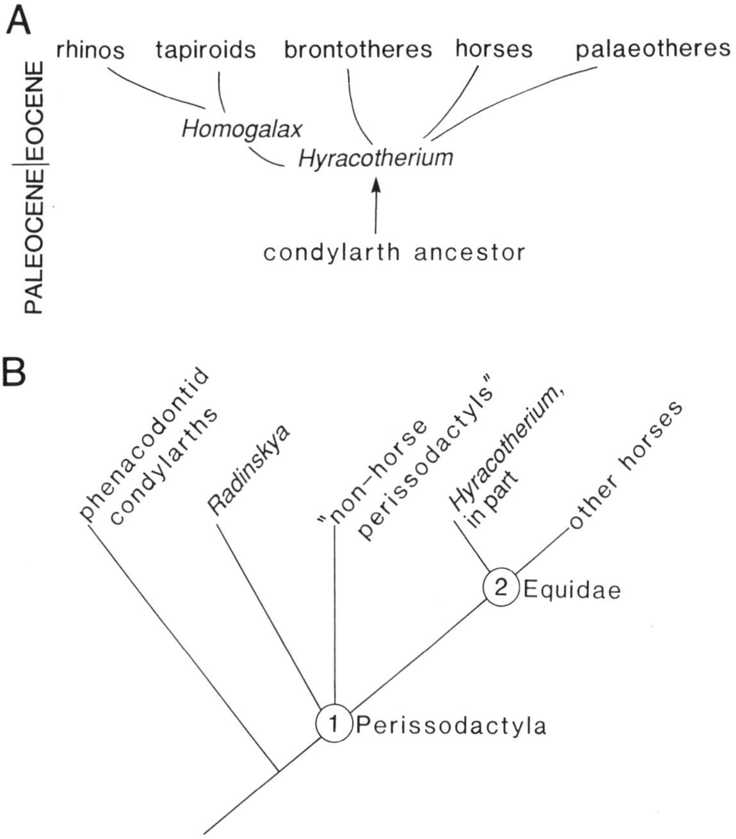

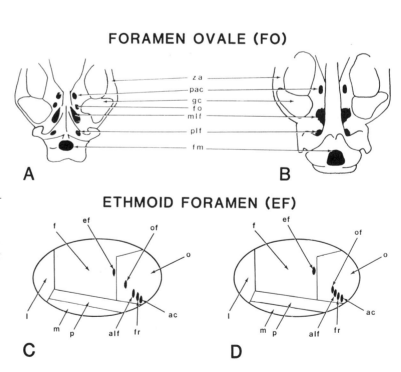

Figure 5.11. *(A) Traditional view of* Hyracotherium *as the stem genus from which all other perissodactyls descended. (B) Cladogram depicting revised interrelationships of primitive perissodactyls. Here the recently described* Radinskya *from Mongolia (McKenna et al. 1989) is close to the origin of the Perissodactyla, an order defined by several synapomorphies. In contrast to primitive perissodactyls, a revised concept of* Hyracotherium *suggests that part of this genus shares derived characters with other members of the Equidae. Modified from Mac-Fadden (1985a); reproduced with permission of Plenum Press.*

Figure 5.12. *Derived characters that unite the Equidae, modified from MacFadden (1976), showing primitive (A, C) and derived (B, D) states in the basicranium and left orbital regions. In the Equidae, a separate foramen ovale (fo) is lost in the basicranium, and the optic foramen (of) is positioned more posteroventrally in the orbit relative to primitive mammals. Other abbreviations; ac, anterior opening of the alisphenoid canal; alf, anterior lacerate foramen; ef, ethmoid foramen; f, frontal bone; fm, foramen magnum; fr, foramen rotundum; gc, glenoid cavity; l, lacrimal; m, maxilla; mlf, middle lacerate foramen; o, orbitosphenoid; p, palatine; pac, posterior opening of the alisphenoid canal; plf, posterior lacerate foramen; za, zygomatic arch. Not to scale.*

Grades, clades, and horses

The genus Merychippus *is extremely difficult to diagnose, since its species intergrade almost imperceptibly with those of* Parahippus *on the one hand and those of* Pliohippus, Calippus, Nannippus, Hipparion, *and* Neohipparion *on the other.*   – Stirton (1940:178)

By the turn of the twentieth century it was becoming clear that the diversity of fossil horses in North America was far greater than had been portrayed in the earlier, simple orthogenetic schemes (see Chapter 3). That diversity was greatest during the Miocene, at the base of the adaptive radiation of grazers that lived contemporaneously with browsing forms. The classic studies around the turn of the century did much to add many newly described species to the literature, but phylogenetic groupings were difficult to identify (e.g., Cope 1889; Gidley 1907; Osborn 1918). This is understandable, because workers during that time had to rely principally on dental characters, which are now known to be highly plastic and to have evolved in parallel in several contemporaneous lineages. That state of knowledge led paleontologists to recognize horizontal groupings of morphologically similar forms, which in many cases were embodied in the concepts of the genera described before the turn of the century. Thus, Matthew (1926) proposed 10 stages in horse evolution spanning their entire Cenozoic record. Those stages corresponded to the then-recognized generic groups (*Eohippus, Orohippus, Epihippus, Mesohippus, Miohippus, Parahippus, Merychippus, Pliohippus, Plesippus,* and *Equus*). Stirton's classic synthesis (1940) of North American equid phylogeny (see Figure 3.12), which contains many polyphyletic taxa, is certainly understandable for its time. As exemplified in the foregoing quotation, the most vexing part of Tertiary equid phylogeny is the problematical merychippines, which are very similar in dental morphology. This is in contrast to later equids, which had diversified their dentitions, so that by the later Miocene individual lineages demonstrated distinctive character states. Nevertheless, despite the excellent hindsight that we can bring to this problem, Stirton's phylogeny rightfully will continue to be an important reference in equid paleontology that represents the state of scientific knowledge a half-century ago.

As with many aspects of mammalian phylogeny and classification in recent years, there has been a trend toward pulling apart horizontal, polyphyletic, wastebasket taxa in favor of redefining groups based on monophyletic descent, which results in a more nearly "vertical classification." With regard to equid systematics and phylogeny, Quinn (1955) was the first to propose a radical vertical regrouping of the then-recognized genera of Miocene horses based on collections that he examined from the Gulf Coast plain of Texas (Figure 5.13). For example, Quinn

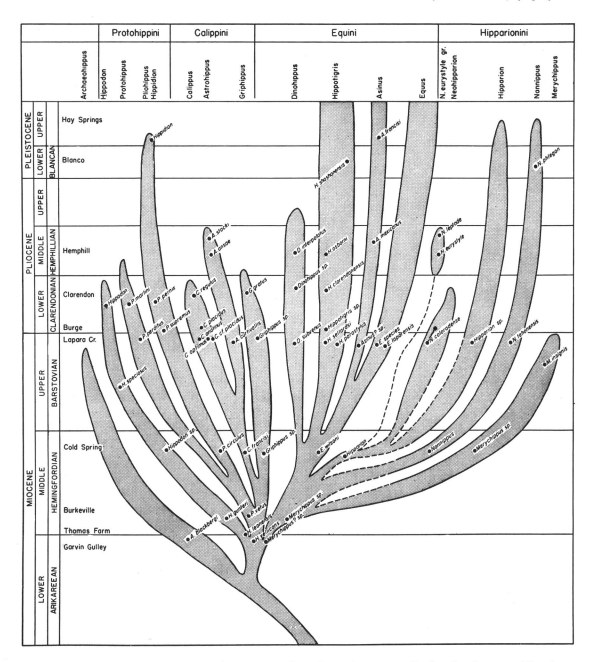

Figure 5.13. *Quinn's phylogeny (1955) of late Cenozoic horses based on his work on collections from the Texas Gulf Coast plain. Although this was an early attempt to split up the equid generic grades of previous authors, many of Quinn's phylogenetic decisions were quite radical (e.g., a 17–18 Ma "Eoequus" wilsoni) and were not followed by most subsequent workers. Reproduced with permission of the Bureau of Economic Geology, Austin, Texas.*

believed that he could recognize a Miocene ancestor of *Equus*, and he called it *Eoequus wilsoni* [which Hulbert (1988a) considers a junior synonym of *Protohippus perditus*]. That decision and many others contained in Quinn's publication have since met with disfavor (e.g., Webb 1969a), and his phylogenetic schemes, although predating the current trend toward recognition of monophyletic clades, were rarely followed in the subsequent literature on equid systematics.

Skinner and MacFadden (1977) proposed the genus *Cormohipparion* to include species previously allocated to the merychippine and hipparion grades of equid evolution. That paper was a radical departure from most prior taxonomic practice, because rather than using overall dental morphology, in that study we recognized this generic clade based principally on the derived state of the preorbital facial fossa. This practice initially met with strong opposition. For example, although she used facial fossae in her discussion of *Sinohippus* (Forsten 1982a), Forsten (1982b) believed that we had merely redescribed merychippines, and Eisenmann et al. (1987) believed that we had placed too much emphasis on this single character complex [see the response (MacFadden 1987b)]. Nevertheless, in many subsequent papers the condition of preorbital facial fossae is considered, along with other, more traditional characters, in order to understand equid systematics, whether in the North American radiation (e.g., Bennett 1980; MacFadden 1984a,b) or the Old World radiation (e.g., Woodburne & Bernor 1980; Forsten 1982a).

In summary, during the past few decades there have been many studies that have updated and revised various aspects of equid systematics. Although the general pattern of the phylogeny of the Equidae (Figure 5.14) is superficially similar to those produced during the middle of this century (e.g., Figures 3.11–3.13), there has been a decided trend toward more emphasis on vertical classification and recognition of strictly monophyletic clades.

Major clades of New World Equidae

Figure 5.15 is a cladogram showing the currently recognized genera of Equidae. The derived character states that unite the important (numbered) nodes are summarized in Table 5.2. Detailed discussions of other specific portions of this cladogram are available within the primary literature cited in Table 5.2.

Node 1: Equidae

The family Equidae is defined by the character states of the basicranium described earlier (MacFadden 1976) and the presence of a postprotocrista on P3 (Hooker 1989) (Table 5.2). Relative to *Hyracotherium* Owen 1840, however, all other hyracotheres,

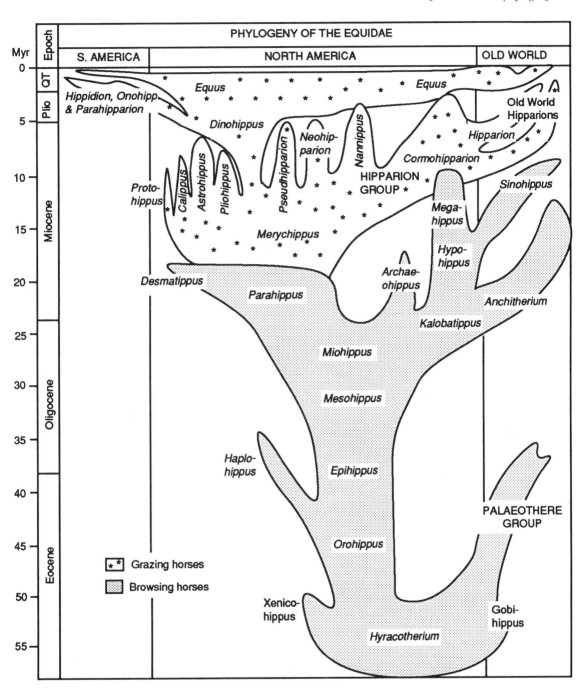

Figure 5.14. *Interrelationships of the currently recognized genera of horses. Note that in the Old World, in addi-* tion to the genera Hipparion *and Cor-*mohipparion, *the interrelationships of the "Old World Hipparions" are poorly* *understood and include the endemic* Stylohipparion, Cremohipparion, Sivalhippus *group, and* Proboscidipparion.

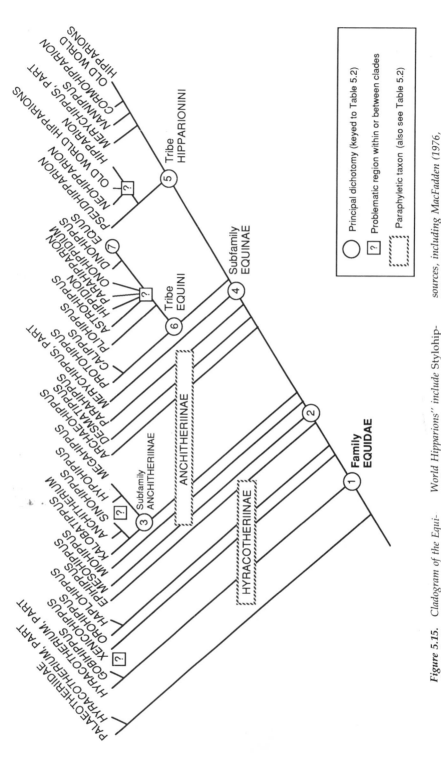

Figure 5.15. Cladogram of the Equidae, including all recognized genera (see Appendix II). Because the exact interrelationships are uncertain, the "Old

World Hipparions" include Stylohipparion, Cremohipparion, Proboscidipparion, and the Sivalhippus group. Compiled from numerous

sources, including MacFadden (1976, 1988a), Hooker (1989) for Hyracotherium; Hulbert (1989), and work in progress.

Table 5.2. *Shared-derived character states (and source references)
used to corroborate nodes in cladogram of North American Equidae
(Figure 5.15)*

Node 1: Family Equidae Gray 1821
Foramen ovale absent (or confluent with medial lacerate foramen);
optic foramen located posteroventrally in orbit (MacFadden 1976); P3
postprotocrista present (Hooker 1989).

Node 2: Subfamily †"Anchitheriinae"[a] Osborn 1910 and subfamily
  Equinae Gray 1821
Completely molarized upper and lower P2–M3 (see Figure 11.4);
tridactyl manus and pes (MacFadden 1976); metacarpal V reduced;
I1–3 with pitted crowns, extended premaxilla, longer diastema; angle
of jaw posteriorly rounded, without notch (Prothero & Shubin 1989).
Note: R. J. Emry (personal communication) has one specimen of a
Chadronian *Mesohippus* that has a fourth metacarpal preserved;
however, that is the only example for which this primitive
(hyracothere) character state is known to exist for this genus.

Node 3: †"Anchitheriinae" *sensu stricto*
Greatly increased tooth crown area and estimated body size
(MacFadden 1987a, 1988b); relatively well developed cingula, and
loss of ribs between styles on cheek teeth (Stirton 1940; Forsten
1975).

Node 4: Subfamily Equinae Gray 1821
Cement formed on deciduous and permanent cheek teeth; degree of
development of pli caballin on premolars and molars; presence of pli
entoflexid; moderately deep ectoflexid on p2; unworn M12 molar
crown heights greater than about 23–28 mm (Hulbert 1989; Hulbert
& MacFadden 1991).

Node 5: †Tribe Hipparionini Quinn 1955
Well-developed and persistent pli caballin on molars; MC V
articulates primarily with MC IV (Hulbert & MacFadden 1991).

Node 6: Tribe Equini Gray 1821
Dorsal preorbital fossa (DPOF) depth moderate; DPOF with shallow
posterior pocket; protocone–protoloph connection during early wear
stages on P34; hypoconid lake on p34; p34 metastylid much smaller
and located more lingually than metaconid; m12 metastylid much
smaller and located more labially than metaconid (Hulbert &
MacFadden 1991).

Node 7: Genus *Equus* Linnaeus 1758
DPOF poorly developed or absent; very high crowned and relatively
straight teeth; complex enamel plications and protocones (Bennett
1980; Azzaroli 1982; MacFadden 1984a; MacFadden & Azzaroli 1987);
elongated and either robust or gracile metapodials (Bennett 1980;
Forsten 1986); well-developed intermediate tubercle (INT) on distal
humerus (Hermanson & MacFadden 1992).

[a]Paraphyletic taxa are listed with quotation marks.

namely, the genera in the subfamily Hyracotheriinae, are more
derived on the basis of showing an advanced stage of molarization of the premolars. It is important to note that in the hyracotheres, so far as is known, there are four metacarpals in the

manus and three metatarsals in the pes; these are the primitive characters for the family, (Kitts 1956, 1957; MacFadden 1976). As mentioned earlier, *Hyracotherium* is used here in the restricted sense to include only those species currently allocated to this genus with equid affinities, *sensu* Hooker (1989). The genera *Xenicohippus* Bown & Kihm 1981, from the early Eocene of the western United States, and *Gobihippus* Dashzeveg 1979, from the Eocene of Mongolia, may not be distinct, but they are included here as valid pending further study (see Appendix II). *Haplohippus* McGrew 1953, from the early Oligocene of western Texas, is known only from a few lower dentitions and therefore is the poorest-known of all New World equid genera. As currently constituted, Hyracotheriinae is a paraphyletic taxon because it does not include within it members of the Anchitheriinae and Equinae.

Node 2. Mesohippus through Parahippus

Mesohippus Marsh 1875 and the other genera within the subfamily Anchitheriinae are set apart from the hyracotheres by at least seven shared-derived character states of the skull, jaws, dentition, and metapodials (MacFadden 1976; Prothero & Shubin 1989) (Table 5.2). Although *Miohippus* Marsh 1874 has shared-derived characters that set it apart from *Mesohippus* (Prothero & Shubin 1989), the interrelationships of the other taxa within this subfamily are not well established. One of the more interesting genera within this subfamily is *Archaeohippus* Gidley 1906, a clade of dwarf forms that evolved a unique metacarpal morphology (Sondaar 1968).

The genus *Parahippus* Leidy 1858 of the older literature (e.g., Stirton 1940) is a heterogeneous, horizontal assemblage. One lineage normally included within this genus, but here referred to *Desmatippus* Scott 1893, remained low-crowned and yet demonstrated increased size, apparently in parallel with anchitheres. Another lineage, including advanced *Parahippus,* had cheek teeth that were incipiently hypsodont and were partially covered with cement. Based on a PAUP analysis of cranial, dental, and postcranial characters, *Parahippus leonensis* is the closest outgroup of the Equinae (MacFadden & Hulbert 1988; Hulbert & MacFadden 1991).

Node 3: Anchitheriinae sensu stricto

This apparently monophyletic taxon, which is among the most poorly studied of the Equidae, is united by shared-derived characters of dentition and increased body size (Table 5.2). In North America this clade consists of three genera, *Kalobatippus* Osborn 1915 (= *Anchitherium* von Meyer 1844, *s.l.,* of some literature;

however, relative to Old World *Anchitherium, Kalobatippus* has greatly elongated metapodials), *Hypohippus* Leidy 1858, and *Megahippus* McGrew 1937, and, in the Old World, *Anchitherium sensu stricto* (e.g., Abusch-Siewert 1983) and *Sinohippus* Zhai 1962 (Forsten 1982a). In North America this is an extremely interesting group because later forms within *Hypohippus* and *Megahippus* evolved highly specialized dentitions, apparently for browsing.

Node 4: Equinae

This monophyletic clade consists of 18 currently recognized genera and represents the major adaptive radiation of hypsodont fossil Equidae, including the extant genus *Equus* Linnaeus 1758. This node is well corroborated by five shared-derived character states of the cheek teeth (Hulbert 1989; Hulbert & MacFadden 1991) (Table 5.2).

Like several other late Tertiary horse genera, *"Merychippus"* Leidy 1857, *s.l.* (i.e., including *Protohippus* Leidy 1858) has traditionally been a large, polyphyletic assemblage with many species [e.g., 36 listed by Stirton (1940)]. Recent studies, however, have separated the merychippine grade into monophyletic clades (e.g., Evander 1989; Hulbert 1989; Hulbert & MacFadden 1991). This has resulted in far fewer species currently included in the "core" concept of *Merychippus*, with the other species now being reassigned to other genera within the tribes Equini or Hipparionini. Furthermore, *Protohippus*, usually placed as a subgenus of *Merychippus* (e.g., Stirton 1940), is now considered generically distinct (Hulbert 1988a). Interestingly, cladistic analysis of *Protohippus* (Hulbert 1988a) corroborates a close relationship with equines, a hypothesis traditionally stated in the literature (e.g., Stirton 1940). Much of the taxonomic reorganization of equine horses in recent years has resulted from the use of new character complexes in addition to those of teeth and limbs, most notably preorbital facial fossae.

Preorbital facial fossae in equid systematics

Many mammals, including primitive eutherians (Novacek 1986), have preorbital facial depressions in the cheek region that serve as sites for origin of the snout and lip musculature (respectively, levator alae nasi and levator labii superioris). In many taxa of fossil horses these depressions are greatly elaborated into pits, or fossae. The dorsal one usually is called the lacrimal, naso-maxillary, or dorsal preorbital fossa, whereas the ventral one usually is called either the malar or maxillary fossa. For over a century (e.g., Cope 1889) it has been realized that these pits are distinctive and may be of possible systematic utility. However, in recent years the growing use of these structures to discern monophyletic equid clades and their phylogenetic interrelationships has been controversial. Although this topic has been dealt with extensively elsewhere in the literature (Webb 1969a; Skinner & MacFadden 1977; Bennett 1980; Forsten 1982a,b; MacFadden & Skinner 1982; MacFadden 1984b, 1987b; Eisenmann et al. 1987), because of its systematic utility it will be briefly addressed here.

Two criticisms of the use of the facial fossa complex have been that (1) it may be sexually dimorphic or otherwise variable, and (2) its function is poorly known. Critics have stated that if one or both of these assertions are correct, then facial fossae should not be used in horse systematics (e.g., Forsten 1982b; Eisenmann et al. 1987).[4]

SEXUAL DIMORPHISM. Osborn (1918) stated that within *Pliohippus* (*Dinohippus* of some more recent studies) the development of facial fossae was sexually dimorphic; yet elsewhere in that same monograph he used the morphology of the fossae to diagnose certain hypsodont horse species. In particular, workers such as Skinner and MacFadden (1977; MacFadden 1984a) have investigated this general question for late Tertiary hypsodont horses with their studies of hipparions (Figure 5.16). They have analyzed the range of morphology represented in large quarry samples of *Cormohipparion occidentale* from the late Miocene of North America. As shown in Figure 5.17, variations in the development of the dorsal preorbital fossa (DPOF) were similar in males and females. Other studies that potentially could corroborate Skinner and MacFadden (1977) have been hampered by several factors: (1) There is a paucity of well-preserved fossil equid quarry samples in which the sexes of individuals can be determined.[5] (2) In many fossil equid species the morphology of preorbital facial fossae is difficult to quantify (the teardrop-shaped DPOF bounded by a well-defined rim in *Cormohipparion* is a notable exception). (3) There is no homologous structure in living horses, and topographically similar structures in other outgroups (e.g., the lacrimal fossa and facial gland in ruminant artiodactyls) seem to have different functions. Nevertheless, so far as is known, quarry samples of fossil horses do not show sexual dimorphism in preorbital facial fossae, and thus these are valid characters in equid systematics.

FUNCTIONAL CONSIDERATIONS. Some systematists have suggested that if the function of a structure is not known with certainty, then it should not be used in taxonomy. That criticism has been leveled at the use of preorbital facial fossae in fossil horses (e.g., Forsten 1982b; Eisenmann et al. 1987). Though certainly it is better to know the function of a morphological struc-

[4]Despite Forsten's criticism of the systematic utility of facial fossae (Forsten 1982b), in the same year she (Forsten 1982a:673) stated that "the most marked difference between [*Sinohippus*] *zitteli* and *Hypohippus* relate [*sic*] to the development of the preorbital fossa."

[5]Many mammals exhibit secondary sexual characters such as dimorphism, including extant *Equus*. The sex of a fossil horse is determined by reference to modern analogy, in which individuals with large canines are interpreted to be males, and those with smaller canines to be females.

ture, I do not agree that uncertainty regarding its function invalidates its systematic utility. Furthermore, despite claims to the contrary, there are plausible hypotheses to explain the function of preorbital facial fossae in fossil horses. Even so, I shall discuss this challenge to their use.

As mentioned earlier, in modern *Equus* the facial fossae are poorly developed or virtually absent (except for some fossil species, e.g., *E. simplicidens*), and therefore we have no extant homologues for comparison. The prominent anatomist W. K. Gregory (1920) addressed the problem of their function using dissections of Grevy's zebra, *Equus grevyi*, in which the preorbital facial fossae, although rudimentary, are slightly better developed than in domesticated horses. He concluded that the malar (ventral) fossa in fossil horses was the site of origin of lip musculature (maxillo-labialis superior = levator labii superioris). However, he believed that the principal function of the dorsal (lacrimal) preorbital fossa was to house a greatly enlarged nasal diverticulum located beneath the overlying musculature. Although the exact function of the nasal diverticulum was uncertain (e.g., Matthew 1926), it may have housed a small sac that opened just within the nostrils. MacFadden (1984b) suggested that the DPOF and nasal diverticulum functioned in vocalization, although there are other possible functions. For example, a plausible hypothesis to explain the function of cheek musculature within the facial fossae can be addressed by morphological evidence from the fossil record and an understanding of the behavioral ecology of extant *Equus*.

The maximum morphological diversity of facial fossae occurred during the maximum taxonomic diversity of fossil horses in the late Miocene of North America. At that time, at least 12–15 species lived contemporaneously, and many of these were also sympatric. With that large diversity of apparently coexisting horses, they would have finely partitioned their niches. This seems to be corroborated by feeding adaptations, where, for example, the shape of the incisor arcade and muzzle, which form the cropping mechanism, varied morphologically in different clades of Miocene hypsodont horses (see Figure 11.8), not unlike the morphological partitioning of this same functional complex in modern ungulates of the African savannas (Owen-Smith 1985; Janis & Ehrhardt 1988). Thus, the diversified facial fossae and associated muscles at that time may have functioned in different feeding specializations. Another possible hypothesis relates to equid social behavior. We know that modern horses have evolved a complex social behavior in which the lips and nasal region are frequently flared during display communication (e.g., Feist & McCullough 1975; Berger 1986). We also know that the preorbital region of extant *Equus* houses muscles related to

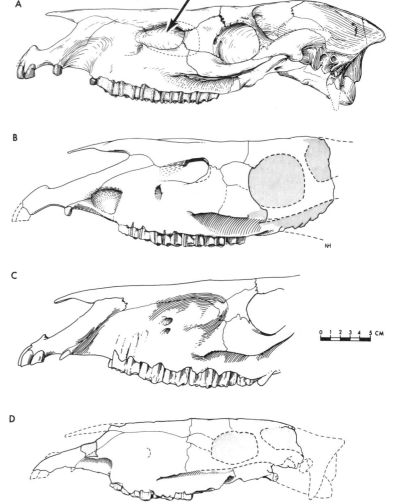

DORSAL PREORBITAL FOSSA

A

B

C

D

Figure 5.16. *The four genera of North American hipparions, showing the development of the dorsal preorbital fossa (DPOF). Modified from MacFadden (1984a; 1985a). (A) Cormohipparion is characterized by a well-developed DPOF with a well-defined rim and posterior pocket (dashed line). (B) Hipparion s.s. also has a relatively distinct DPOF, although the anterior region is less well defined than in Cormohipparion. (C) Neohipparion is characterized by a poorly defined DPOF. (D) In advanced, "typical" species of Nannippus, the DPOF is absent, although current research (R. C. Hulbert, Jr., personal communication) indicates that primitive forms within this clade have a relatively well developed DPOF similar to Cormohipparion.*

movement of the lip and nasal region. Thus, different preorbital facial musculatures could also have functioned in different patterns of social display.

In summary, the two major objections to the systematic utility of preorbital facial fossae in fossil horses (i.e., they are highly variable and are of unknown function) are without strong empirical justification. Traditionally these objections resulted from an emphasis on the use of the more ubiquitous dentitions and postcranial elements. The teeth are more practical because they are the most frequently encountered fossilized elements. The postcranial elements, although widely used, are of limited utility in diverse equid faunas, where associations with skulls and dentitions often are ambiguous. Furthermore, the great evolutionary

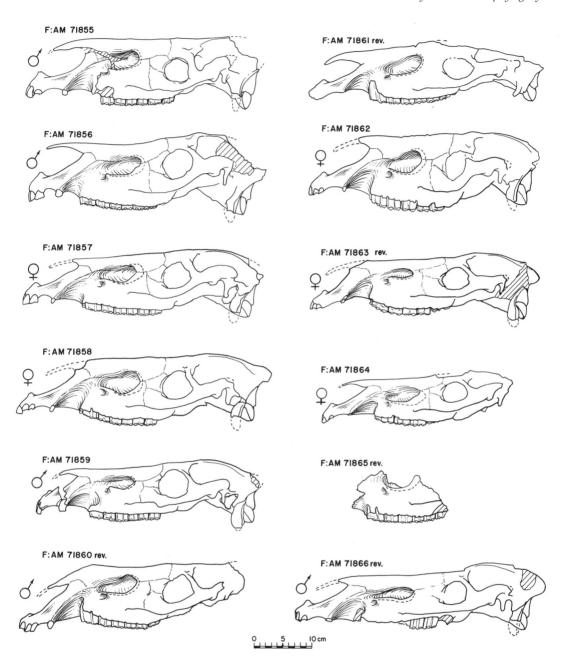

F:AM 71855

F:AM 71856

F:AM 71857

F:AM 71858

F:AM 71859

F:AM 71860 rev.

F:AM 71861 rev.

F:AM 71862

F:AM 71863 rev.

F:AM 71864

F:AM 71865 rev.

F:AM 71866 rev.

0   5   10 cm

Figure 5.17. *Quarry sample of Cor-mohipparion occidentale from the late Miocene (late Clarendonian) Hans* *Johnson quarry of north-central Ne-braska, showing males and females based on relative size of the upper ca-* *nine. So far as can be seen from this sample, the development and variation of the DPOF are not sexually dimorphic.*

plasticity of these characters requires that they be used with caution. In practice, the best systematic methodology is to use as many different characters as possible, and this is particularly so with the advent of computer-based phylogenetic algorithms.

For fossil horses this includes a combination of cranial, dental, and postcranial characters.

Node 5: Tribe Hipparionini

This clade is united by at least two shared-derived character states of the dentition and metacarpus (Table 5.2). The core taxa within this clade from North America include *Hipparion* de Cristol 1832, *Neohipparion* Gidley 1903, *Nannippus* Matthew 1926, and *Cormohipparion* Skinner & MacFadden 1977. Old World derivatives, namely, "hipparions" of previous workers (e.g., MacFadden 1984b), include *Hipparion s.s.*, *Cormohipparion*, *Proboscidipparion* Sefve 1927, "*Sivalhippus*," *Cremohipparion* Qiu et al. 1987, and *Stylohipparion* van Heopen 1932. Advanced hipparion taxa are further united by several derived characters, including persistently isolated protocones in the upper cheek teeth (Figure 5.18). This distinctive character has been of prime importance in recognition of hipparions in the traditional literature. However, starting with Skinner and MacFadden (1977), and as discussed previously, characters of the DPOF (the malar fossa is consistently absent from most hipparions) also have become critical to an understanding of hipparion systematics and phylogenetic interrelationships. As such, distinctive facial morphologies characterize the North American (e.g., MacFadden 1984a) (Figure 5.16) and Old World hipparion clades (e.g., Woodburne & Bernor 1980).

Since the nineteenth century, paleontologists have realized that hipparions were Holarctic in distribution. Some workers at that time made the distinction that *Hipparion* should be reserved principally for Old World forms, whereas *Neohipparion* or *Hippotherium* were more common in the New World (e.g., Cope 1889; Gidley 1907; Stirton 1940). These distinctions, Prothero and Schoch (1989b) notwithstanding, seem less valid in more recent literature; for example, both *Hipparion s.s.* and *Cormohipparion* have been recognized from throughout Holarctica (e.g., MacFadden 1984a) (see Chapter 7).

Node 6: Tribe Equini

This clade is united by at least six shared-derived character states of the DPOF, dentition, and limbs (Table 5.2). In addition to forms that were endemic to North America, this clade includes some morphologically derived forms that were predominantly in South America: *Hippidion* Owen 1869 and *Onohippidium* Moreno 1891 [recent work has revealed the presence of these genera in the late Cenozoic of North America (MacFadden & Skinner 1979)]. Both of these are characterized by large body size and a

A. Hipparionine   B. Equine

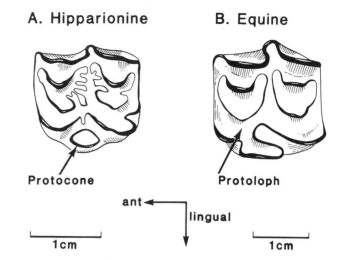

Protocone   Protoloph

ant ◄──
│
lingual
▼

1cm   1cm

Figure 5.18. *Protocone morphology in the two clades of hypsodont horses. In general, the hipparionines (A) (e.g., Nannippus minor) are characterized by an isolated protocone, whereas the equines (B) (e.g., Astrohippus stocki) are characterized by a protocone that is connected to the protoloph. Exceptions to this pattern include primitive species within these clades and individuals of early and late wear stages.*

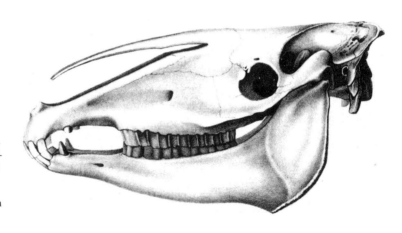

Figure 5.19. *Lateral view of beautifully preserved cranium of* Hippidion *"neogaeum" described by Burmeister (1875) [= H. bonaerense C. Ameghino of Sefve (1912)] and based on specimen from the Pleistocene ("Pampean") of Argentina. Note greatly retracted nasal notch and elongated nasal bones ("splints") characteristic of* Hippidion *and* Onohippidium.

greatly retracted nasal notch, in which the nasal bones are delicate, elongated splints (Figure 5.19). Equini is important because it contains the only extant genus within the Equidae, namely, *Equus*.

Primitive taxa within the Equini (e.g., *Protohippus*) that are dentally primitive are essentially at a merychippus grade of evolution [it is understandable that this genus was previously considered a subgenus of *Merychippus* by many workers, e.g., Stirton (1940)]. Advanced taxa have many derived characters, particularly in the dentition. One character that has been used to distinguish the tribes Hipparionini and Equini is protocone morphology, specifically, whether it is persistently isolated, as in hipparionines (Figure 5.18A), or connected to the protoloph, as in equines (Figure 5.18B). Ontogenetic differences that occur during wear obscure many distinctive characters in hypsodont horses; in many cases, allocation of an isolated tooth or other fragmentary dental material to a particular taxon is difficult or impossible. Usually a comprehensive sample that displays the range of character variation is needed for unambiguous alpha-level taxonomy of fossil horses.

Despite what is traditionally stated in the literature, primitive taxa within the Equini were tridactyl, whereas advanced forms of *Pliohippus* Marsh 1874, *Astrohippus* Matthew & Stirton 1930, and, so far as is known, *Dinohippus* Quinn 1955 were monodactyl. Many species within the Equini increased their body sizes. The genus *Calippus* Matthew & Stirton 1930 has traditionally been thought to represent a lineage of dwarf forms within the Equidae. More recent work, however, shows that trends in *Calippus* were more complex, and both size increases and decreases occurred in separate clades within this genus (Hulbert 1988a).

There has been considerable discussion in the literature concerning the origin of *Equus*. In contrast to the ideas of many nineteenth-century European paleontologists, who derived *Equus* from *Hipparion* (e.g., Gaudry 1867) (see Figure 3.2), today there is universal agreement that the closest relative of *Equus* is within the Equini. However, beyond this consensus, the exact phylogenetic interrelationships within this clade have been matters of considerable debate, particularly with the recent emphasis on new suites of characters.

It is traditionally stated in the literature that *Equus* originated from some advanced species of *Pliohippus s.l.*, namely, within *Astrohippus* (e.g., Osborn 1918; Stirton 1940; Lance 1950). More recently, however, it has been recognized that on the basis of preorbital facial fossae there are two distinct lineages within this prior concept of *Pliohippus*. One, with well-developed dorsal and malar preorbital facial fossae, includes *Pliohippus* and *Astrohippus* (Figure 5.20A,B). The other, for which the genus *Dinohippus* (Figure 5.20C) was erected (Quinn 1955), includes "pliohippines" (of the older literature) with poorly developed preorbital facial fossae. Given the observation that in *Equus* (Figure 5.20D) these structures are also poorly developed, rudimentary, or absent, recent studies suggest *Dinohippus* (not *Pliohippus s.s.* or *Astrohippus*) to be the closest outgroup to the modern genus (Quinn 1955; Webb 1969a; Bennett 1980; Azzaroli 1982; MacFadden 1984b). In fact, one advanced species of *Dinohippus*, *D. mexicanus* [= *Pliohippus mexicanus* of Lance (1950)], from the latest Miocene–early Pliocene (Hemphillian) of North America, has highly derived dental characters [including elongated protocones, complex enamel plications, and straight, high crowns (MacFadden 1984b, 1986)] that are not morphologically far re-

A PLIOHIPPUS

DORSAL PREORBITAL FOSSA

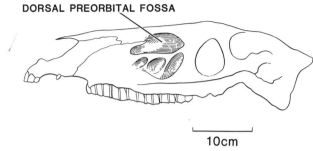

10cm

B ASTROHIPPUS

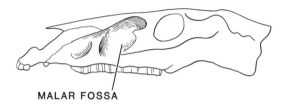

MALAR FOSSA

C DINOHIPPUS

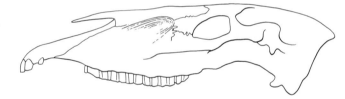

D EQUUS

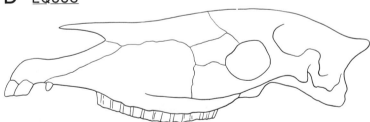

Figure 5.20. Configurations of preorbital facial fossae in the advanced equine genera Pliohippus *(A),* Astrohippus *(B),* Dinohippus *(C), and* Equus *(D). From MacFadden (1988a); reproduced courtesy of Plenum Press.*

moved from primitive *Equus* from the Pliocene (Blancan) of North America (Figure 5.21).

On the basis of distinctive zebrine versus caballine patterns of the lower dentitions (Figure 5.22A,B), certain workers have

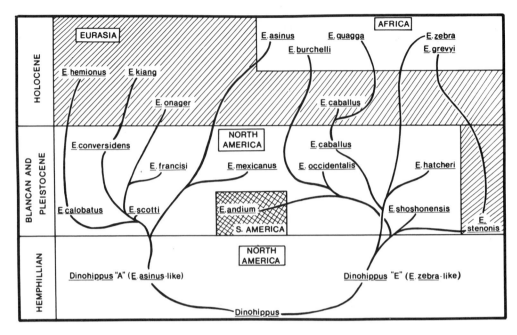

Figure 5.21. *Phylogeny of fossil and Recent species of* Equus, *and origin of* this genus *from* Dinohippus. *From Bennett (1980); reproduced with permission of the Society of Systematic Biology and author.*

Figure 5.22. *Morphology of cheek teeth in different groups of* Equus. *(A) In the zebrine pattern there is a deep ectoflexid and V-shaped linguaflexid. (B) In the caballine pattern there is a shallow ectoflexid (resulting in an isthmus) and deep, U-shaped linguaflexid. (C) In the hemionine pattern there is a shallow ectoflexid and shallow V- or sometimes U-shaped linguaflexid. It should be noted, however, that these characters are only generally applicable, and as discussed in the text, there are many exceptions to them. Not to scale. Modified from Skinner (1972) and Eisenmann (1986).*

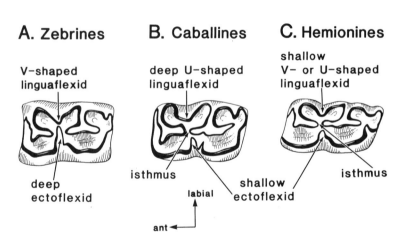

A. Zebrines

V-shaped linguaflexid

deep ectoflexid

B. Caballines

deep U-shaped linguaflexid

isthmus

labial

shallow ectoflexid

ant ←

C. Hemionines

shallow V- or U-shaped linguaflexid

shallow ectoflexid

isthmus

suggested that *Equus* had a polyphyletic origin from two or more lineages of pliohippines *s.l.* (McGrew 1944; Dalquest 1978, 1988). These dental characters, however, are known to be highly plastic, because they have undergone parallel evolution in several lineages of hypsodont horses (e.g., Eisenmann 1977), and there are few other characters that can be used to corroborate this hypothesis. Thus, arguments for the polyphyletic origin of *Equus* represent a minority viewpoint that has not received much support in subsequent literature.

Node 7: **Equus**

> *The paleontologist working with bones and the neomammalogist working primarily with external characters (hide) and geographic distribution must reach a compromise if the recently extinct (Pleistocene) and living equids are to have taxonomic stability.*
>
> – Skinner (1972:117)

Equus is obviously central to the concept of horse systematics. As Morris Skinner so aptly noted, this genus also is exemplary of the problems confronting neontologists and paleontologists in synthesizing consensus phylogenetic schemes for taxa that contain fossil and extant representatives using different sets of characters. To compound the problems surrounding this already complex genus, the taxonomic literature on *Equus* contains about 230 named species (Winans 1989), many of which are based on very fragmentary material or have inadequate descriptions. Of these, 58 have been allocated for North American fossil species, which Winans (1985, 1989) synonymizes into five valid taxa. Neontologists currently recognize six to eight (e.g., Walker 1975) valid extant species of *Equus,* and although various distinct genera have been proposed (e.g. *Zebra, Onager, Asinus, Hemionus, Hippotigris, Dolichohippus*), in the more conservative approach all of these can be considered groups (usually subgenera) within the genus *Equus* (e.g., Simpson 1945; Walker 1975).[6]

The array of characters that have been used to interrelate various species of *Equus* is astounding, to the point that there have been few studies that have attempted to integrate all of them into a single scheme. With regard to osteology, although groupings have been made using characters of the cranium and upper dentitions (Bennett 1980; Eisenmann 1980), many workers, including both neontologists and paleontologists, have focused on the importance of lower-molar characters in the taxonomy of *Equus* (Skinner 1972; Dalquest 1978, 1988; Eisenmann 1981; Forsten 1986). Although the situation is significantly more complicated than portrayed here (Groves & Willoughby 1981; Forsten 1986), generally three subgroups of *Equus* can be discerned from a few simple characters of the lower cheek teeth (Figure 5.22): (1) Old World zebrines (= stenonids) exhibit deep ectoflexids and V-shaped linguaflexids, (2) caballines exhibit shallow ectoflexids and deep U-shaped linguaflexids, and (3) asinines and hemionines exhibit shallow ectoflexids and shallow V- or U-shaped linguaflexids. Certain workers place stronger emphasis on the development of the ectoflexid (e.g., Skinner 1972), whereas others consider the linguaflexid to be more important (e.g., Forsten 1986). It should be emphasized, however, that like most characters, these rarely provide a panacea for prior problems in systematics. As pointed out by Dalquest (1988), whether a particular ectoflexid is shallow or deep, or a linguaflexid is V-

[6]On the basis of osteological characters, Eisenmann (1980, 1981) recognizes 10 valid species of modern *Equus;* however, that is a slightly higher figure than is recognized by most neontologists.

or U-shaped, sometimes depends on the subjective judgment of the investigator. Furthermore, these characters are almost certainly ontogenetically and geographically variable. For example, in northern populations of the extant hemionine, *Equus kiang*, the linguaflexid tends toward being U-shaped, whereas in southern populations it tends toward being V-shaped (Groves & Willoughby 1981). In a somewhat disparaging conclusion to the exclusive or predominant use of these characters in *Equus* systematics, Groves and Willoughby (1981:341) state that "the belief that characters of the dentition (especially the cheekteeth) are completely diagnostic of species-groups in the Equidae, is unfortunately, a myth: largely sustained, we suspect, by the wishful thinking of palaeontologists who often have nothing else to work with!"

No one would argue against using as many characters as possible to construct phylogenies and classifications, but given the considerable magnitude of this task in the case of *Equus*, such analyses have had only limited success. For six extant species of *Equus*, Groves and Willoughby (1981) presented a multivariate phenetic analysis of 65 characters involving the cranium, dentition, postcranium, external anatomy, karyotype, reproduction, and behavior. Following that comprehensive synthesis of different kinds of characters, their findings confirmed six subgeneric groupings (*Equus s.s.*, *Asinus*, *Hemionus*, *Hippotigris*, *Quagga*, and *Dolichohippus*). However, with regard to phylogenetic interrelationships, those authors concluded that it was difficult to trace any kinship among these subgenera without fossil evidence.

Various biochemical techniques developed over the past few decades have broadened the potential array of characters available for discerning the systematics of extant taxa. For example, mitochondrial DNA (mtDNA) sequencing has been used as an indication of phylogenetic interrelationships, divergence times, and evolutionary rates of particular taxa, such as humans and great apes (Miyamoto, Slightom, & Goodman 1987) and artiodactyls (Kraus & Miyamoto 1991). Not surprisingly, phylogenetic interrelationships and divergence times calculated using biochemical data have been compared with paleontological evidence, though not without controversy; see, for example, Radinsky (1978a) for one point of view. Of relevance here, some interesting systematic studies using biochemical data have recently been presented for *Equus*.

Using tissues from the placenta, liver, spleen, heart, and testes, George and Ryder (1986) analyzed mtDNA sequences in six species of *Equus*, including *E. przewalskii* (Figure 5.23), which is considered to be the ancestor of *E. caballus*. Their data produced a nested hierarchy in which the zebrines seem to form a group separate from hemionines, asinines, and *E. przewalskii*,

Figure 5.23. Przewalski's horse (Equus przewalskii) in the San Diego Zoological Park. Wild stocks of Przewalski's horse from the steppes of Asia are considered by many workers to be the ancestors of the modern domesticated horse. Photograph courtesy of Michael O. Woodburne.

Figure 5.24. Interrelationships of six extant species of Equus derived using mitochondrial DNA. Adapted from George and Ryder (1986, Fig. 4a,b combined). Although their data did not resolve the interrelationships of E. africanus and E. hemionus, they did set these species apart from the three zebrines E. burchelli, E. greyvi, and E. zebra. On the basis of other data, the most primitive species, E. przewalskii, is closest to E. caballus. The data presented by George and Ryder (1986) suggest that the extant species of Equus diverged from a common ancestor of about 3.9 Ma, which gives surprisingly close agreement with paleontological data (Lindsay et al. 1980).

Six extant *Equus* species

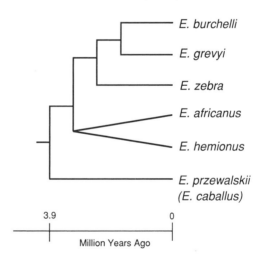

the latter being confirmed to be close to *E. caballus* based on other data (Figure 5.24). Of interest here, assuming a rate of nucleotide substitution of 2% per million years, their data suggest that the modern species of *Equus* diverged from a common ancestor at about 3.9 Ma, which gives surprisingly good agreement with evidence from the fossil record (e.g., Lindsay, Opdyke, & Johnson 1980).

Higuchi et al. (1984) investigated mtDNA in the quagga (*Equus quagga*), a South African equid that became extinct near the end of the nineteenth century (Figure 5.25). A small piece of muscle

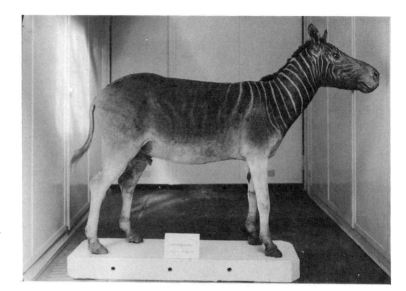

Figure 5.25. Specimen of Equus quagga *on exhibit in the Netherlands. Originally endemic to southern Africa, quaggas were hunted to extinction during the middle of the nineteenth century. Photograph courtesy of the National Museum of Natural History, Leiden.*

Figure 5.26. Interrelationships of the quagga, using reconstituted hide from a museum specimen. Adapted from Higuchi et al. (1984). These data suggest that E. quagga diverged from other Equus (e.g., E. zebra) at about 4–3 Ma; this is of similar magnitude to other biochemical (see Figure 5.24) and paleontological results for the timing of diversification of this genus. The values at the nodes (and the lengths of the branches) indicate the numbers of nucleotide sequences that differ between the taxa. Reprinted (slightly modified) by permission from Nature, 312:284; copyright © 1984 Macmillan Magazines Ltd.

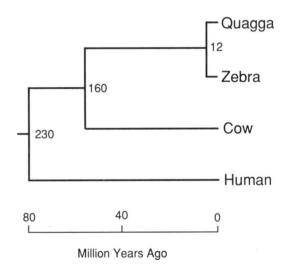

was obtained from a salt-preserved hide in the Museum of Natural History in Mainz, West Germany, from an individual that had died some 140 years earlier. From 0.7 g of tissue they extracted mtDNA, which was then replicated and compared with the mountain zebra (*E. zebra*) and two reference outgroups: the cow (*Bos taurus*) and the human (*Homo sapiens*) (Figure 5.26). Their results indicate that the zebra and quagga differ by only 5% (12 of 229 positions compared) in their mtDNA sequences and that they apparently diverged from a common ancestor at 3–4 Ma, which is consistent with the results presented earlier for the extant species of *Equus* (George & Ryder 1986) and the relevant paleontological data for this genus (Lindsay et al. 1980). More recent mtDNA studies on the quagga (Rau 1986; Hughes 1988) have shown its affinities to the zebra, *E. burchelli*, rather than the domesticated horse, *E. caballus*, the latter having been suggested as a close relative of the quagga based on osteological characters (Bennett 1980).

Those studies of the quagga were among the first reported uses in systematics of biochemical data obtained from an extinct species. In what was without question an understatement, Higuchi et al. (1984:284) concluded by saying that "if the long-term survival of DNA proves to be a general phenomenon, several fields including palaeontology, evolutionary biology, archaeology, and forensic science would benefit."

Comments on the systematics of the Equidae

For a variety of reasons, the systematics of the family Equidae illustrate several general issues that confront many practicing systematists, regardless of the group being studied.

Groups that contain both extinct and living representatives pose exceptional challenges for the systematist, for several reasons. From a practical standpoint, the paramount problem is that there will be only a limited number of characters (i.e., mostly osteological) to assess the systematics of the entire group. In fossil horses, teeth have played an immense role in systematics, and yet these structures by no means provide a panacea. Particularly in hypsodont taxa, teeth are highly prone to ontogenetic variation, and as was pointed out earlier for *E. kiang*, important characters of the lower dentition can be geographically variable. Although not an insurmountable problem, such variation must be carefully considered when constructing general morphological descriptions of populations or when using these or any other characters in systematics.

Even with a suite of potentially fossilized characters, some may not be preserved in certain taxa, and the effects of these missing data on the topology of resulting cladistic phylogenies are important topics in current research (Novacek 1988). Furthermore, some neontologists have balked at the idea of inserting newly discovered, usually extinct, primitive taxa into a phylogeny in which, in a strict cladistic approach, the resultant classification would be highly unstable (because of shifted hierarchies). Thus, some neontologists have called for abandonment of classifications that combine fossil and living taxa. It would be absurd to separate fossil and modern taxa based on these criteria, because to do so would obscure the diversity within particular clades. Of the 34 recognized genera of Equidae (see Appendix II), 33 are extinct, making this family a "taxonomic relict," *sensu* Simpson (1953a), of a once far greater diversity.

The history of systematic thought regarding all taxonomic levels, from Ungulata, to Perissodactyla, to Equidae, to the various subfamilies, to the genus *Equus*, teaches us an important principle: Systematic hypotheses evolve, and classifications are in a constant state of flux. There never will be, nor should there be, a "final" phylogeny or classification for any group. "Final" implies absolute knowledge; however, what we really do in the science (*sensu* Popper) of systematics is construct testable hypotheses of relationships using available characters in the operational taxonomic units (OTUs) currently being considered. Most classifications can represent no more than the current state of knowledge. Particularly for large groups, the systematist rarely has the same degree of confidence about all hypotheses presented in a classification. For example, in the cladogram of the Equidae (Figure 5.15), the phylogenetic interrelationships of some taxa that are represented by dichotomies have been well studied, whereas some others that are represented by unresolved multichotomies are less well known. Similarly, for classification, the validities of certain genera have been corroborated by recent studies, whereas others are less well known, and regarding the latter, future studies may or may not demonstrate that they are indeed distinct.

Another lesson to be learned is that the "old-timers" were not so far off the mark (by our present-day standards) in many cases. For example, based on dentitions, traditional studies had proposed to split up the polyphyletic merychippine grade into two natural clades, represented by the genera (or subgenera) *Merychippus s.s.* and *Protohippus*. Using other characters, in addition to those of the dentition, recent studies have corroborated at least two clades of hypsodont equine horses descended from the merychippine grade, namely, the Hipparionini (including certain species of *Merychippus s.s.*) and the Equini (including certain species of *Protohippus*). Furthermore, some of the "new" characters that we currently use, such as preorbital facial fossae, have been recognized to be of potential systematic importance for over a century. Truly novel discoveries and concepts in science are indeed rare.

The recent work on the mtDNA of the extant and recently extinct species of *Equus* demonstrates what probably will become an entirely new class of characters in fossils based on biochemical data. From a practical view, for the near future these techniques would seem to be limited mostly to Holocene or extraordinary late Pleistocene occurrences in which dried tissue is preserved, for example, in frozen proboscideans or dried sloth "mummies." However, as other biochemical data are extracted from normally fossilized skeletal tissues (e.g., Glimcher et al. 1990) and their modern counterparts, then these characters, along with those of traditional osteology, will play important roles in elucidating the systematics and evolution of groups containing both extinct and extant taxa. Thus, in the future, the artificial barrier between the paleontologist and neontologist will disappear.

6

Isotopes, magnetic reversals, fossils, and geological time

*There are philosophical topics other than time. Some of them engage our
attention for extended periods. But sooner or later each is sucked down
into the vortex of that one problem that is truly permanent: time.*
— Wood (1989:xi)

As historians of nature, paleontologists are keenly aware of the
importance of time for an understanding of the evolution of
organic diversity. No other discipline in the natural sciences can
provide insight into this question without recourse to the fossil
record. As discussed in Chapter 2, the ability of geologists and
paleontologists to discern temporal events in the past has become increasingly more precise, particularly during the second
half of the twentieth century. The purpose of this chapter is,
first, to discuss the kinds of geological time and how they have
developed over the past two centuries. There is an immense
body of literature on this subject, and more detailed discussions
of the calibration of geological time and the development of the
geological time scale are available elsewhere (e.g., Eicher 1976;
Berry 1987). The second half of this chapter will focus on the
development of the Cenozoic land-mammal geochronology of
North America, particularly as it relates to fossil horses.

Some workers use the term "geochronology" in a restricted
sense to include only radioisotopic decay techniques (e.g.,
U–Pb, Rb–Sr, and K–Ar) applied to determining geological time.
"Geochronology" is used here in the broad sense (one that connotes the etymology "study of the time of the earth") to include
all principles, techniques, and applications, whether they be
radiogenic or stable-isotopic, paleomagnetic, or paleontological.
This is conceptually similar to Woodburne's definition (1987:xiv)
of geochronology as "the discipline of discerning and ordering
events or other phenomena of Earth history with respect to
geological time."

Relative geological time

All introductory students of historical geology read in textbooks about two different kinds of geological time, namely, relative and absolute, and the fundamental differences between the two. Until the eighteenth century, although most scientists understood that fossils represented the remains of extinct taxa, few wondered about any further hierarchical arrangement of geological time. The principal reason for that was because most Western philosophers believed that the earth and its inhabitants had existed for only several thousand years, and that concept was underscored by prominent scholars. In the middle of the seventeenth century, John Lightfoot, vice-chancellor of Cambridge University, deduced that the earth had been created at 9 a.m. on September 17 in 3928 B.C. Theologians such as Archbishop Ussher of Ireland confirmed that thinking; in the 1650s Ussher stated that the earth had been created on October 23 in 4004 B.C. (Dunbar & Waage 1969; Gould 1991).

Comte Buffon was a highly influential figure in natural history during the eighteenth century. He was one of the first to advocate that the six days of creation should not be taken literally, that each "day" may have spanned long epochs of time (Dunbar & Waage 1969). Buffon performed classic experiments on the rates of cooling of metal spheres as scaled-down analogues of the earth. From those studies he estimated that about 35,000 years had elapsed since the time that the earth became cool enough to sustain life. Although originally he believed that the age of the earth could have been as much as 75,000 years, he later speculated in unpublished manuscripts that it might be much older, on the order of 3 myr (Albritton 1980; Buffetaut 1987).

Buffon also was one of the first scholars to believe that different extinct forms had lived at different times in the past (Buffetaut 1987). Conceptually that provided the foundation for faunal succession and the practice of biostratigraphy, both of which are fundamental principles in the assessment of relative geological time. In this context, two pioneering individuals are usually discussed: William Smith and Georges Cuvier. William Smith (1769–1839) was a British surveyor and geologist who made important geological maps of England and Wales during the early nineteenth century. While doing his fieldwork and examining relevant fossil collections, Smith realized that distinctive fossil faunas characterized rocks of different ages and that such variations provided useful indices to determine relative geological time. Those observations and principles were embodied in his *The Geological Map of England and Wales*, completed in 1815. On the maps and accompanying sheets, all rock formations that contained a characteristic set of fossils were given the same color. Smith's observations of patterns of fossil occurrences in sediments provided the essence of the "principle of faunal succession." The importance of that, however, was not immediately grasped by the scientific community, probably because Smith was not a learned scholar and did not publish his ideas (Berry 1987). Only later were Smith's accomplishments recognized in scientific circles. As Berry (1987:58) noted, "probably the high point in this recognition was Adam Sedgewick's address to the Geological Society of London, given in February 1831 when he presented the Society's first award of the Wollaston Medal to Smith. He stated in that address that there could be no doubt but that it was the Society's duty 'to place our first honour on the brow of the Father of English Geology.' "

At the same time that Smith was mapping the geology in Great Britain, Cuvier and his geological colleague, Alexandre Brongniart (1770–1847), were studying the succession of Cenozoic faunas in the Paris Basin. Cuvier reached several conclusions that are of significance to relative dating. First, he believed, like Buffon, that there were distinct faunas of different ages. In a lecture presented in 1801, Cuvier noted that "the older the beds in which the bones are found, the more they differ from those of animals we know today" (Rudwick 1972:127). Being a catastrophist, Cuvier also believed that each successive fauna was unique and resulted from repeated series of divine creations, followed by mass extinctions (that was different from his earlier belief in a single "world anterior to ours") (Rudwick 1972). It should be noted that neither Cuvier nor Smith was an evolutionist, and the principles of relative dating do not require evolutionary theory as part of the framework.

As a result of the observation of faunal correlation, two related concepts were born during the nineteenth century: synchrony and homotaxis. Both of these terms imply similarities in faunas, but synchrony also connotes the further assumption of age equivalence, whereas homotaxis merely implies ecological or taxonomic similarity, not necessarily synchrony. It was impossible to test those ideas when they were first formulated, but subsequent geochronological calibrations have confirmed that in all known cases, similar mammalian faunas are both synchronous and homotaxial.

Thus, by the first half of the nineteenth century, relative dating, faunal succession, correlation, and synchrony had been incorporated into the intellectual framework of geology, paleontology, and natural history. These principles and estimates of the great antiquity of the earth were to provide fundamental support for Darwin's ideas on evolution during the second half of that century. Darwin believed that the earth was 200–400 myr old. He lamented that even that amount of time was not sufficient to account for all the observed diversity of extinct and extant organisms (Patterson 1987). Though not fundamentally altering the *patterns* among extinct faunas observed by Smith, Cuvier, Owen, and other nonevolutionists or antievolutionists, Darwinian selection did provide a new paradigm to explain the underlying *process* of organic change.

"Absolute" geological time: isotopes and paleomagnetism

Early attempts to determine absolute geological time were centered around estimates of the age of the earth. Those estimates were made on the basis of observations of factors such as rates of deposition and erosion, thicknesses of ancient sediments, and the amount of salt in the oceans. They were greatly divergent, ranging from 3 myr to 1.5 billion years, considerably shorter than the current estimate of about 4.5 billion years based on studies of moon rocks, meteorites, and radioisotopes (Eicher 1976; Faure 1986). At the turn of the twentieth century, the durations of the Cenozoic and Pleistocene were thought to be, respectively, about 4.2–6.4 myr and 0.5–1.0 myr (Osborn 1910). Because of their inherent great uncertainty, those techniques and estimates still did not provide a general framework for determining "exactly" the ages of certain geological, paleontological, and evolutionary phenomena.

Although the term "absolute geological time" is firmly entrenched in the textbooks, modern geochronologists cringe at this concept because of the implication of final or absolute knowledge. Radiometric (also termed isotopic or radioisotopic) time scales are known to change, for a variety of reasons, such as (1) revisions of radioisotopic decay constants [e.g, Dalrymple (1979) for $^{40}K-^{40}Ar$] or (2) new data that recalibrate certain parts of the time scale [e.g., Swisher & Prothero (1990) for the Eocene–Oligocene boundary].

Within the past few decades, paleomagnetism has gained widespread acceptance as an "absolute" or numerical dating technique. However, it never produces an age without additional chronological data from other disciplines, most frequently radioisotopy or biostratigraphy, as discussed later. "Paleomagnetic dating" is frequently misunderstood by nonspecialists. Sometimes paleomagnetists are sent oriented rock samples from colleagues, in the hope that they can produce a "date" for an important locality or fossil. In almost all cases (except for the Quaternary) this is impossible. Paleomagnetism, or, more specifically, magnetic-polarity stratigraphy (also termed magnetostratigraphy), produces a *pattern* of normal- and reversed-polarity zones that in conjunction with interbedded radioisotopic or biostratigraphic data is then correlated to the standard, global reference, the magnetic-polarity time scale (MPTS).

Isotope geochronology

The discovery of radioactivity in the mid-1890s opened an entirely new field of scientific inquiry that has had immense consequences for geological time. In 1911 the British geologist Arthur Holmes published the first paper in which uranium-series radioactive decay was applied to dating geological samples (Eicher 1976). The principles and methodologies described in Holmes's early works were to provide the foundation for radioisotopic ("absolute") geochronology for the next several decades.

Of the many different isotopic systems that have been applied to the determination of geological time since the early twentieth century [see Faure (1986) for a synthesis], two have been employed most frequently in calibrating the Cenozoic, namely, ^{40}K–^{40}Ar (Dalrymple & Lanphere 1969) and the fission-track method (Naeser 1979). Two others, ^{40}Ar/^{39}Ar (McDougall & Harrison 1988) and ^{87}Sr/^{86}Sr (Elderfield 1986; Koepnick, Denison, & Dahl 1988), although newer on the scene, have important potential for calibrating sequences containing fossil mammals.

There is no one best geochronological technique; each has its strengths and limitations in particular situations. Since the classic article by Evernden et al. (1964), the ^{40}K–^{40}Ar method has been the mainstay of land-mammal age calibration, but it requires particular geological settings in which volcanic units, usually flows or air-fall ashes, are interbedded with fossiliferous sediments. This technique has been immensely useful in providing a chronological framework for North and South America, but because of the lack of suitable sediments, it will continue to be of limited application in the important sections in Europe and Australia. The ^{40}K–^{40}Ar method is susceptible to analytical error because two different laboratory techniques are required for age determinations, one to assess the K content, and another to assess the Ar. Furthermore, radiogenic Ar can be lost in certain sedimentary systems (e.g., during mountain-building and thermal events), and that will produce anomalous ages in the laboratory. Although these problems can be controlled by analysis of replicate samples and the use of different minerals, a new technique is now available that has also produced high-quality radioisotopic age determinations: The ^{40}Ar/^{39}Ar method, which recently has been applied to Cenozoic mammal-bearing sediments (e.g., Deino et al. 1990; MacFadden et al. 1990), does not involve many of the problems of the conventional ^{40}K–^{40}Ar method and undoubtedly will become more commonplace in future studies.

The fission-track method, also based on radioactive decay, analyzes defects or "tracks" (observable under the microscope) in the crystal lattice of a mineral (usually zircon) caused by spontaneous, time-dependent fission of ^{238}U. In the external-detector method (Naeser 1979), these naturally occurring tracks are compared with others resulting from fission of ^{235}U induced in nuclear reactors. Like the ^{40}K–^{40}Ar method, the fission-track method is prone to error if the sediments have been heated, resulting in annealing (shortening or erasure) of tracks.

Another recently developed technique analyzes the ratio $^{87}Sr/^{86}Sr$ in seawater (e.g., Elderfield 1986; Koepnick et al. 1988). In contrast to the isotopes used in the techniques described earlier, these are stable isotopes, rather than radiogenic decay series. The principle of the $^{87}Sr/^{86}Sr$ method is that this ratio has varied in the oceans over geological time in a systematic manner (Figure 6.1). Unless subsequent alteration (diagenesis) has occurred, the $^{87}Sr/^{86}Sr$ ratio at a given time will remain preserved in marine sediments and fossils deposited at that time. Thus, at any given point in the past, this ratio was the same in the oceans worldwide, and therefore such ratios can be used to determine the ages of sediments in which Sr-bearing minerals are found. For dating land mammals, this technique is primarily limited to continental margins, where marine/nonmarine sections are preserved. Recent studies have already shown the potential of this technique. MacFadden, Bryant, and Mueller (1991) used Sr isotopic ratios to date the base of the adaptive radiation of Miocene grazing horses in Florida (Figure 6.2). Webb et al. (1989) used Sr isotopes in conjunction with magnetostratigraphy to date the very rich middle Pleistocene (Irvingtonian) mammal-bearing Leisey Shell Pit in central Florida. In suitable paleoenvironments the use of the Sr isotopic method is likely to increase, particularly in situations in which the geological prerequisites (e.g., volcanic units or long sections) for other dating techniques are lacking.

Paleomagnetism

Like radioisotopic dating, paleomagnetism, more specifically magnetostratigraphy, has been widely used in recent years as a geochronological tool. During the early part of the twentieth century, geologists and geophysicists began examining the earth's ancient magnetic fields preserved as natural remanence (NRM) in rocks, which is the essence of paleomagnetism. In 1929 the Japanese scientist Matuyama realized that all late Quaternary rocks were of normal polarity (like that of the present-day field), whereas early Quaternary rocks had reversed polarity (Tarling 1983). That observation laid the foundation for magnetostratigraphy, because the earth's magnetic field has had irregular periods of normal and reversed polarities that have produced unique, worldwide patterns that can be used for correlation to the magnetic-polarity time scale (MPTS).

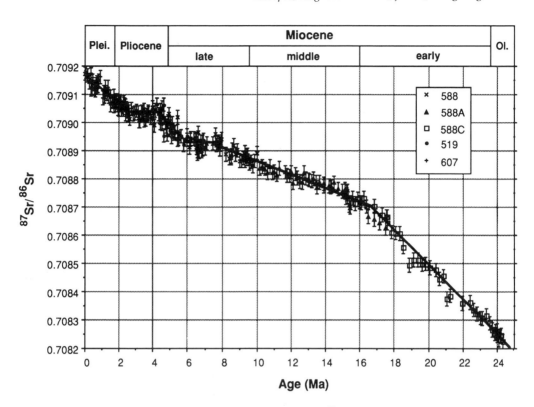

Figure 6.1. *Sr-isotopic seawater composition and variation for the Neogene. This curve allows age determinations for sediments or fossils containing $^{87}Sr/^{86}Sr$. From Hodell, Mueller, and Garrido (1991); reproduced with permission of the Geological Society of America and senior author.*

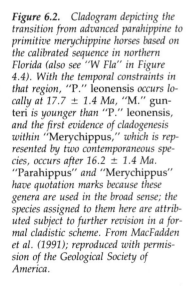

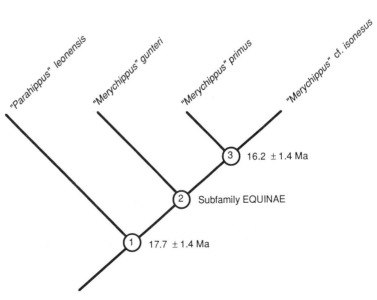

Figure 6.2. *Cladogram depicting the transition from advanced parahippine to primitive merychippine horses based on the calibrated sequence in northern Florida (also see "W Fla" in Figure 4.4). With the temporal constraints in that region, "P." leonensis occurs locally at 17.7 ± 1.4 Ma, "M." gunteri is younger than "P." leonensis, and the first evidence of cladogenesis within "Merychippus," which is represented by two contemporaneous species, occurs after 16.2 ± 1.4 Ma. "Parahippus" and "Merychippus" have quotation marks because these genera are used in the broad sense; the species assigned to them here are attributed subject to further revision in a formal cladistic scheme. From MacFadden et al. (1991); reproduced with permission of the Geological Society of America.*

Magnetostratigraphy is suitable for use with a wide variety of fine-grained sediments, but it cannot be used in coarse sands and conglomerates. The technique begins with collection of paleomagnetic horizons (sites) at regular intervals throughout stratigraphic sections that are to be dated (Figure 6.3). Sometimes this sampling regimen can result in hundreds of sites, involving literally thousands of hours of subsequent preparation and analysis (Lindsay et al. 1987; MacFadden 1993). In the laboratory, the samples are treated (demagnetized) with either alternating current or heat to remove secondary, post-depositional magnetization in order to isolate the primary magnetization and interpret the polarity that was acquired at the time of deposition. Once the polarities of the individual sites are determined, then the resultant magnetic-polarity pattern of normal and reversed zones is correlated to the MPTS. In almost all instances this correlation is done in conjunction with other interbedded geochronological data, usually radioisotopic or biostratigraphic age determinations.

Evolution of precision and improved calibrations

An important point from the preceding discussion is that no single geochronological technique or method can be considered as the best. Obviously, studies that use several techniques as independent chronological checks are to be preferred. Different techniques provide different levels of precision, that is, the size of the error (the familiar "±") associated with an age determination (Dalrymple & Lanphere 1969). In radioisotopic age determinations, the error, which, depending upon the technique or the laboratory, can represent a standard deviation or standard error, usually is a percentage of the mean determination. Accordingly, geologically older dates obviously will have larger errors in terms of number of million years. For example, assuming a 5% error, an early Pliocene date of 5.0 ± 0.25 Ma has a 0.5-myr uncertainty, whereas an Eocene date of 50.0 ± 2.5 Ma has a 5-myr uncertainty. In contrast, if a particular magnetostratigraphic event can be unambiguously identified, then the uncertainty error is constant at about 0.33 myr for the Cenozoic (Flynn, MacFadden, & McKenna 1984) (Figure 6.4). Recent work on paleomagnetic transition stratigraphy has been concerned with geological events that occurred within 10,000-year intervals, which is the time required for the earth's geomagnetic field to change from fully normal (or reversed) to fully reversed (or normal) polarity (Badgley & Tauxe 1990). This is one of the few geochronological techniques currently available that can determine pre-Pleistocene time to a precision of tens of thousands of years. Unfortunately for paleontologists, there is the requirement that the fossil to be dated by this method be contained

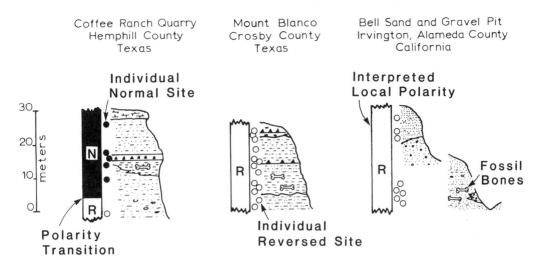

Figure 6.3. *Paleomagnetic sampling at three important sections (i.e., containing typical Hemphillian, Blancan, and Irvingtonian mammals) in western North America. At each locality, individual sites are represented by filled (normal) and open (reversed) circles to the left of the stratigraphic columns. From a sequence of these superposed sites, polarities and transitions are interpreted. Modified from Lindsay et al. (1975).*

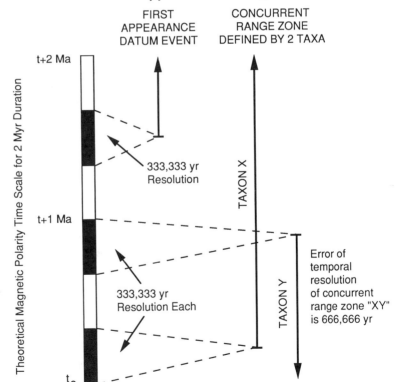

Figure 6.4. *Theoretical example of temporal resolution of a single biostratigraphic datum plane (A) and overlapping (concurrent) range zone "XY" (B). Assuming that the uncertainty of an average Cenozenic polarity chron is about 0.33 myr, then for a single event that lies within it, the uncertainty of magnetostratigraphic age determination is within 333,333 years. For an overlapping range zone bounded within two polarity chrons, the uncertainty is 666,666 years. Modified from Flynn et al. (1984).*

within the zone of polarity transition. Although transition stratigraphy has exciting potential in certain situations, its requirement of capturing a polarity transition "in the act" limits its applicability to only a small portion ($\ll 1\%$) of the time scale.

Until recent years, radioisotopic errors using the $^{40}K-^{40}Ar$ or fission-track method usually were about 5%, although some were as low as about 1% (Figure 6.5). With an uncertainty of about 0.33 myr (Flynn et al. 1984), magnetostratigraphy provided a more precise resolution of geological time during the early Cenozoic. However, because of recent advances in analytical instrumentation (i.e., mass spectrometers), as well as the use of different radioisotopic techniques, such as $^{40}Ar/^{39}Ar$, errors are now one or two orders of magnitude smaller than before. For example, recent age determinations for the Cretaceous–Tertiary boundary have errors of less than 100,000 years (Swisher 1989).

Development of Cenozoic land-mammal chronology and fossil horses

The slow stages in the attainment of perfection in the grinding teeth of the Eocene horses are of great value as time-keepers.
– Osborn (1910:45)

The principle of faunal succession was embodied in the thinking of Cuvier and was codified by Smith, and it provided the foundation for local, regional, and worldwide correlations. As fossil equid collections increased during the middle and late nineteenth century, it was realized that their widespread abundance made them important guide fossils. Thus, because of their relatively rapid evolution, along with foraminifera and rodents, horses are viewed by many paleontologists as powerful biochronological indices of Cenozoic time.

Life zones and mammal ages

Orthogenetic diagrams for extinct mammals such as fossil horses were conceived during the late nineteenth century as a means of understanding the evolutionary history of ancestors and descendants. However, these were subsequently modified, and their essence was applied in determining the geochronology of certain parts of the Cenozoic. The mammalian "life zones" or "faunal zones" of the early twentieth century (e.g., Osborn & Matthew 1909; Osborn 1910; Scott 1913) (Figure 6.6) were thought to represent roughly equal intervals of time that spanned the entire Cenozoic, although there were no techniques then available to test that assertion; it was based on the general "amount" of faunal evolution observed in each biochron. Each of these zones was thought to be unique (i.e., nonoverlapping), and their names were carried by diagnostic fossils. Although they varied, depending upon the worker, for horses the most

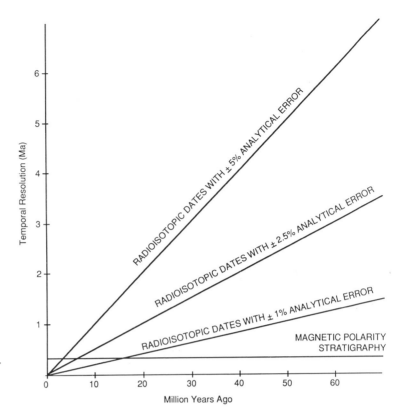

Figure 6.5. *Plot of temporal resolution for radioisotopic age determinations and magnetic-polarity stratigraphy throughout the Cenozoic. In radioisotopic ages the analytical error (which is a percentage of the age determination) increases with increasing time before the present. If the average duration of any given polarity chron is about 0.33 myr (see Figure 6.4), and if a particular chron can be identified, then there is a constant error in age determination throughout the Cenozoic. Modified from Flynn et al. (1984).*

common biochronological name-bearers for life zones included genera such as *Eohippus* (or *Hyracotherium*), *Orohippus*, *Merychippus* (or sometimes *Protohippus*), *Plesippus,* and *Equus* (Figure 6.6). Scott (1913:131) exemplified that importance when he stated that "over the Great Plains the principal Pleistocene formation is that known as the *Sheridan,* or, from the abundance of horse-remains which are entombed in it, the *Equus* Beds."

Various problems with North American land-mammal biochronology arose during the early twentieth century, including inconsistent usage, poor faunal characterization, and local terminology (Tedford 1970). Accordingly, in 1937 the Wood Committee (Wood et al. 1941) was formed to codify land-mammal nomenclature and correlation for Tertiary continental deposits of North America. In their summary article (Wood et al. 1941) they formalized the usage of many mammalian biochrons and proposed 18 "provincial ages," which subsequently have also been called "North American Land Mammal Ages" (NALMAs). Like the life zones of previous workers, these were supposed to represent a nonoverlapping succession of widespread mammalian faunas and were depicted to span roughly equivalent portions of geological time. As in previous schemes, horses were prominent in the biochronological characterization of the

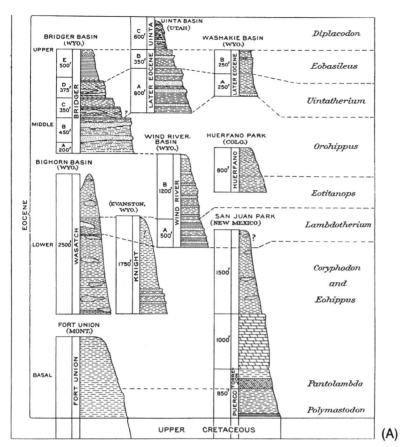

(A)

Diplacodon
Eobasileus
Uintatherium
Orohippus
Eotitanops
Lambdotherium
Coryphodon and Eohippus
Pantolambda
Polymastodon

UINTA BASIN (UTAH) — C 600' — Later Eocene Uinta — B 350' — A 800'
BRIDGER BASIN (WYO.) — UPPER — E 500' — D 375' — C 350' — B 450' — A 200' — MIDDLE — Bridger
WASHAKIE BASIN (WYO.) — B 250' — A 250' — Later Eocene
WIND RIVER BASIN (WYO.) — B 1200' — A 500' — Wind River
HUERFANO PARK (COLO.) — 800' — Huerfano
SAN JUAN PARK (NEW MEXICO) — 1500' — 1000' — 850' — Puerco Torrejon
BIGHORN BASIN (WYO.) — Wasatch 2500' — LOWER
(EVANSTON, WYO.) — Knight 1750'
FORT UNION (MONT.) — Fort Union — BASAL
EOCENE

UPPER CRETACEOUS

AGE		GENERAL DIVISIONS MTS & PACIFIC	PLAINS	HORSES
PLEISTOCENE				EQUUS
PLIOCENE	U	SAN PEDRO		PLESIPPUS
PLIOCENE	M		BLANCO	PLIOHIPPUS / HIPPARION / PROTOHIPPUS
PLIOCENE	L	SANTA FE / RATTLESNAKE / RICARDO	REPUBLICAN R. / VALENTINE / OGALALLA "LOUP FORK"	
MIOCENE	U	DEEP RIVER / BARSTOW / MASCALL	PAWNEE	MERYCHIPPUS
MIOCENE	M		SHEEP CREEK / "ARIKAREE"	PARAHIPPUS
MIOCENE	L	FT. LOGAN / JOHN DAY	HARRISON	
OLIGOCENE	U			MIOHIPPUS
OLIGOCENE	M		BRULÉ / WHITE RIVER	MESOHIPPUS
OLIGOCENE	L		CHADRON	
EOCENE	U	DUCHESNE / UINTA		EPIHIPPUS
EOCENE	M	BRIDGER		OROHIPPUS
EOCENE	L	"WIND RIVER" / "WASATCH"		HYRACOTHERIUM
PALEOCENE	U	CLARK FORK / TIFFANY		NO HORSES
PALEOCENE	M	TORREJON		
PALEOCENE	L	PUERCO		

(B)

EPOCHS	PROVINCIAL AGES	FAUNAL ZONES
PLEISTOCENE		EQUUS
PLIOCENE	BLANCAN	PLESIPPUS
PLIOCENE	HEMPHILLIAN	DIPOIDES
MIOCENE	CLARENDONIAN	EUCASTOR
MIOCENE	BARSTOVIAN	MONOSAULAX
MIOCENE	HEMINGFORDIAN	MERYCHIPPUS
MIOCENE	ARIKAREEAN	DICERATHERIUM
OLIGOCENE	WHITNEYAN	PROTOCERAS
OLIGOCENE	ORELLAN	OREODON
OLIGOCENE	CHADRONIAN	BRONTOTHERIUM
EOCENE	DUCHESNEAN	TELEODUS
EOCENE	UINTAN	AMYNODON
EOCENE	BRIDGERIAN	OROHIPPUS
EOCENE	WASATCHIAN	HYRACOTHERIUM
PALEOCENE	CLARKFORKIAN	PLESIADAPIS
PALEOCENE	TIFFANIAN	PLESIADAPIS
PALEOCENE	TORREJONIAN	PANTOLAMBDA
PALEOCENE	DRAGONIAN	DRACOCLENUS
PALEOCENE	PUERCAN	TÆNIOLABIS
UPPER CRETACEOUS		TRICERATOPS CERATOPSIA

(C)

NALMAs, with *Hyracotherium, Orohippus, Merychippus,* and *Plesippus* being name-bearers for the corresponding faunal zones (Figure 6.6). [*Equus* was associated with the Pleistocene; mammal ages for that time were not formalized until Savage (1951)]. The Wood Committee explicitly stated the means by which these ages were characterized, using four kinds of biochronological information: (1) index fossils, those taxa restricted to the biochron (e.g., *Hyracotherium* for the Wasatchian), (2) characteristic fossils, which although common in a particular NALMA also occurred in either younger or older deposits (e.g., *Mesohippus* during the Chadronian, but it was also known from later deposits), (3) first appearances (e.g., *Parahippus* in the Arikareean), and (4) last appearances (e.g, *Nannippus* in the Blancan). Of the 13 NALMAs (Wasatchian to Blancan) during which horses are known to have existed in the North American Tertiary, the Wood Committee used horses to characterize 12 of them. Only the biochronological characterization of the Orellan (Oligocene) does not mention horses (Wood et al. 1941).

In summary, several fundamental points about the NALMAs proposed by the Wood Committee were assumed to be correct, principally because there were no independent geochronological techniques then available to test them: (1) The NALMAs were depicted to be of roughly equal durations (based principally on their amounts of faunal evolution); (2) they were both homotaxial and synchronous, because everywhere they indicated the same age (relict faunas were not recognized); (3) they were believed to be nonoverlapping, and all portions of the Tertiary were represented; (4) the first- and last-appearance datum planes were synchronous throughout North America; and (5) the boundaries of the Cenozoic epochs and relevant NALMAs were depicted as coincident (e.g., the Miocene–Pliocene and Barstovian–Clarendonian). We shall return to these assumptions later in a context of improved geochronological techniques that were developed after the Wood Committee report.

Figure 6.6. Life zones and faunal stages based on the terrestrial succession in western North America, as envisioned during the first half of this century. (A) Composite stratigraphic section from Paleocene and Eocene showing faunal stages used for correlation, including the equids Eohippus and Orohippus. From Osborn and Matthew (1909). (B) Tertiary correlations using fossil horses. From Simpson (1933). (C) Part of composite correlation chart published by the Wood Committee (Wood et al. 1941) showing faunal zones, including those whose name-bearers are equids, namely, Hyracotherium, Orohippus, Merychippus, Plesippus (= Equus) and Equus.

Intercontinental correlations, datum planes, and horses: FADs and LADs

Here we see a striking example of the importance of invading types in determining the age of a fauna.   – Colbert (1935:6)

The correlation of widely distant formations is so intimately bound up with problems of geographic distribution and migration that the two series of problems must be studied and solved together.   – Matthew 1915 (in 1939:24)

Given the observation of earlier workers that most extinct mammals within faunas were recognizable only within North America, it was difficult to use NALMAs for intercontinental correlations. However, some extinct taxa, including the horses *Hyracotherium, Anchitherium,* and *Hipparion,* were known to have occurred throughout Holarctica. Thus, it was soon realized that one of the best means to achieve intercontinental correlations in different biogeographic realms was to use first-appearance datum planes, which are interpreted to represent dispersal events.[1]

The concept of a datum plane is widely used in the marine realm for the geologically instantaneous appearance or extinction of biostratigraphically important taxa, such as foraminifera during the Cenozoic, and this concept has been extended to terrestrial mammals. Thus, first-appearance (also immigration) and last-appearance (also extinction) datum planes are well entrenched in the literature. Because of the problems of ascertaining that a particular biochronological event was actually the first (FAD, first-appearance datum plane) or last (LAD, last-appearance datum plane), some workers have recently suggested the less inferential equivalents, lowest and highest stratigraphic datum planes (LSD and HSD, respectively) (e.g., Lindsay et al. 1987), particularly when discussing local or regional range zones. However, for continental and global correlations, FADs and LADs are still widely used.

The *Hipparion* and *Equus* FADs have been among the most frequently cited examples in mammalian biochronology. Workers such as Colbert (1935) have stated that both of these horses arose in North America during the late Cenozoic and subsequently dispersed into the Old World. Furthermore, those dispersal waves were rapid, on the order of 1,000 years (Kurtén 1957), certainly beyond the range of any available geochronological resolution, and thus were synchronous throughout much of Eurasia. The first appearances of *Hipparion* and *Equus* were viewed as roughly coincident with, respectively, the Miocene–Pliocene and Pliocene–Pleistocene boundaries, thus giving them further significance in the calibration of Cenozoic epochs and marine/nonmarine correlations. These equid genera continue to be biochronologically important, although their utility as guide fossils has been modified in light of more recent geochronological advances.

Geochronology during the second half of the twentieth century

As mentioned earlier, radioisotopic and paleomagnetic techniques had their origins in the early part of this century. However, they were not extensively applied to the calibration of fossil-bearing sediments until the 1960s. Although the Reynolds mass spectrometer, which has produced many of the Cenozoic K–Ar age determinations, had been developed after World War

[1] Although many workers use the term "migration," in a biochronological and paleobiogeographic context "dispersal" is preferred. As used here, a migration is a seasonal expansion or contraction of the local range of a species, whereas a dispersal is the long-term expansion of a species range.

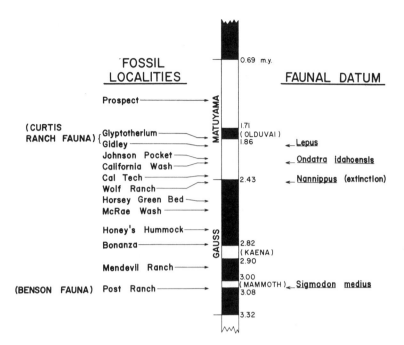

**FOSSIL
LOCALITIES**

FAUNAL DATUM

0.69 m.y.

Prospect

(CURTIS
RANCH FAUNA) {
Glyptotherlum
Gidley

Johnson Pocket
California Wash
Cal Tech
Wolf Ranch
Horsey Green Bed
McRae Wash

Honey's Hummock
Bonanza

Mendevil Ranch

(BENSON FAUNA) Post Ranch

MATUYAMA

1.71
(OLDUVAI)
1.86

2.43

GAUSS

2.82
(KAENA)
2.90

3.00
(MAMMOTH)
3.08

3.32

Lepus

Ondatra idahoensis

Nannippus (extinction)

Sigmodon medius

Figure 6.7. *Fossil mammals and magnetostratigraphy. The paper by Johnson et al. (1975) on the Pliocene–Pleistocene sequence in the San Pedro Valley, Arizona, was the first to integrate mammalian biostratigraphy with reversal chronologies and set the stage for an entirely new field of interdisciplinary research. Reproduced with permission of the Geological Society of America.*

II (Glen 1982), the magnetometers available at that time were not sufficiently sensitive to measure the magnetic intensities encountered in most mammal-bearing sediments. Although originally developed in the 1960s for use with deep-sea sediments, the spinner magnetometer was to become the mainstay of pioneering magnetostratigraphic studies on land. Since the early 1970s these two geochronological methods have been used successfully in tandem to date many fossil-bearing deposits throughout the world, including those that contain fossil horses.

Evernden et al. (1964) published a landmark paper in which they compiled 92 ^{40}K–^{40}Ar age determinations for volcanic units collected from sediments containing most of the NALMAs and spanning much of the Cenozoic era. A decade later, two important papers on magnetostratigraphy were published: (1) Johnson, Opdyke, and Lindsay (1975) used this technique to calibrate Pliocene–Pleistocene mammal-bearing sediments in the San Pedro Valley, Arizona (Figure 6.7). (2) Lindsay, Johnson, and Opdyke (1975) presented the first correlation of paleomagnetically determined late Cenozoic NALMAs for the western United States (Figure 6.3). Those papers provided the foundation for the newly expanding field of geochronology. Now virtually every important mammal-bearing locality in North America amenable to these techniques has been studied, as have many critical sites in Eurasia (most notably the hominoid-bearing Siwaliks in Pakistan) and South America (Lindsay et al. 1987; Prothero 1988; Opdyke 1990; MacFadden in press). These techniques and the vast new influx of relevant data have revolutionized geochronology and the kinds of interesting questions that can be asked about Cenozoic mammalian paleobiology and evolution. However, this new phase in geochronology has also forced us to rethink some of the basic premises of the NALMAs.

NALMAs, 50 years later

There were several assumptions on which the NALMAs were based, but they could not be tested because of the low level of geochronological resolution available at the time of publication of the Wood Committee report (1941). A half-century later, we can now ask which of those premises are still valid, which have to be modified in light of current knowledge, and how such changes will impact the biochronology of fossil horses. Five topics related to those premises are discussed next. The recent synthesis of North American land-mammal biochronology in the Woodburne (1987) volume provides an excellent, detailed view of this subject and how things have changed over the past half-century.

DURATION OF NALMA AGES. The NALMAs were originally depicted to span roughly equal increments of faunal evolution, and therefore each was considered to have about the same duration (Figure 6.6). Radioisotopic and paleomagnetic data since that time have shown that to be generally correct, although there is a quite a range of variation. With 19 currently recognized NALMAs spanning the entire Cenozoic, if all had equal durations, then each would be 3.5 myr long. Although many are close to this figure, several are widely divergent (Figure 6.8). For example, the Oligocene Orellan and Whitneyan are now thought to have been quite short, each spanning about 1 myr.[2] In contrast, the late Oligocene–early Miocene Arikareean, spanning 9 myr, is the longest NALMA (Berggren et al. 1985; Emry, Russell, & Bjork 1987; Tedford et al. 1987). As will be discussed in later chapters, these changes in land-mammal age durations have affected our interpretations of the patterns of evolutionary change in fossil horses. For example, traditionally the Oligocene was viewed as a time of "normal" rates of evolution in the genera *Mesohippus* and *Miohippus*. However, with the Orellan and Whitneyan together spanning a relatively short time of about 3 myr, the speciation of horses in that interval is now seen to have been considerably more rapid than was previously believed.

HOMOTAXY, SYNCHRONY, AND HETEROCHRONY. So far as the available geochronological data indicate, all land-mammal ages were both similar in their essential faunal composition (homotaxial) and synchronous throughout the regions in which

[2]The Cenozoic time scale continues in a state of flux as more geochronological data become available for critical stratigraphic sequences. For consistency, I use the time scale of Woodburne (1987) for NALMAs and that of Berggren et al. (1985) for the MPTS throughout this book, even though I realize that there have been and will continue to be modifications to various portions of those scales [e.g., see Swisher & Prothero (1990) for the Eocene–Oligocene boundary].

Figure 6.8. *Histograms of durations of NALMAs, using the time scale of Woodburne (1987): RLB, Rancholabrean; IRV, Irvingtonian; WHI, Whitneyan; ORE, Orellan; TOR, Torrejonian; PUE, Puercan; BLA, Blancan; HEM1, Hemphillian; CLA, Clarendonian; DUC, Duchesnean; BRI, Bridgerian; CLC, Clarkforkian; TIF, Tiffanian; BAR, Barstovian; HEM2, Hemingfordian; CHA, Chadronian; UIN, Uintan; WAS, Wasatchian; ARI, Arikareean.*

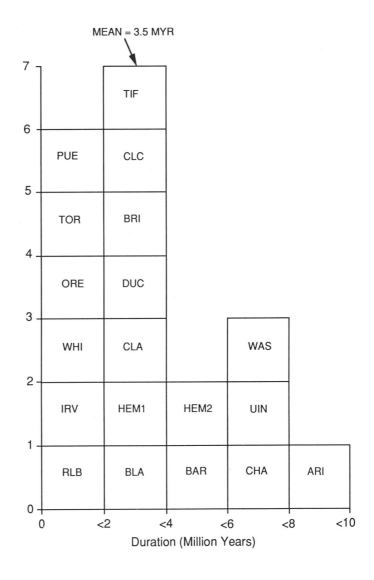

[3]The term "heterochrony," which literally means "different time," has been used in several senses. One current use in the evolutionary literature is developmental or ontogenetic heterochrony (McKinney 1988), in reference to the different timing of acquisition of characters or traits in successive populations, lineages, or clades. In contrast, as used here, heterochrony implies different ages for faunas of similar taxonomic compositions.

have been characterized. In recent years, a challenge to these important assumptions has been presented by Hickey et al. (1983). On the basis of magnetostratigraphic correlations they suggested that high-latitude Wasatchian faunas from the Canadian Arctic existed some 2–4 myr earlier than those of the well-known Rocky Mountain basins in the western United States. Obviously, this faunal heterochrony,[3] or nonequivalence of geological time, if correct, would have serious consequences for biochronology, including William Smith's principle of faunal succession. The original paleomagnetic data and correlations, however, were tenuous (Flynn et al. 1984; Kent et al. 1984), because recollecting, reanalysis, and revised paleomagnetic correlations to chrons C25 and C26 on the MPTS now indicate synchrony of the Arctic and lower-latitude Wasatchian faunas (Tauxe & Clark 1987). As will be discussed in Chapter 7, more precise geochronological calibrations of *individual taxa* of some fossil horses do have varied temporal ranges in different regions. Thus, there are examples of "relictual" taxa that persist in certain regions; for example, *Cormohipparion* persisted into the Pliocene (ca. 2 Ma) in the southeastern United States, whereas it had become extinct in the midcontinent about 6–7 Ma (Hulbert 1987, 1988b). However, these same calibrations confirm the long-held view that equid *faunas* are valid biochronological indices for Cenozoic terrestrial sequences.

NONOVERLAP OF NALMAS. When the Wood Committee (1941) formalized the NALMAs, their intent was to have a sequence of nonoverlapping biochrons that spanned the entire Tertiary, with no temporal gaps. Of the 18 NALMAs originally proposed by the Wood Committee, 17 are still valid. Only the Dragonian is now considered to represent part of the early Torrejonian (Archibald et al. 1987; *contra* Tomida & Butler 1980; Tomida 1981).

After the Wood Committee report, some workers proposed additional NALMAs. Because Wood et al. (1941) addressed only the biozonation of the North American Tertiary, Savage (1951) proposed the terms Irvingtonian and Rancholabrean for the Pleistocene. Unlike these names, which are now universally accepted, other NALMAs proposed subsequent to the Wood Committee met a different fate. The most common of these was the Valentinian, as a local, and faunally distinct, NALMA for the Great Plains and High Plains region. However, it overlapped the late Barstovian and early Clarendonian and thus has not been used in recent synthetic compilations and discussions (e.g., Tedford et al. 1987). Other suggestions, such as Geringian (Martin 1975) and Harrisonian (Wilson 1960) for, respectively, the late Oligocene and early Miocene parts of the Arikareean, were poorly characterized and locally conceived, and they overlapped existing NALMAs. These Tertiary biochrons proposed subsequent to Wood et al. (1941) have therefore not been widely adopted and are not considered valid (e.g., Tedford et al. 1987). Thus, the original intent that the NALMAs be sequential and nonoverlapping has been confirmed by a half-century of subsequent geochronological advances.

SYNCHRONOUS DATUM PLANES. Dispersal events are considered by many biostratigraphers to be powerful indicators of time equivalence, or synchrony. Given the rates of passive dispersal (mixing of foraminifera) in the oceans and geologically instantaneous dispersal events on land (Kurtén 1957), this has

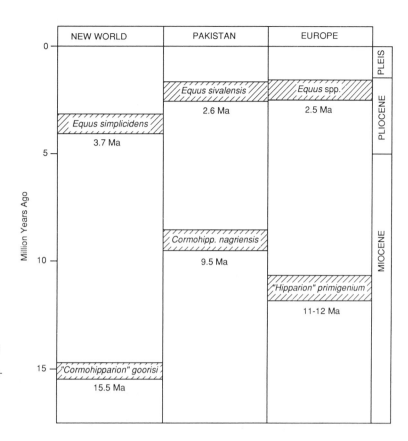

Figure 6.9. *Calibration of the "Hipparion" and* Equus *FADs throughout Holarctica. Although these two events were thought to have been roughly synchronous by earlier workers, they are now known to have been time-transgressive. Modified from Flynn et al. (1984) and Lindsay et al. (1980).*

not been challenged. However, because of our powers of resolution, this premise has recently come into question.

One of the best examples of how increased precision requires rethinking of dispersal events concerns the *Hipparion* and *Equus* FADs throughout Holarctica that Colbert (1935) believed to be roughly synchronous. It now seems that the primitive hipparion *"Cormohipparion" goorisi* was close to the base of the clade that dispersed into the Old World (MacFadden & Skinner 1981; Woodburne, MacFadden, & Skinner 1981). On the basis of faunal correlations of radiometrically calibrated sites it seems that the first appearance of *"C." goorisi* in North America was at about 15.5 Ma (Figure 6.9). In contrast, the first appearance of close relatives in the Old World seems to have been at 11–12 Ma in Europe, and then about 9.5 Ma (Badgley, Tauxe, & Bookstein 1986) in Pakistan, the latter probably representing a regional paleobiogeographic isolation within the Indo-Pakistan subcontinent.

Similarly, the origin and dispersal of *Equus* are now perceived to be time-transgressive, rather than synchronous as reported in previous studies (Colbert 1935). As discussed in Chapter 5, the genus *Equus* descended from an advanced species of *Dino-hippus*, close to the late Hemphillian *D. mexicanus*. On the basis of calibrations of relevant stratigraphic sections it seems that the most primitive *Equus*, *E. simplicidens*, first occurred in North America at about 3.7 Ma (Figure 6.9). Its first occurrence through-out the Old World was significantly later. By about 2.5–2.6 Ma, *Equus* consistently occurred at many calibrated sites throughout the Old World (Lindsay et al. 1980; Azzaroli 1982, 1990). In summary, these two examples of datum planes that involve fossil horses are now time-transgressive between North America and the Old World. Thus, given the current state of geochron-ological resolution, dispersal events between continents can be identified, and the potential exists for future recognition of re-gional dispersal events within continents.

COINCIDENCE OF EPOCH AND AGE BOUNDARIES AND EQUID DISPERSALS.   In the absence of evidence to the contrary, it was certainly understandable that earlier workers believed that the boundaries of the NALMAs approximated the Cenozoic ep-och boundaries. Thus, the following were considered coincident in the Wood Committee (1941) report (Figures 6.6 and 6.10): (1) base of Puercan = base of Paleocene; (2) Clarkforkian–Wasatch-ian = Paleocene–Eocene; (3) Duchesnean–Chadronian = Eocene–Oligocene; (4) Whitneyan–Arikareean = Oligocene–Miocene; (5) Barstovian–Clarendonian = Miocene–Pliocene; (6) end of Blancan = end of Pliocene. Several lines of evidence, most notably from marine/nonmarine intercontinental correla-tions, as well as radioisotopic and paleomagnetic calibrations, require significant revisions in our current thinking. Of the six NALMA/epoch boundaries listed, none are now considered coincident. In perhaps the most divergent example for Cenozoic land-mammal biochronology, the Barstovian–Clarendonian boundary, now taken at 12.5 Ma, is some 7.5 myr older than the Miocene–Pliocene boundary at 5.0 Ma (Berggren et al. 1985; Tedford et al. 1987).

The FADs for horses have also been thought to be equivalent to important biochronological and epoch boundaries. Thus ac-cording to Wood et al. (1941), *Hyracotherium* and *Orohippus* first appear, respectively, at the bases of the Wasatchian and Bridg-erian (both Eocene) (Figures 6.6 and 6.10). Similarly, the *Mery-chippus* FAD is depicted at the base of the Hemingfordian (Miocene), the *Plesippus* FAD at the base of the Blancan, and the *Equus* FAD at the base of the Pleistocene. As in the case of the other revisions to our concept of the NALMAs described earlier, these coincidences of equid FADs and epoch or age boundaries have become obsolete. This has resulted from numerous factors, most notably revised calibrations and revised taxonomy. For

A.
Wood et al. (1941)

B.
Woodburne (1987)

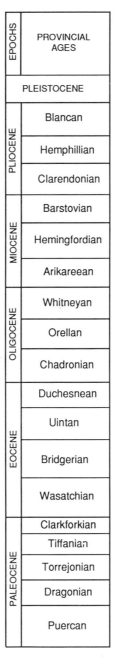

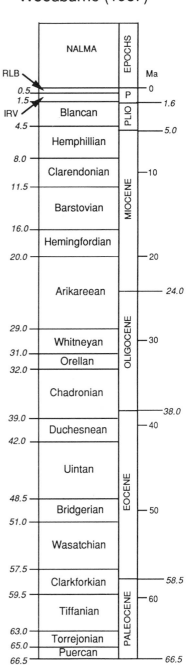

Figure 6.10. *NALMAs, then and now: (A) the original proposal of the Wood Committee (1941) and (B) the current scheme. Modified from Woodburne (1987, Fig. 10.1). The current scheme is calibrated in millions of years ago (Ma) on the radioisotopic time scale.*

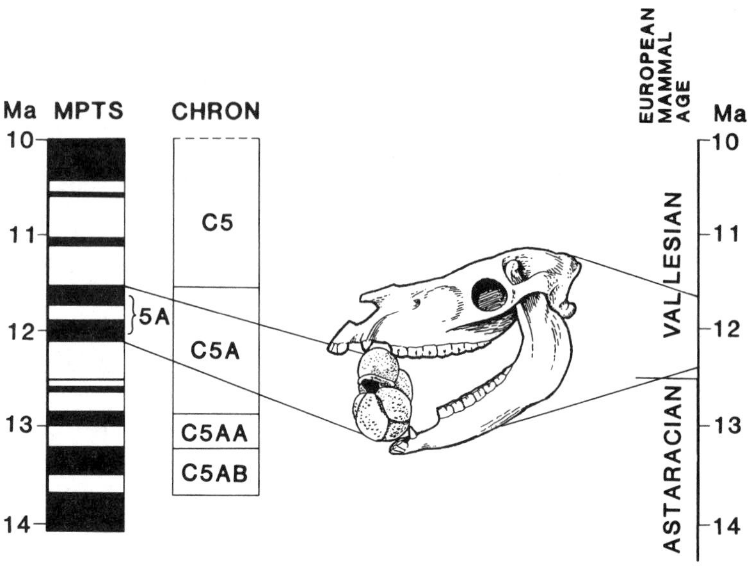

Figure 6.11. *Illustration symbolizing the interdisciplinary nature of correlations in the European Neogene:* Hipparion s.l. *with the planktonic foraminiferan* Globigerina nepenthes *in its mouth (not to scale). Adapted from Berggren and Van Couvering (1974), with more recent time scale of Berggren et al. (1985), central figure reproduced with permission of Elsevier Scientific Publishing Company.*

example, the *Equus* (which now includes *Plesippus*) FAD ranges from 3.7 to 2.5 Ma throughout Holarctica (Figure 6.9); it therefore is significantly older than the base of the Pleistocene, now taken at about 1.6 Ma.

Concluding remarks

The land-mammal ages in North America (NALMAs) and on other continents form the temporal framework for an understanding of the distribution of fossil horses. These biochronological units have undergone periodic refinements and revisions over the years, principally as a result of new geochronological methods and techniques. The precise calibrations and correlations now available are interdisciplinary in scope and combine research resulting from previously disparate fields, such as vertebrate paleontology, micropaleontology, radioisotopes, and paleomagnetism (Figure 6.11).

Without a well-developed understanding of systematics (Chapter 5) and biochronology, studies seeking to interpret the paleobiogeographic distribution of extinct Equidae would be far less informative, and studies of evolutionary patterns and processes would be either impossible or invalid. All succeeding chapters in this book build upon this framework. Furthermore, as we can see from the development of more precise geochronologies, scientific thought also evolves. This allows us to ask finer-scale questions in paleobiology, questions that previously would have been outside the realm of scientific inquiry.

7

Ancient geography, changing climates, dispersal, and vicariance

It is hardly more than a pardonable exaggeration to say that the deter-
mination of the meaning of nature reduces itself principally to the dis-
cussion of the character of time and the character of space.
                                                    – Whitehead (1957:33)

Even before the family Equidae achieved prominence as evidence for evolution, the ancient distributions of fossil horses had captured the imagination of nineteenth-century natural historians. During the voyage of the H.M.S. *Beagle* in the 1830s, when Darwin went ashore in the region of Rio La Plata (in what is now Argentina), he collected fossil equid specimens in association with other extinct Pleistocene mammals. He was so taken with that occurrence that he remarked in his later writings (Darwin 1852:130) that "certainly it is a marvellous fact in the history of the Mammalia, that in South America a native horse should have lived and disappeared, to be succeeded in after ages by countless herds descended from the few introduced with the Spanish colonists." Sir Richard Owen (1840b) described the fossils brought back on the *Beagle*, including specimens of Pleistocene *Equus*, which were later thought to be closely related to the then-extant South African quagga (*E. quagga*). Those similarities caused interesting speculation about possible ancient South Atlantic or Antarctic land bridges (Murray 1866).

In the second half of the nineteenth century, collections of fossil horses began to be accumulated from Cenozoic deposits throughout the Northern Hemisphere. Judging from the names given to those specimens, paleontologists understood that biogeographic interchange had occurred between the Old World and the New World. For example, the use of the genus name *Hyracotherium* (originally described by Owen in 1840 from the London Clay) by most North American paleontologists of that time (except, not surprisingly, Marsh, who gave it the junior synonym *Eohippus* in 1876), as well as Leidy's description of *Palaeotherium* in 1851 from the *mauvaises terres* of the Dakota Territory, implied intercontinental connections and mammalian dispersal. Likewise, in Europe, the use of the name *Hippotherium* (by numerous workers), which was originally described from North America, implied the same notion.

The purpose of this chapter is twofold: In the first part I briefly review the development of biogeographic thought as it relates to Cenozoic land mammals. This will be done first in a context of stable continental configurations and then in light of the plate-tectonic paradigm. In the second part I present a discussion of fossil horses as a case study in light of modern paleobiogeographic theory.

Biogeography before plate tectonics

As far back as the time of the Greek philosophers, particularly Aristotle, it was realized that the distribution of life forms on earth had changed through time. During the Middle Ages it was realized that the presence of marine invertebrate fossils in areas no longer covered by the seas indicated changing geography. Buffon and Cuvier are generally regarded as the founders of the science of biogeography. Lull (1917) clearly described the essence of biogeography, which he divided into three types of distributions: (1) geographic, horizontal or surficial, (2) bathymetric, vertical or altitudinal, and (3) geological, time or duration.

Of relevance to vertebrate paleontology, Buffon observed, as did Benjamin Franklin and Darwin (Buffetaut 1987), the remains of large, extinct herbivorous mastodons that presumably had fed upon lush, "tropical" vegetation entombed in deposits at high latitudes, whereas today their relatives are restricted to a more equatorial distribution. To Buffon, that ancient distribution indicated significant climate changes and fit his model of a cooling earth. Such observations by early natural historians initiated the close interconnection between inquiries about climate change (paleoclimatology) and the distribution of ancient life (paleobiogeography).

During the middle and late nineteenth century, several important books were written on the biogeography of extinct and living mammals (Murray 1866; Sclater 1874; Wallace 1876; Allen 1878; Lydekker 1896; Sclater & Sclater 1899). Of those, perhaps Wallace's has been the most popular because it included much of Darwin's thinking in a biogeographic context. During that time the now-familiar "biogeographic regions" were proposed by Sclater (1858) based on modern-day distributions of avifaunas (Figure 7.1): (1) Australian, (2) Neotropic, (3) Ethiopian, (4) Oriental (or Indian), (5) Nearctic, and (6) Palaearctic [the term "Holarctic" was proposed by Heilprin (1887) to encompass the

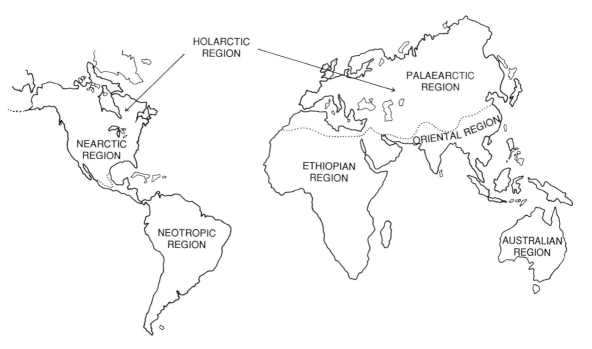

Figure 7.1. *Biogeographic regions of the world, based on modern distributions. Modified from Simpson (1965).*

latter two regions]. In all of those essays, ancient distributions were discussed in light of the currently available paleontological evidence. However, as in the case of Darwin discussing evolution, the fossil record tended to muddle rather than elucidate biogeographic theory. Thus, it was soon realized that biogeographic principles based solely on modern distributions would have to be modified in light of the fossil record. By the beginning of the twentieth century, as the first wave of the great paleontological expeditions produced extraordinary and important collections of Cenozoic land mammals, particularly from the New World (those from the Old World had been known for some time), the paleobiogeographic significance of those discoveries was readily discussed by the prominent paleomammalogists of that time (Osborn 1910; Scott 1913).

Determinants of land-mammal distribution

The known facts point distinctly to a general permanency of continental outlines during the later epochs of geologic time.
— Matthew 1915 (in 1939:3)

Biogeography is the study of observed distributions of life as well as the processes that resulted in those patterns. Thus, natural historians have been keenly interested in understanding the geological, climatic, and organic influences on the distribution of life, all of which are interrelated.

Geological framework

Until only a few decades ago, much biogeographic thought was predicated on the assumption of stable continental configurations, and that was espoused by many influential scientists of this century (e.g., Matthew 1939; Darlington 1957; Simpson 1965). Even in his later years, as the weight of plate-tectonic evidence undeniably supported continental drift, Simpson still maintained that it was not a great factor in understanding the distribution of Cenozoic land mammals. His reasoning was that 65 myr of plate tectonism had not been sufficient time for drift to be as important as other geological factors.

In a stabilist context, the dominant mechanism to explain changing continental configurations was global sea-level fluctuation (obviously related to climate, as discussed later), which could give rise to marine barriers during transgressions and emergent bridges during regressions, particularly in areas of shallow continental shelves and margins. Although over-water dispersal was considered plausible, land bridges frequently were invoked as mechanisms for ready intercontinental dispersal of land mammals. Perhaps the most celebrated (or notorious, depending upon one's perspective) of these, for which there is little geological evidence, was "Atlantis," the late Tertiary land bridge postulated to have connected the Old World and the New World (e.g., Murray 1866). Although Atlantis was originally proposed based on plant distributions, Leidy (Murray 1866) favored such a land bridge to explain the similarities of Miocene horses from Nebraska and Europe. In discussing land-mammal dispersal mechanisms, Simpson (1980:215) lamented that "reckless biogeographers late in the 19th and early in the 20th century wound up giving land bridges a bad name. Land bridges were constructed by undisciplined minds to explain any and all occurrences of related animals in now separated regions of the earth."

In a series of writings, Simpson (e.g., 1940b, 1953b, 1965) did much to clarify the processes of land-mammal dispersal. He coined the now-familiar terms "corridor," "filter," and "sweepstakes" to describe the ease of dispersal, or lack of barriers, and the amount of faunal similarity in regions being compared (now called the Simpson index) (e.g., Flynn 1986). At one end of the spectrum, a "corridor" implies the fewest barriers to dispersal and the highest faunal similarity, for example, the distribution of extant mammals in Florida and New Mexico (Simpson 1936, 1940b). In contrast, "sweepstakes" implies very low chances of dispersal and corresponding low faunal similarity, for example, extant land-vertebrate faunas in northern South America versus the Caribbean Antilles, where the more distant the island in the archipelago, the less similar the fauna to that on the mainland.

Sweepstakes dispersal implies the presence of significant barriers, and those taxa that "win" the dispersal lottery usually are viewed as doing so via island-hopping or rafting. Although certain biogeographic events may seem highly improbable, such as the over-water dispersal of rodents and primates into South America at about 25 Ma (MacFadden et al. 1985; MacFadden 1990), Simpson was an advocate of the "power" of geological time. Hence, if given enough time, the improbable becomes probable, thereby making chance events plausible (e.g., Simpson 1952b, 1980). "Filters" imply effects, usually climatic (but they also can be altitudinal), that tend to restrict or filter out the dispersal of certain taxa, whereas others are relatively unaffected by this condition. Examples of this include the lack of primates at high latitudes throughout most of the Cenozoic, and the filtering of certain mammals by the isthmus of Panama since about 4 Ma. As will be seen later, Simpson's corridor-filter-sweepstakes triad has been modified in light of plate-tectonic theory.

Climate change

Since the late eighteenth century it has been recognized that distributions of land mammals have been greatly influenced by climate change (e.g, the example cited earlier concerning extinct proboscideans used by Buffon, Franklin, Darwin, and others). Thus, it is generally stated in the literature (and has been confirmed by paleotemperature studies, as discussed later) that the early Tertiary earth was generally warmer, with more equable climates at relatively high latitudes (up to 70°S and 70°N). This has been confirmed by relevant fossil evidence, such as the presence of cold-sensitive crocodilians and turtles in the Canadian Arctic (West, Dawson, & Hutchison 1977) and cool-temperature paleofloras in the Antarctic (Case 1988). That climatic regime changed during the Eocene, resulting in a 10°C decrease in mean global temperature, increased aridity, and contraction of the tropical and temperate regional belts (Wolfe 1971, 1978; Hickey 1980). Although that global climatic change was originally recognized from paleobotanical evidence, for fossil mammals it also was a time of increased extinctions and faunal shifts worldwide (e.g., Prothero 1989). (Because of the major taxonomic turnover in Eocene–Oligocene mammal faunas in France, that time is also referred to as the "Grande Coupure," i.e., "the big cut.") Scott (1913:116) summarized the prevalent thinking during the early twentieth century based on qualitative observations: "There are many reasons for believing that the Oligocene climate marked the beginning of the very long and gradual process of refrigeration which culminated in the glacial conditions of the Pleistocene epoch."

Although temperature is one of the most important factors related to climate change, many others have the potential to influence the distribution of land mammals. Most notably, the increased global aridity during the late Cenozoic seems to have affected the adaptations and distribution of ungulate mammals, particularly fossil horses (Gregory 1971; Webb 1987). Some workers have cited other climatic factors affecting paleobiogeography and evolution, such as wind directions, rainfall patterns, planetary albedo, and even orbital cycles, but up to now these usually have been discussed in the context of pre-Cenozoic life or within the marine realm.

Organic interactions

Changes in the distributions of certain species within a particular region can have a profound effect on other life that exists within the fragile webs of ecosystems. For example, the spread of primary producers (i.e, the photosynthetic plants) on land, both in theory and from empirical observations, affects all dependent levels within the food chain, including the primary consumers, herbivorous mammals, and the "top carnivores." This is exemplified by the spread of grasslands in the middle to late Tertiary of the New World. Grasses first occur in the South American fossil record no later than about 35 Ma, and presumably they affected the rapid evolution and spread of relatively high crowned indigenous notoungulates and litopterns throughout that continent (Patterson & Pascual 1972). Paleobotanical evidence (Elias 1942; MacGinitie 1962; Retallack 1982) indicates that grasslands spread throughout North America some 15 myr later, and that newly available biome had a profound influence on the evolution and distribution of grazing ungulates, including horses.

A complex interaction: case study from the late Tertiary of the Great Basin

Several factors that influenced ancient distributions of life have been mentioned separately. However, to examine only one of these determinants in isolation from the others belies the complexity of natural ecosystems and communities. In reality, changing geology, geography, climate, and biotas are all interrelated. For fossil horses this complexity has been realized since the classic studies (e.g., Kowalevsky 1873) and is nicely exemplified by Shotwell's insightful work (1961) on fossil horses from the Great Basin of the western United States (see Figure 4.4). His study is somewhat dated in terms of current taxonomy and geochronology. However, the general patterns of change that Shotwell observed represent a valid case history using fossil horses.

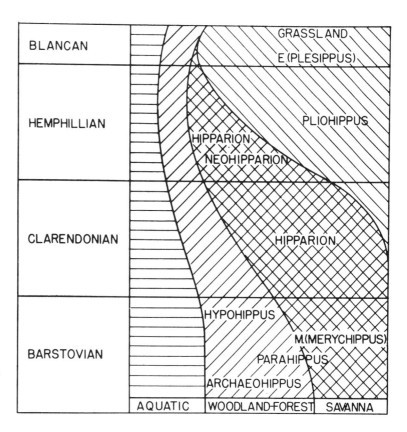

Figure 7.2. *Changing proportions of equid faunas from the Great Basin, Nevada, during the Miocene. From Shotwell (1961). Although the systematics of some of the equid genera portrayed here are now out of date, the general patterns and changes suggested by Shotwell for the horses of that region still seem valid. Reproduced with permission of the Paleontological Society.*

Shotwell collected a sequence of mammalian faunas from the Great Basin that ranged from Barstovian to Blancan in age and represent a duration of about 12–13 myr (Tedford et al. 1987). Not unexpectedly, horses were a dominant component of those faunas. Shotwell noted significant differences in the taxonomic compositions of those horses (Figure 7.2): Browsing taxa (e.g., *Archaeohippus* and *Hypohippus*) were predominant during the Barstovian; they were then replaced by grazing equines, namely, the three-toed hipparions, followed by one-toed "pliohippines" (Equini in the usage here).

Although undoubtedly there were evolutionary interactions within the mammalian community that shaped the composition of the Great Basin faunal sequence, Shotwell understood that other climatic and biotic factors were also important. Numerous paleobotanical studies [cited by Shotwell (1961)] from that region had indicated a predominantly woodland/forested biome during the middle Miocene (Barstovian) that changed to a more open, savanna-grassland biome during the late Miocene–Pliocene (Clarendonian–Blancan). Those floral changes seem to have resulted from increasing global aridity during the late Tertiary. Thus, Shotwell (1961) interpreted the shift from browsing to grazing adaptations, as well as from forested to more open country equid faunas in the Great Basin, to have resulted from a complex interaction of ancient organismic, ecological, and climatic determinants.

Centers of origin and patterns of dispersal

The source of the world's supply of mammals, the great homes, centers, or continents from which the orders evolved and took their distinctive form, still remains as one of the great problems of the Age of Mammals.
— Osborn (1910:64–5)

Were I a historical biogeographer, I would most likely pursue two great tasks: (1) to find the place of origin, traced through the wanderings, of the taxon or taxa of interest to me; and (2) to find the place of origin of the taxa comprising the fauna (or biota) of interest to me.
— Udvardy (1981:17)

As indicated by these quotations, much discussion in the literature has been aimed at determining the centers of origin and subsequent dispersal patterns for many different kinds of organisms. At the same time, however, many biogeographers have realized that such pursuits frequently have yielded unsatisfactory results. A complete discussion of centers of origin and patterns of dispersal would be far too complex to present here, and many such discussions are already available in the literature (e.g., Darlington 1957; Udvardy 1969; Nelson & Platnick 1981; Nelson & Rosen 1981).

Systematists working on predominantly extant taxa with poor or nonexistent fossil records have developed a series of criteria to recognize centers of origin and subsequent dispersal (e.g., Cain 1944). Of these criteria, diversity gradients and abundances have frequently been used to identify centers of phylogenetic origin for various taxa. Thus, regions with the greatest taxonomic diversity and abundance are closest to the center of origin. In a classic example, the southern high-latitude extant distribution of *Nothofagus*, the southern beech, indicates the region of origin and diversification of this genus (Darlington 1965).

Although these kinds of assumptions may be correct in some instances, there are many exceptions to such a "rule," and therefore they are rendered of dubious value in biogeography. Take, for example, the Equidae, and assume that there was no fossil record of this group. Its present-day distribution would argue for an Old World origin for this group, but with the addition of fossil data, the origin of the Equidae becomes a more complex question. Although perplexing at times, the fossil record has been greatly relied upon for elucidating centers of origin for taxa that either are entirely extinct or contain both extinct and living representatives.

The recognition of ancient centers of origin requires that first appearances (FADs, to biostratigraphers; see Chapter 6) and subsequent dispersal be discernible from the fossil record. However, this practice must be scrutinized in all cases because of the problem of potential biases in poorly sampled sedimentary sequences. With well-dated and well-collected stratigraphic sections, FADs are becoming increasingly more valuable for recognition of centers of origin. The *"Hipparion"* and *Equus* FADs throughout Holarctica are prime examples of how the paleontological record, coupled with precise geochronology, can resolve dispersal events. Most recorded dispersals of land mammals throughout continents have taken from hundreds of years to about 10,000 years; consider, for example, North American Pleistocene mammals (Kurtén 1957), or the historical introduction of *Equus* into the Americas (Berger 1986, 1987). Although the currently available geochronological precision is unable to resolve events within these orders of magnitude, this may change as future analytical advances are made in dating techniques.

In addition to investigating individual taxonomic events, systematists for over a century have been discussing the paleobiogeographic origins of higher taxonomic categories and even entire faunas. Thus, Sclater's biogeographic regions, although originally used to describe distinct faunal entities, were later used as centers of origin, as, for example, in Osborn's concept (1910) of "creative centers of origin." Not surprisingly, many workers in North America and Eurasia have postulated, respectively, Nearctica or Palaearctica (or Holarctica) as centers of origin for many mammal taxa, with these groups subsequently radiating to more equatorial regions. The notion of Hickey et al. (1983) that Wasatchian mammals were 2–4 myr older in the Canadian Arctic islands than in the lower latitudes of North America would have corroborated that idea; however, as mentioned in Chapter 6, the chronological framework for that study has since been shown to have been incorrect (Tauxe & Clark 1987), and there is no demonstrable faunal heterochrony exhibited in Wasatchian mammals (Kent et al. 1984; Flynn et al. 1984). The Argentinian paleontologist Florentino Ameghino (1854?–1911) believed that southern South America (principally Patagonia) was the center of origin for many modern mammals, including horses and humans; that was not surprising, because many of his important fossils were collected from that region.

Some workers believe that the task of determining centers of origin is so fraught with methodological problems and poor empirical data that this question should not be one of scientific interest to biogeographers. This has led to the formulation of the "vicariance" method or school of biogeography, which had its origins in the work of the botanist Croizat. Vicariance has since been applied to studying the distributions of diverse suites of plants and animals, both fossil and extant. The literature on this subject is far too large to cite here, and it is treated at length elsewhere (e.g., Rosen 1976; Nelson & Rosen 1981). The essence of vicariance is that, in contrast to a concern about centers of origin, the more important question is the separation of once-continuous spatial distributions of taxa through the formation of barriers, also called vicariance events.

With regard to land mammals, although there are many kinds of barriers (e.g., climatic, or epicontinental seaways) that could result in vicariance, continental drift (resulting from plate tectonics) is now viewed as a dominant, active biogeographic mechanism that caused separation of previously continuous distributions. The majority of fossil horses date from the late Cenozoic, which, as Simpson noted, was a time when there were few major changes in continental positions. Therefore, most recent discussions of equid biogeographic origins have used dispersal models, although vicariance also probably occurred during the earlier part of their history. Thus, studies of fossil horses (discussed later) and other groups have shown that dispersal is not completely incompatible with vicariance (MacFadden 1981). In a context of plate tectonics, both can be integrated into modern biogeographic theory.

Plate tectonics, Noah's Arks, vicariance, and isotopes

Viewed against a plate tectonic background, Simpson's corridors, filters, and sweepstakes actually take on more importance than ever as methods of dispersal. – McKenna (1973:295)

At the beginning of this century, a few brave scientists challenged the time-honored paradigm of the permanence of continents. Workers such as Taylor and Wegener believed that the continents had drifted across the surface of the earth. In the 1960s, further evidence from sea-floor spreading showed that the continents had drifted, but as passive riders along with the dynamic ocean basins. The new paradigm of plate tectonics, which for biogeography includes the prior notion of continental drift, has revolutionized thinking in the natural sciences. Furthermore, recent developments in stable-isotope geochemistry have revolutionized our understanding of ancient temperatures and climate change. Accordingly, these are important starting points for discussing modern biogeographic theories developed during the past few decades.

Prior to the acceptance of plate-tectonic theory in the 1960s, it was reasonable for biogeographers to assume a stable continental framework. As the revolution in earth sciences occurred and the weight of evidence was tipping in favor of drift, it was realized that modifications to previous models were necessary. Without throwing out Simpson's ideas of corridors, filters, and

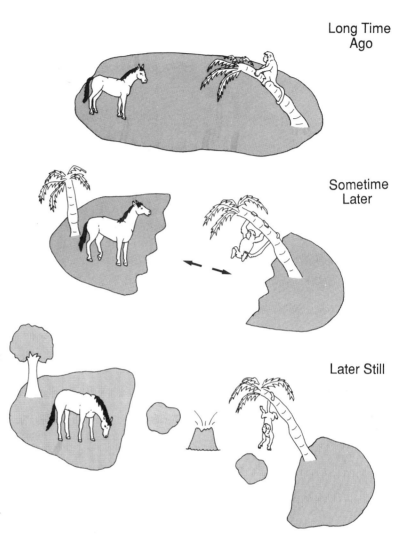

Long Time
Ago

Sometime
Later

Later Still

Figure 7.3. *Hypothetical example of continent and indigenous biota (top) that subsequently become fragmented (a vicariance event) as a result of an active oceanic spreading center (middle). After plate tectonics continue (bottom), the two fragmented continents and their biotas move farther apart, and both the oceanic and climatic barriers become greater. Thus, dispersal of land mammals, particularly those that are climatically sensitive, between these two landmasses becomes further impeded. In this scenario, over time a corridor changes into a filter, which then changes into a sweepstakes dispersal route.*

sweepstakes, in a series of papers McKenna (e.g., 1973, 1975a, 1983) modified those concepts in light of plate-tectonic theory. Thus, Simpson's three terms are not completely distinct from one another, and one can change into the other over time. This is illustrated by the following scenario:

In a hypothetical model, assume that a landmass with a generally homogeneous fauna resulting from a dispersal corridor lies over a newly created and tectonically active spreading center (Figure 7.3, top). Over time, the spreading causes rifting, and then the formation of a sea, the latter resulting in a barrier to dispersal (Figure 7.3, middle). As further time elapses, the faunas on the two new landmasses evolve their own distinctiveness through isolation (Figure 7.3, bottom). In advanced stages of this scenario, the seaway evolves into a considerable oceanic barrier, and then the only potential means of dispersal between the two parts of the formerly continuous landmass is by island-hopping sweepstakes. In summary, McKenna's reinterpretation (1975a, 1983) of Simpson's corridors, filters, and sweepstakes (or the reverse) results in a dynamic continuum that changes with plate activity.

For instances in which there is a one-way dispersal of an entire biota as a result of plate tectonics, McKenna (1975a) coined the term "Noah's Ark," examples of which include India during the Tertiary, and Australia today. When the ark docks, its biotic passengers "off-load" and intermingle with the endemic biota. This is not as farfetched as it may seem; the geological foundation of this idea was suggested earlier by geophysicists (e.g., Wilson 1968) for ancient exotic terranes (i.e., crustal blocks that became accreted to continents). An example of this is the suturing of Florida to North America during the late Paleozoic (Opdyke et al. 1987). McKenna (1973:304–5) also noted that "if one allows two such arks per 100 m.y., one might expect evidence of perhaps a dozen similar situations on the same scale since the beginning of the Palaeozoic." Given lithospheric spreading rates of a few centimeters per year, then over geological time frames of millions or billions of years, voyages of hundreds or thousands of kilometers are possible. In summary, over long periods of geological time, "passive" dispersal of organisms, traditionally thought to have been caused by such factors as wind and water currents (Udvardy 1981), now is seen as having resulted from plate-tectonic activity, producing continents drifting with their resident biotas.

A general notion in the literature has been that for most of the Cenozoic the dominant land bridge for mammalian dispersal has been across "Beringia." Although it is currently submerged, during periods of lowered sea levels the shallow Barents Shelf between Siberia and Alaska has been exposed as dry land (Figure 7.4A). In addition to that route, McKenna (1972, 1975a) has proposed, on the basis of geophysical reconstructions and faunal similarities, the DeGeer dispersal route (Figure 7.4B), a region of faunal interchange across the North Atlantic until the end of the early Eocene (Wasatchian/Sparnacian). This hypothesis is corroborated by a high degree of faunal similarity for Holarctic mammals during that time. After the early Eocene, sea-floor spreading continued long enough to create a significant oceanic barrier to the interchange of terrestrial faunas across the North Atlantic, and the DeGeer route ceased to exist as a Holarctic corridor. Thereafter, Beringia became the dominant Holarctic intercontinental connection for the remainder of the Cenozoic.

Along with drifting landmasses, changing climates have had profound influences on past biotic distributions. Prior to the 1960s, studies of ancient temperatures had been based mostly

A. BERING DISPERSAL ROUTE

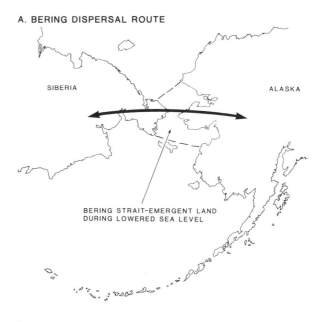

B. DEGEER DISPERSAL ROUTE

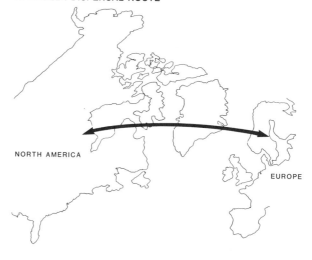

Figure 7.4. *Principal Cenozoic Holarctic dispersal routes for mammals. For most of the Cenozoic, the Bering land bridge (A) has been the dominant dispersal route, particularly during times of lowered sea level, when a shallow shelf has been exposed. However, recent evidence suggests that prior to the middle Eocene the DeGeer route (B) was also active across the North Atlantic. Modified from MacFadden (1988a); reproduced with permission of Plenum Press.*

on qualitative observations of cold- and warm-adapted plants (e.g., palms) and animals (e.g., crocodilians and turtles) in the fossil record. However, because of developments in stable-isotope geochemistry, that was to change. During the 1950s it was realized that ratios of ^{18}O to ^{16}O, usually measured in carbonates and expressed as $\delta^{18}O$ (Appendix I), were temperature-dependent and therefore could be used as paleothermometers. In a series of classic papers, Shackleton and Opdyke (e.g., 1973, 1977) correlated deep-sea magnetostratigraphy and $\delta^{18}O$ fluctuations in Pliocene–Pleistocene pelagic deep-sea cores (Figure

7.5). From those cores they interpreted the observed cycles in the ratios of oxygen isotopes to represent shifts in global paleo-temperature and climate. That discovery had important ramifications, because instead of the traditional notion of four major glacial stages during the Pleistocene, it was seen that there had been about 23 cold/warm ancient global temperature cycles during that time. This observation has since been confirmed in many deep-sea cores worldwide, and $\delta^{18}O$ paleotemperature studies have been extended to older marine sediments (e.g., Shackleton & Kennett 1975). The global water budget implies that times of colder climate will also be times of lowered sea level. In a novel synthesis, Opdyke (1990) found coincidence between high $\delta^{18}O$ values and 12 of 13 land-mammal age boundaries (Figure 7.6). As discussed in Chapter 6, mammalian dispersal events (FADs) are biostratigraphically very useful and therefore are embodied in many of the definitions of land-mammal age boundaries. It is therefore not surprising that the colder intervals during the Cenozoic also would have been times of enhanced interchange of mammals because of lowered sea levels and the opening of land bridges.

Distribution of fossil horses throughout the Cenozoic

Now that the various determinants and principles of land-mammal paleobiogeography have been presented, we can discuss the distribution of fossil horses. It is generally stated in the literature that the family Equidae has predominantly been a North American group. Thus, the presence of horses on other continents traditionally has been interpreted to have been a result of dispersal by land bridges, and that occurred several times during the Cenozoic. Although this idea may be correct in its essential form, there are many details that make the actual biogeographic history of the Equidae considerably more complex than the generally accepted first-order approximation.

The Eocene–Oligocene: DeGeer dispersal, vicariance, and Old World extinction

> *The simultaneous appearance of the earliest of the modernized mammals in Europe (lat. 50° N.) and North America (lat. 40° N.) points to some contiguous land-mass as the original home of these creatures.*
> – Lull (1921:560)

The earliest known members of the Equidae are early Eocene in age, and they had a distribution that extended throughout Holarctica, including occurrences in Europe, Asia, the Canadian Arctic, and western North America (see Figure 4.4). There is no discernible time-transgressive sequence of those occurrences that could be used to determine the center of origin and sub-

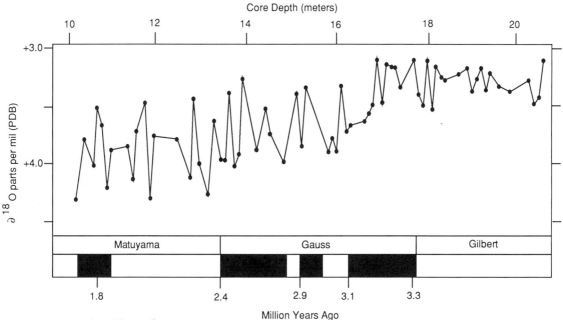

Figure 7.5. Data for a Pliocene deep-sea core showing variations in oxygen isotopes representing climate cycles, calibrated using the magnetic-polarity time scale (Shackleton & Opdyke 1977). The shifts at the top and bottom of the plot are interpreted to represent, respectively, relatively warmer and colder times, and these are also observed to continue into the Pleistocene. Reproduced (slightly modified) with permission from **Nature**, 270:218; copyright © 1977 Macmillan Magazines Ltd.

Figure 7.6. Correlation between Cenozoic climate changes, as interpreted from oxygen-isotope ratios (δ^{18}O) in benthonic foraminiferans, and NALMA boundaries. Modified from Opdyke (1990), "Magnetic stratigraphy of Cenozoic terrestrial sediments and mammalian dispersal," The Journal of Geology, 98:621–37; © by and reproduced with permission of the University of Chicago Press.

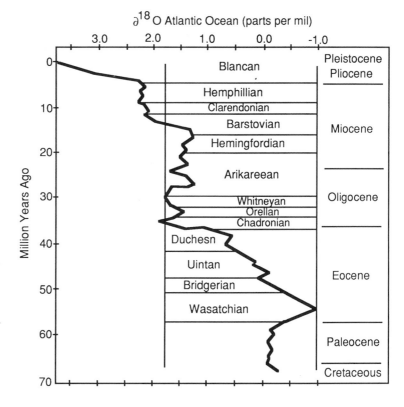

sequent dispersal of this family. Although the question of the biogeographic origin of the Equidae is currently unresolved, there certainly is a diversity of possibilities. If the closest outgroup of the Equidae was the late Paleocene *Radinskya* from China (McKenna et al. 1989), then this family may have originated in Asia. Following the traditional notion of a close relationship between phenacodontid condylarths and *Hyracotherium* (Radinsky 1966; MacFadden 1976), then North America may have been the center of origin of the Equidae. Recently, Gingerich (1989) suggested that several groups of early Tertiary mammals (including perissodactyls and therefore, by inference, horses) originated in Africa from taxa as yet unknown. Regardless of which scenario is correct, during the early Eocene (Wasatchian of North America, Sparnacian of Europe) the family Equidae was represented by *Hyracotherium* throughout Holarctica. This generic-level similarity is observed in many other contemporaneous Holarctic mammals, and if McKenna (1975a, 1983) is correct, as discussed earlier, this implies relatively unimpeded interchange across the North Atlantic DeGeer dispersal route (Figure 7.7A).

At the beginning of the middle Eocene (Bridgerian of North America, Ypresian of Europe) the mammalian faunal similarity between Palaearctica and Nearctica decreased dramatically, indicating a closing of the DeGeer connection because of continued sea-floor spreading in the North Atlantic (McKenna 1975a, 1983). This is also apparent in early perissodactyls, in which there was a split resulting in Equidae (*sensu stricto*) in North America and the Palaeotheriidae in Eurasia. This can be interpreted as a vicariant separation into two perissodactyl families of a previously continuous biogeographic distribution (MacFadden 1988a). The Palaeotheriidae became extinct during the late Oligocene, and Radinsky (1969) attributed their demise to competition from other sympatric perissodactyls in the Old World.

In North America, the Equidae evolved in isolation from the Old World during the middle Eocene to the late Oligocene. The fossil record indicates that the first dispersal of horses after that time occurred at the beginning of the Miocene, at about 24 Ma, with the presence of the browsing anchitheres [including *Kalobatippus* and *Anchitherium s.s.;* see Abusch-Siewert (1983) for Old World forms] in Holarctica and Africa (Figure 7.7B). Paleoclimatological studies indicate that to have been a relatively cold time, and presumably a time of lowered sea levels, which opened up previously inundated land bridges for enhanced dispersal (Opdyke 1990). With the closure of the DeGeer route after the end of the early Eocene, the shallow Barents Shelf between Siberia and the Aleutian Islands periodically emerged during times of lowered sea levels, forming the Bering land bridge, the dominant pathway of Holarctic dispersal throughout the remainder of the Cenozoic.

The Miocene: hipparions and Holarctica

The next equid dispersal between North America and the Old World involved the hipparions (Figure 7.7C), a complex group of grazing, three-toed horses (see Figures 5.14 and 5.15 and Appendix II). As mentioned in Chapter 6, the origin of this clade and its subsequent range expansion, resulting in the *"Hipparion"* datum, provide a classic example of combined geochronology and mammalian paleobiogeography (See Figure 6.9). To briefly summarize here, the first occurrence of this group was recorded from localities in Texas dating to about 15 Ma, and thereafter hipparions were fairly common throughout North America until the late Miocene. The first occurrences of this group in the Old World were dated at about 11.5 Ma in Europe and 9.5 Ma in the well-sampled Siwalik sequence of India and Pakistan. The latter event seems to have represented a biogeographic isolation of the Indian subcontinent, possibly related to the uplifting Himalayas. Thereafter, hipparions underwent an extensive adaptive radiation in the Old World, rivaling that seen in North America (Woodburne & Bernor 1980). They also dispersed into Africa during the late Miocene.

Also during the Miocene, the other dominant clade of grazing horses, the tribe Equini, diversified in North America and Central America. In contrast to the three-toed hipparions, later equine horses were monodactyl. Prior to the Pliocene (as discussed later), the Equini did not disperse into the Old World. Along with the hipparions, they were the dominant equids at the height of the diversity of the family Equidae during the Miocene, between about 15 and 8 Ma.

The Pliocene and Pleistocene: the maximum spread of horses

Although equine horses were predominantly a North American group, some members of this clade dispersed into South America during the great faunal interchange that occurred about 2–3 Ma as a result of the dry-land connection formed by the isthmus of Panama (Figure 7.7D). These horses are represented at many late Cenozoic (predominantly Pleistocene) sites in South America by the two relatively common genera *Onohippidium* and *Hippidion* (see Figure 5.19), and possibly *Parahipparion* (although the latter is less well known). Until recently, these South American "endemics" were known only from that continent. However, MacFadden and Skinner (1979) described a primitive species of *Onohippidium*, *O. galushai*, and possibly also *Hippidion*, from the late Miocene of, respectively, Arizona and Texas. Thereafter, these genera are not recorded in North American deposits, except possibly for a "back-dispersal" of *Hippidion* represented by a fragmentary specimen from the late Pliocene to early Pleisto-

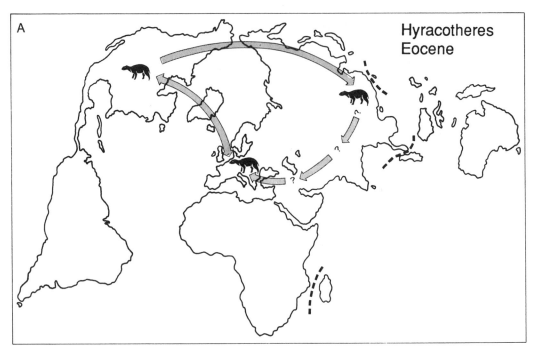

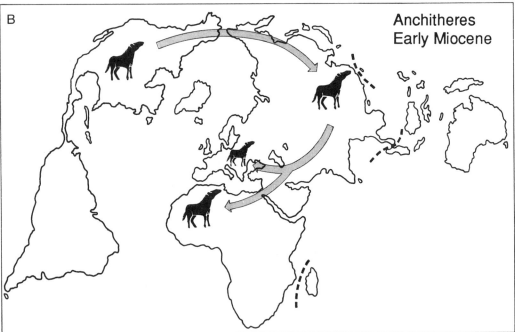

Figure 7.7. *Dispersal patterns for fossil horses throughout the Cenozoic. As mentioned in Figure 7.4, the dominant dispersal route during the early Eocene (A) seems to have been the DeGeer or North Atlantic route, followed in the middle and late Eocene by the Bering route during the early Miocene (B) the three-toed browsing anchitheres (Anchitherium, Kalobatippus, and Sinohippus) had a combined distribution that included Holarctica and Africa, as*

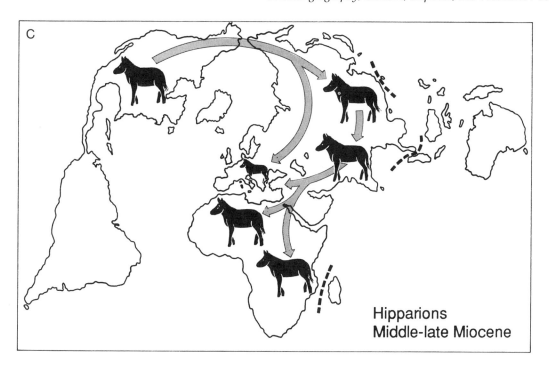

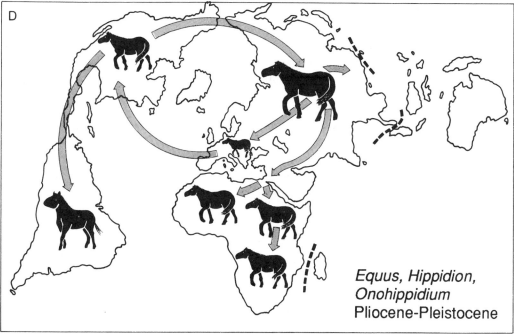

did the hipparions during the middle and late Miocene (C). During the Pliocene and Pleistocene (D), horses attained their maximum geographic distribution, represented by Equus throughout their range (Holarctica, Africa, and South America), as well as Hippidion *and* Onohippidium *in* southern North America and widespread in South America. *Modified from Franzen (1984).*

cene (Irvingtonian) sequence of southern California (MacFadden & Skinner 1979).

Equus

The genus *Equus* has had a widespread and complex biogeographic history. The closest recognized outgroup of *Equus*, *Dinohippus mexicanus*, had a distribution in southern North America and Mexico during the late Miocene (late Hemphillian), at about 5 Ma. The earliest fossil *Equus* found in the New World dates to about 3.7 Ma (Lindsay et al. 1980) (see Figure 6.9).

During the late Pliocene, *Equus* dispersed into the Old World, including both Eurasia and Africa, and it is commonly encountered at many terrestrial sites dating after about 2.5 Ma (Lindsay et al. 1980). *Equus* is also common at numerous Pleistocene localities in South America, again representing a southward dispersal event across the Panamanian land bridge. Thus, within a period of 1–1.5 myr, *Equus* had dispersed into every continent and biogeographic region, with the exceptions of Australia and Antarctica. Like many other large mammals, *Equus* became extinct in many parts of its previous biogeographic range about 10,000 years ago as part of the late Pleistocene megafaunal extinction. The causes of this event have been hotly debated, but they probably included a combination of climatic and vegetative changes, as well as "prehistoric overkill" resulting from hunting by humans, who had rapidly dispersed throughout the New World by 10,000 years ago. Today, the extant genus *Equus* has a considerably contracted biogeographic range and is restricted in the natural state to parts of the Old World (Figure 7.8). Further extinctions and range contraction have continued almost up to the present day. For example, as a result of extensive hunting, the quagga *E. quagga* was extirpated from southern Africa in the middle of the nineteenth century. Many introduced populations of horses that have secondarily become feral, such as *E. caballus* in the Great Basin of the western United States (Berger 1986), have flourished, although they too will be in peril if conservation efforts are not carefully implemented.

Concluding comments

Since the 1960s, the science of biogeography has undergone a major theoretical and methodological revolution as a result of studies in plate tectonics, cladistics, and vicariance. McKenna has shown that some of the fundamental biogeographic concepts pertaining to land mammals, such as Simpson's corridors, filters, and sweepstakes, though not to be abandoned, must be modified in light of continental drift. For those interested in discerning dispersals on the basis of paleontological data, new

EXTANT *EQUUS*

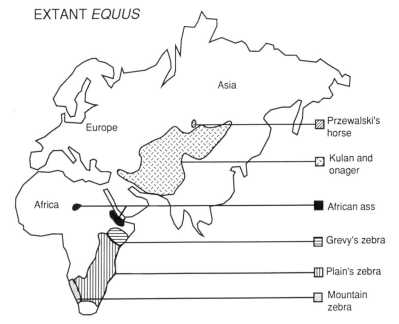

Figure 7.8. *Biogeographic distribution of extant* Equus. *Modified from Berger (1980),* Wild Horses of the Great Basin; © 1986 by and reproduced with permission of the University of Chicago Press.

geochronological techniques and the excellent fossil record of horses allow more precise timings of ancient distributions than were previously possible. A final note is that modern biogeographic theory does not have to be an either/or phenomenon. Thus, as has been noted previously (e.g., MacFadden 1981), vicariance and dispersal can both be integrated into a comprehensive model to explain ancient and modern distributional histories of taxa. A modern view of the paleobiogeography of horses is one in which the center and timing of origin of the Equidae can be constrained only to the Northern Hemisphere, and probably to the late Paleocene to early Eocene. After the early Eocene, the pan-Holarctic distribution of equoids was divided by a vicariance event. Thereafter, the dominant biogeographic pattern was isolation of the New World Equidae until about 24 Ma, followed by several active dispersal events across the Bering land bridge as a result of lowered global sea level. The most widespread distribution of horses occurred during the Pleistocene with *Equus*, followed by dramatic extinctions and rapid range contraction for this family.

In addition to both passive and active dispersal, either during times of continental drift or via land bridges, the distribution of horses has been affected by climatic gradients, resulting in diversity shifts, and organic interactions with other members of their successive biotas (e.g., grasses). No single factor is totally responsible for the observed patterns of distribution. The chapters that follow will show how, in addition to biogeographic determinants, the history of fossil horses has been greatly influenced by evolution and ecology, both within species and within coexisting biotic communities through time.

8

Evolutionary processes: variation, speciation, and extinction

> *The paleontologist, indeed, generally comes much too late to find anything but bones. Instead he finds something denied to the neontologist: the time element; and the crowning biological achievement of paleontology has been the demonstration, from the history of life, of the validity of the evolution theory.*
>
> – Kurtén (1953:5)

Most scholars agree that the concept of evolution is the single most comprehensive and unifying principle in biology. Several fundamental questions center around our interest in a better understanding of organic diversity. These focus principally on evolutionary *processes* and the resulting evolutionary or phylogenetic *patterns*, or, as Simpson (1944) titled his landmark book, *Tempo and Mode in Evolution*. Despite the fact that it is far from perfect, as even Darwin lamented, the fossil record is continually examined for long-term evidence of this theory.

Evolution is a dominant and pervasive theme of this book. Whereas earlier chapters have dealt with horse evolution from a historical perspective (e.g., orthogenesis), or have provided the necessary systematic, temporal, and biogeographic framework, the purpose of the next three chapters is to show how fossil horses illustrate various evolutionary concepts. This chapter is devoted to processes, whereas Chapters 9 and 10 discuss rates and patterns of evolution.

Phenotypic variation and selection: the raw material and evolutionary catalyst

Genes and chromosomes do not fossilize, so most of the evidence that paleontologists can offer for an understanding of the genetic basis for evolution comes from phenotypic variability and change (e.g., Kurtén 1953, 1955). In the fossil record, this is most often expressed as variation in morphology. Variation is usually quantified in terms of the coefficient of variation (V)

for a given character in a given population, where V = mean/ standard deviation. There have been few rigorous discussions of the biological meaning of V's (Lande 1977; Sokal & Braumann 1980). Mayr (1969) stated that V's of taxonomic relevance usually are less than 30%, although this does not take into consideration the various sources of variation, including genetic, ontogenetic, and geographic. Simpson, Roe, and Lewontin (1960:91) stated that "as a matter of observation, the great majority of them [V's] lie between 4 and 10, and 5 and 6 are good average points. Much lower values usually indicate that the sample was not adequate to show the variability. Much higher values usually indicate that the sample was not pure, for instance, that it included animals of decidedly different ages or of different minor taxonomic divisions."

Because of its potential importance in understanding evolutionary mechanisms, there have been numerous studies of the amount of phenotypic character variation in many extant vertebrate groups, including mammals [e.g., the detailed synthesis by Yablokov (1974); Penguilly (1984)]. Kurtén's landmark paper (1953) analyzed variation and its evolutionary consequences in several fossil groups, including the Miocene horses *Merychippus* and Old World *Hipparion*. Other studies of fossil horses have found that "normal" variation (i.e., not related to sexual dimorphism) for what are interpreted to represent fossil species is usually between 4% and 10% [e.g., *Hyracotherium* (Gingerich 1981); *Mesohippus* (Forsten 1970); *Parahippus* (Bader 1956); *Merychippus* (Downs 1961); hipparions (MacFadden 1984a, 1989)], which are values within the limits of paleontologically "homogeneous" samples (Simpson et al. 1960).

Comparison of V's for different characters both within and between samples has several pitfalls that must be considered. Yablokov (1974) found that smaller mean character values had larger V's, a phenomenon that he termed "drift." MacFadden (1984a, 1989) found drift of characters (other than those related to sexual dimorphism) between about 5% and 10% in fossil and Recent hypsodont horses (Figure 8.1). Thus, in certain cases, drift of variation can be significant, and this must be taken into account when comparing V's, particularly those whose means span different orders of magnitude.

In order to better understand certain microevolutionary processes, it has been a common, although usually untested, assumption among paleontologists that fossil quarry samples represent ancient demes, or "paleopopulations," that is, individuals that lived and died together, or a local assemblage that appears to have been geologically instantaneous. MacFadden (1989) compared the amounts of dental- and cranial-character variations in fossil quarry samples and extant equid population samples and found that V's usually were less than 10%. Thus, in terms of amount of variation, fossil horse quarry samples

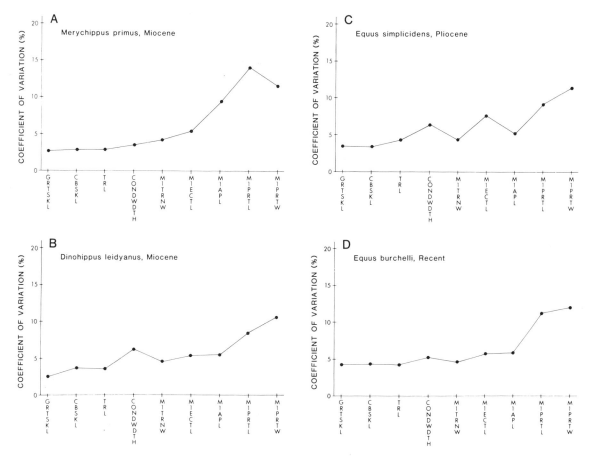

Figure 8.1. *Variability profiles showing character variation in samples of fossil horses (A–C) and an extant population of* Equus burchelli *(D). Along the x-axis the characters are arranged in order of mean size, from largest (left) to smallest (right); these show the "drift" (sensu Yablokov 1974), or increased V's with decreasing size. Characters are abbreviated as follows: GRTSKL, greatest skull length; CBSKL, condylar basilar length; TRL, cheek-tooth row length (P2–M3); CONDWDTH, condylar width; M1TRNW, greatest transverse width of M1; M1ECTL, greatest ectoloph length of M1; M1APL, greatest anteroposterior length of M1; M1PRTL, greatest protocone length; M1PRTW, greatest protocone width. From MacFadden (1989); reproduced with permission of Oxford University Press.*

(particularly those that are catastrophic assemblages) often seem to represent "paleopopulations," or analogues of modern demes, and therefore are valid samples for comparison with extant populations in testing microevolutionary theory.

Among the studies of character variation in fossil horses, two from the Miocene of North America are insightful because they promote a better understanding of evolutionary processes.

Merychippus *from California and the Great Basin*

Downs (1961) presented a detailed study of variation in a 1–2-myr sequence of samples of the three-toed grazers *Merychippus seversus* from the Columbia River plateau and Great Basin of

Oregon and Nevada and *M. californicus* from Coalinga in southern California (see Figure 4.4). By modern standards, Downs's study had several flaws, including (1) the use of samples from different localities (and without demonstrable superposition), (2) its lack of direct geochronological calibrations, and (3) its use of presumed ancestral and descendant species that may have been only distantly related. Nevertheless, his study was exemplary in terms of its thoroughness and the interesting evolutionary questions that it sought to answer.

Although such analysis is commonplace in more recent studies, Downs analyzed in great detail the amounts of variation in 14 measured and counted (meristic) characters of the upper dentition (P4–M2) in those two species. For the continuous variates he found V's between 3.7% and 20.5%, with most of them falling below 10%. On the basis of more recent work (e.g., MacFadden 1984a, 1989) it would seem that the higher V values may have represented drift of smaller mean size (e.g., protocone dimensions) or ontogenetic variation (crown height). He also found very high V's for meristic characters, namely, counts of the numbers of enamel plications (this was also found in hipparions) (MacFadden 1984a). Downs (1961, Table 4) concluded that some characters served better than others for microevolutionary and taxonomic interpretations, and he ranked them as "Excellent" (e.g., crown height), "Good," "Fair," and "Poor" (e.g., pli postfossette enamel lengths). He also compared the amounts of variation in his fossil samples with the amounts for extant *Equus burchelli* and *E. zebra* from Africa. He concluded that the amounts of variation in the fossil and Recent samples were similar, and thus the concepts of demes and biological species also applied to extinct species.

On the basis of the data he had collected, Downs (1961) concluded that he could interpret two general evolutionary processes. Within *M. severus* and *M. californicus* he saw microevolutionary trends in measured characters, including increased crown heights, transition from oval to elongated protocones and from rectangular to square occlusal surfaces, increased enamel, and decreased curvature of crowns. He noted that certain characters, such as crown height, evolved faster than others, an observation reported elsewhere for horses (e.g., Simpson 1953a; MacFadden 1988b) and, as a general evolutionary phenomenon, termed "mosaic evolution" (de Beer 1954). Downs also concluded that these observed trends were not reversible (Dollo's "law," see Chapter 10).

A small sample from the Virgin Valley of Nevada, although referred to *M. seversus*, was significantly more advanced than other populations from the Mascall Fauna of Oregon. Downs suggested that the Virgin Valley sample was transitional between *M. seversus* and *M. californicus* and therefore provided an example of relatively rapid evolution between ancestral and descendant species. Downs (1961) concluded that for the most part, those merychippines had been chronoclinal species that had demonstrated gradual evolutionary change.

Merychippus *and* Pseudhipparion *from Nebraska*

Van Valen (1963, 1964, 1965) wrote a series of papers on samples of *Merychippus primus* from Thomson quarry (see Chapter 4) and *Griphippus* (now *Pseudhipparion*) *gratum* from Burge quarry, both located in Nebraska. Van Valen believed that in quarry samples, geographic variation probably was not an issue, and ontogenetic variation could be factored out. If so, then any resulting variation would be a measure of the intensity of natural selection acting upon the population (Haldane 1954; also see Marcus 1964). Van Valen compared variations and calculated selection intensities in different age cohorts (determined by tooth-wear classes) and different dental characters for these fossil horses. Of his conclusions (also described in later chapters), the following are relevant here: (1) Different characters were exposed to different selection intensities. For example, the P2 dimensions exhibited relatively weak selection, whereas selection was strong for crown heights. (2) Different age cohorts had different selection intensities. Thus, the juveniles and old individuals were exposed to relatively strong selection, whereas in the middle-aged individuals, selection was weaker. Intuitively, or because of the great emphasis placed on them in the literature, one might think that horses are prime examples of relatively rapid evolution under a regime of strong selection. On the contrary, Van Valen (1964:106) states that "a quantitative estimate shows that weak natural selection is adequate to account for the most rapid known evolutionary change in the Equidae."

Species recognition in paleontology

A universal premise in paleontology is that extinct species, like extant biological species, are fundamental units in the Linnean hierarchy; that is, individuals of each have been capable of interbreeding and producing viable offspring. Whereas neontologists are able to examine the biological-species concept in action by watching successive generations interbreed in the laboratory or in the field, paleontologists do not have this luxury. Yet in both neontology and paleontology, genetic integrity of species often is assumed to exist because of the overall phenotypic similarities of individuals and populations.

Whereas neontologists have available a spectrum of phenotypic characters for study, paleontologists are limited to studies of those characters that fossilize, which usually include skeletal and dental remains. In recognition of such differences, some workers have used terms such as "paleontospecies" or "paleospecies" (e.g., Mayr 1970) or "morphological species" or "morphospecies" (Cain 1954; George 1956) for extinct taxa, which obviously are constructs that can never be tested in the field or laboratory as can biological species. Despite these limitations, paleobiologists strive to apply principles of evolution and ecology uniformly to species regardless of whether they are extinct or extant.

A primary goal in paleontology is to do alpha-level taxonomy, that is, to determine the number of species represented by a collection of fossil specimens, whether they be in a quarry, in a local sedimentary basin, or from numerous localities within a region. Morphologically similar individuals form a phenon,[1] and these often are considered to represent the same entities as biological species. However, there are several caveats that must be considered before the conclusion can be reached that a phenon represents a fossil species. Of relevance to paleontologists, sibling species, that is, those that are morphologically indistinguishable but have other intrinsic isolating mechanisms, will never be recognized in the fossil record (unless biochemical assays for hard parts are developed in the future). A classic example of this is passerine birds, in which extant species often are distinguished by feather and vocalization patterns. Because their morphology is so uniform, fossil passerines are either impossible or notoriously difficult to distinguish below the familial level (Olson 1985).

Another problem inherent in the study of fossil species is that the converse is not always true; that is, different morphologies do not always imply different species. The most striking examples of this result from sexual and ontogenetic variation. For the former, bimodal size distributions of otherwise very similar morphologies, or other secondary sexual characteristics such as larger canines in males, suggest sexual variation within extinct species (e.g., Kurtén 1969; MacFadden 1984a) (see Chapter 12). Another possible source of variation occurs when different ontogenetic stages within a species exhibit different morphologies. Hypsodont horses offer a prime example of the latter, in which crown heights, occlusal measurements, and dental patterns are ontogenetically variable. There are numerous instances in the earlier literature in which different fossil species were described for what are now interpreted as ontogenetic stages of a single species. However, with the polytypic-species concept and numerous studies of variation, like those presented here, intraspecific variation is now more commonly recognized in paleontology.

[1]"Phenon (pl. phena). A sample of phenotypically similar individuals; a phenotypically reasonably uniform sample." (Mayr 1970:422).

Interpreting anagenesis and cladogenesis from the fossil record: theoretical and practical considerations

Two major modes of species evolution and the speciation process are generally interpreted from the fossil record. "Anagenesis," although frequently used to describe gradual microevolution within a species, has also been associated within macroevolution, or phyletic transformation from ancestral to descendant species (e.g., Gingerich 1976, 1977). Anagenesis has also been termed phyletic speciation (e.g., Simpson 1953a), where the number of species at any given time remains constant because the descendant species replaces the ancestor with no temporal overlap.

Cladogenesis, or branching speciation, results in the multiplication or diversification of species from an ancestral form (e.g., Simpson 1953a; Rensch 1959). This is the dominant mode of speciation in rapidly evolving clades, particularly during an adaptive radiation (MacFadden & Hulbert 1988). Although it was discussed in the older evolutionary literature, cladogenesis has recently achieved new prominence because of its importance in systematic methodologies and evolutionary models. In the cladistic method, it is usually accepted to be the dominant mode in which nested species (or other higher-level operational taxonomic units) have resulted from a known sequence of cladogenetic branching events. Although anagenesis is not incompatible with cladistics, cladogenetic branching usually is considered to be the dominant mode. With regard to evolutionary theory, cladogenesis has also recently gained popularity as the dominant mode during punctuated equilibria (Eldredge & Gould 1972; Gould & Eldredge 1977). In this model, stasis is punctuated by rapid evolutionary shifts (e.g., in morphology), resulting in the origin of new species. Despite the differences between anagenesis and cladogenesis, in certain situations these two modes are compatible even within a given lineage or clade. For example, Simpson (1953a) noted that anagenetically evolving lineages rarely are observed to occur in the fossil record for long periods of time without cladogenetic branching events.

Assuming a well-sampled sequence of species representing their actual biochronological ranges, how can anagenesis and cladogenesis be determined from the fossil record? If ancestral and descendant species do not overlap in their observed ranges (Figure 8.2 left), this event is interpreted to be anagenetic speciation (MacFadden 1985a). On the other hand, if the ancestral and descendant species overlap in their ranges, then cladogenesis is inferred (Figure 8.2, right). There are several obvious problems and limitations to these models. If there is stasis or continuous evolutionary change in anagenesis, as exemplified by early Eocene *Hyracotherium* from Wyoming (Gingerich 1989) (Figure 8.3), then the criteria used to draw the boundary between two species usually are arbitrary or difficult to justify. Sometimes

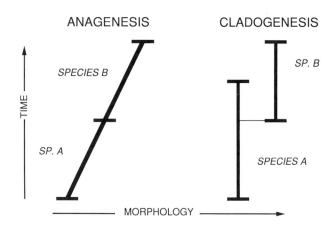

Figure 8.2. *Theoretical examples of speciation from ancestral species A to descendant species B. When a fossil ancestral species and a descendant species do not overlap, then the speciation mode is interpreted to be anagenetic (left); when they coexist for some period of time, then the speciation mode is interpreted to be cladogenetic (right).*

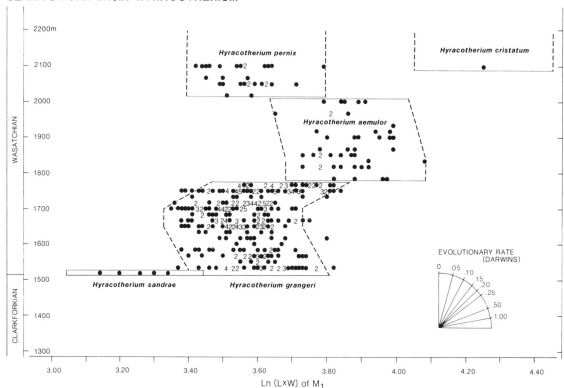

Figure 8.3. *Biostratigraphy of the five species of* Hyracotherium *recognized from the Clark's Fork Basin, Wyoming. From Gingerich (1989); reproduced with permission of the Museum of Paleontology, University of Michigan.*

there are morphological discontinuities; at other times a percentage of change in morphology is taken to indicate a new species. Figure 8.3 also illustrates the example of chronospecies (e.g., Stanley 1979), that is, a continuously evolving species

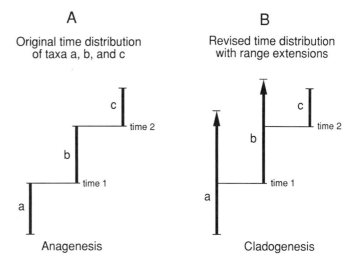

A
Original time distribution
of taxa a, b, and c

B
Revised time distribution
with range extensions

Figure 8.4. *Anagenesis reinterpreted as cladogenesis. (A) The original biostratigraphic distribution of ancestral–descendant species a, b, and c is nonoverlapping, and thus the mode of speciation is interpreted to be anagenesis. (B) New fossil discoveries demonstrate that a and b occur later in time. Thus, although the timing of speciations (times 1 and 2) is the same, the revised distribution is one of overlapping ranges and coexisting ancestral and descendant species, or cladogenesis.*

whose end members are significantly different in morphology but are united in the same taxon because of the presence of a continuum of gradually evolving intermediate forms. The recognition of ancient species in the fossil record is a difficult task. For all the cases described earlier, there were no universally accepted rules or procedures for delimiting species. The final decisions presented in the literature certainly were influenced by the systematic methodology followed by the individual practitioner.

It would be impossible for the fossil record always to preserve the first and last individuals of a particular species, and even if it did, paleontologists would have a difficult time finding them. This has consequences for interpretations of anagenesis. Once cladogenesis is demonstrated by overlap in the ranges of ancestral and descendant species, then further exploration will rarely require reinterpretation of that event as anagenesis. However, the converse, anagenesis reinterpreted to be cladogenesis (in light of later range extensions), is entirely plausible (Figure 8.4). Thus, cladogenesis probably is more common and pervasive than is observed in the fossil record.

There is another limitation of the fossil record that must be emphasized here. In many cases, the number of ancient speciation events observed within a clade undoubtedly represents fewer than those that actually occurred. Others that did not involve discernible shifts in morphology, or were not fossilized, will never be recognized in the fossil record. Studies that compare the numbers of speciation events in extinct and extant taxa must take into account this limitation of the fossil record.

Speciation of fossil horses in North America

Since the 1870s the fossil record of horses in North America has been touted innumerable times as a prime example of gradual evolution. As discussed in Chapter 3, Marsh's orthogenetic depiction of the Yale fossil horses did much to provide evidence for this viewpoint. Since then, many workers have continued to accept the notion of directed evolution and, although not explicitly saying so, the process of anagenesis on both the microevolutionary and macroevolutionary levels.

It is not at all surprising that the traditional view of horse phylogeny is anagenetic. The progressive replacement of one genus with another without temporal overlap is the essence of phyletic evolution. The modern interpretation of horse phylogeny as a complexly branching tree (see Figure 5.14) requires that modes of speciation be reexamined. At first glance it appears that the first half of horse evolution during the Eocene and Oligocene was dominantly anagenetic, with little increase in generic diversity. On the other hand, the second half of horse evolution, represented by the adaptive radiation of hypsodont forms, was decidedly cladogenetic, with numerous generic branches. However, this picture is simplified because it illustrates only patterns on the generic level and thus is an artifact of scale, because genera do not speciate. If one looks at a finer scale within the Linnean hierarchy, namely, at the species level, the mode of evolution tells a different story, one that is much more complex than is traditionally depicted.

Modes of speciation from Hyracotherium *to* Equus

[2]Some workers have previously reported *Hyracotherium* from pre-Eocene deposits in the New World. For example, Morris (1968) described *H. seekinsi* from the supposed late Paleocene of Baja California, and Jepsen and Woodburne (1969) reported the presence of a single specimen of *Hyracotherium* from the Polecat Bench, a classic Clarkforkian sequence of Wyoming. However, since those studies, the age of the Baja locality has been revised to Wasatchian (Novacek et al. 1991), and the exact provenience of the Wyoming find has been challenged (Gingerich 1989). Accordingly, the current consensus is that *Hyracotherium* first occurs in the early Wasatchian faunas of North America (Krishtalka et al. 1987; Gingerich 1989; Novacek et al. 1991). The only possible exception to this datum is the presence of *H. angustidens* from the top of the otherwise late Paleocene (Clarkforkian) Black Peaks Formation of Texas (Schiebout 1974).

In North America the earliest records of horses usually indicate one species of *Hyracotherium* during the earliest Eocene (early Wasatchian),[2] variously referred to as *H. seekinsi* (from Baja California) (Novacek et al. 1991), *H. angustidens* (e.g., New Mexico) (Kitts 1956), and *H. grangeri* (Wyoming) (Gingerich 1981), although a much rarer, contemporaneous *H. sandrae* is also known from the latter sequence (Gingerich 1989). Whether these named species represent distinct taxa evolving in parallel or are synonymous is not known at this time, although some work in progress (Froehlich 1989) may resolve this question in the future. At present it is impossible to determine whether the family Equidae originated by anagenesis or cladogenesis. In Eurasia the stratigraphic record for early hyracotheres and their sister taxa is not sufficiently detailed, although Hooker's work (1989) suggests many cladogenetic events at the base of the perissodactyl radiation (see Figure 5.10). In North America, if phenacodontids are the sister taxon of hyracotheres, then the Equidae probably originated from cladogenesis, because the most perissodactyl-like species of *Phenacodus* (McKenna 1960) co-occur in early Eocene (Wasatchian) deposits in Colorado. Regardless of which perissodactyl outgroup is chosen, no special quantum evolutionary burst or "hopeful monster"[3] need be invoked to account for the change in morphology observed between either possible sister taxa and *Hyracotherium*.

The first abundant evidence for diversification of the Equidae, and, by inference, the process of cladogenesis of horse species, comes from the early Eocene (middle Wasatchian). During that time, two to four species co-occur at many localities in intermontane basinal deposits of western North America, including the classic faunas of New Mexico, Colorado, and Wyoming (Kitts 1956; Gingerich 1981, 1989; Krishtalka et al. 1987; Froehlich 1989). During the middle–late Eocene, horses continued to exhibit low diversity, with one or two species (variously assigned to *Orohippus* or *Epihippus*) found at many western North American sites.

Classically, the Oligocene of North America is also considered a time of low equid diversity. This is usually depicted as a simple anagenetic progression of genera from *Mesohippus* to *Miohippus*, or a single trunk of the phylogenetic tree [e.g., Matthew 1930 (see Figure 3.11); Stirton 1940 (see Figure 3.12); Simpson 1951 (see Figure 3.13)]. This picture has changed dramatically in the past few decades, for several reasons: (1) studies of species biochronology during that time interval (e.g., Clark, Beerbower, & Kietzke 1967; Emry et al. 1987), (2) the description of *Haplohippus* (McGrew 1953), which was contemporaneous (although usually not sympatric) with *Mesohippus*, and (3) a renewed look at the distributions of the two dominant Oligocene genera *Mesohippus* and *Miohippus* (Prothero & Shubin 1989).

Prothero and Shubin's analysis (1989) of the valid species and the temporal ranges for *Mesohippus* and *Miohippus* requires a revised interpretation of the long-term evolutionary patterns of these genera, of which they recognize, respectively, at least five and six valid species. Interestingly, in contrast to the common notion of anagenetic replacement of *Mesohippus* by *Miohippus*, Prothero and Shubin (1989) concluded that these genera were contemporaneous for about 5 myr. Furthermore, at most sites, particularly those in the exceedingly rich White River Group of the northern High Plains (Nebraska, Montana, Wyoming, and South Dakota), three to five coexisting species usually are found; the maximum diversity of six species occurred during the late Chadronian (Figure 8.5). By the middle Oligocene (early Whitneyan) the genus *Mesohippus* became extinct, and species of *Miohippus* continued into the middle and late Oligocene (late Whitneyan to early Arikareean), when they overlapped with a more derived equid fauna that evolved in North America. That assemblage included various contemporaneous species of the genera *Kalobatippus*, *Desmatippus*, *Parahippus*, and *Archaeohippus*.

The advent of high-crowned, grazing horses during the mid-

[3]The concept of the hopeful monster (Goldschmidt 1940) involved the occurrence of a single individual with a dramatic change in genotype (via megamutations) and phenotype that gave rise to a new, higher taxon. Few workers currently accept this notion, and it is now regarded as an interesting curiosity in the history of science.

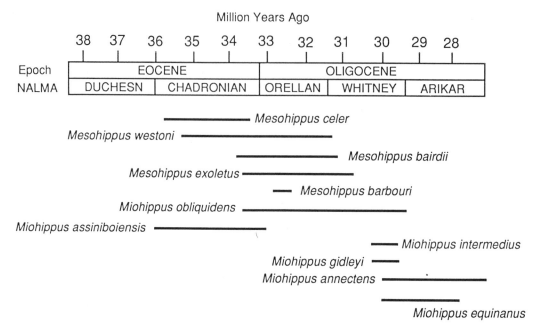

Million Years Ago

Figure 8.5. *Biostratigraphic ranges of the middle Tertiary equid genera* Mesohippus *and* Miohippus. *In contrast to the traditional notion of anagenesis and low diversity for these genera, this chart illustrates the overlapping ranges of the numerous species assigned to these genera, indicating a more complex phylogeny than previously believed. Note that the temporal calibration of this figure is different from that of Woodburne (1987). Modified from Prothero and Shubin (1989).*

dle Miocene at 17–18 Ma is a well-known and often-cited example of an adaptive radiation and diversification of a monophyletic clade. This story is usually depicted on the generic level (e.g., Figure 5.14), with the familiar pattern in the portion of the phylogenetic tree where all the branching begins. We know, however, that adaptive radiations represent the multiplication of species, not higher taxa. Although most neontologists study adaptive radiations of species, given the limitations of the fossil record, the evolutionary patterns of extinct clades often are studied at higher levels, usually genera and families. Patterns discerned from these higher taxa are then considered to be applicable to species.

Recent work has elucidated the species-level systematics of middle Miocene horse phylogeny and has resulted in a revised interpretation of the evolutionary patterns of that important adaptive radiation. MacFadden and Hulbert (1988) and Hulbert and MacFadden (1991) have analyzed the cladistic interrelationships of the equid species at the base of that adaptive radiation. Analysis of their biostratigraphic ranges (Figure 8.6) indicates that early during the radiation, from about 18 Ma to 16 Ma, the dominant mode of speciation was cladogenesis. At that time the rates of speciation were very high, with an intrinsic rate of increase $R = 0.97$, which is among the highest observed for any fossil taxon (Table 8.1). After 16 Ma, R dropped to 0.06, indicating a steady state between origination and extinction (Figure 8.7). Analysis of speciation modes for North American hippa-

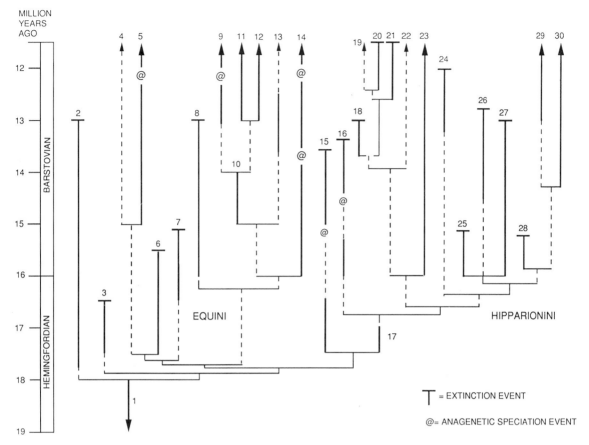

Figure 8.6. *Biostratigraphic distribution, phylogenetic interrelationships, and speciation patterns for 29 equine taxa from the middle Miocene of North America. The outgroup species, "Parahippus" leonensis, is represented by the number 1. The species within the Equinae (including tribes Equini and Hipparionini) are as follows: 2, "Mer-ychippus" gunteri; 3, "M." primus; 4, Astrohippus-Equus-Hippidion clade; 5, Pliohippus mirabilis–P.* pernix *clade; 6, "M."* carrizoensis; *7, "M."* stylodontus; *8, "M."* inter-montanus; *9,* Calippus proplaci-dus–C. placidus *clade; 10,* Calippus sp.; *11,* Calippus sp.–C. elachistus *clade; 12,* C. regulus; *13,* Calippus (Grammohippus) *clade; 14,* Proto-hippus vetus–P. perditus–P. supre-mus *clade; 15, "M."* tertius–"M." isonesus *clade; 16, "M."* sp.–cf. "M." sejunctus *clade; 17, "M."* sp.; *18, "M."* republicanus; *19,* Pseud- hipparion curtivallum–P. gratum *clade; 20,* Pseudhipparion sp.; *21,* P. retrusum; *22,* Neohipparion; *23, "M."* coloradense; *24, "Hipparion"* shirleyae; *25,* M. insignis; *26,* M. californicus; *27,* M. brevidontus; *28, "M."* goorisi; *29,* Nannippus; *30,* Cormohipparion. *Plesions are listed in quotation marks. From Hulbert and MacFadden (1991); reproduced with permission of AMNH.*

rions later during that radiation reveals that anagenesis and clad-ogenesis occurred at about equal frequencies (MacFadden 1985a).

The data and results obtained for the North American grazing equine horses seem consistent with the general patterns of speciation that occur during other documented examples of adaptive radiations, whether in extinct or modern clades. At the beginning of these events, multiplication of species occurs rapidly (high R values), particularly when the clade is exploiting a

Table 8.1. *Values of R^a for species within selected fossil clades[b]*

Taxon	Number of species	Δt (myr)	R	Source
Miocene grazing horses, 18–17 Ma	3, 8	1	0.97	MacFadden & Hulbert (1988)
Miocene grazing horses, 15–13 Ma	14, 16	2	0.06	MacFadden & Hulbert (1988)
All fossil horses	ca. 150	58	0.09	This book
Bovidae	115	31	0.15	Stanley (1979)
Muridae	844	19	0.35	Stanley (1979)
Cretaceous–early Tertiary eutherian mammals			0.12–0.19	Novacek & Norell (1982)
All mammals			0.22	Stanley (1979)
All bivalves			0.06	Stanley (1979)

[a]Intrinsic rate of increase: $R = (\ln N - \ln N_0)/\Delta t$, where N_0 is the initial number of species in a clade, and N is the number of species after elapsed time Δt (Stanley 1979, 1985). For monophyletic clades (e.g., Equinae), $N_0 = 1$, and the equation simplifies to $R = \ln N/\Delta t$.
[b]Also see discussion in Chapter 9.

new resource or environment, or, following Simpson (1944, 1953a), a new "adaptive zone." When the taxonomic "carrying capacity" (i.e., the optimum or equilibrium number of species within the guild or community) is reached, then there is more of a balance between speciation and extinction as faunal turnover occurs. In this later phase of the radiation, although the particular mix of species may change via extinction, "pseudoextinction," or the origin of new species, the total number of species remains constant. In North America there were some 14–17 species of equine horses between about 15 and 9 Ma (MacFadden & Hulbert 1988; Hulbert & MacFadden 1991) (Figure 8.7).

During most of the Cenozoic, South America remained an isolated island continent with a highly endemic mammalian fauna. Between about 3 and 2 Ma there was a major immigration of mammals into that continent from North America during the well-known "great American faunal interchange." The story of horses (before *Equus*) in South America is in itself interesting, and yet it is often neglected, principally because museum collections are so scattered and the systematics of the relevant genera (*Hippidion*, *Onohippidium*, and *Parahipparion*) are greatly in need of a modern revision. Nevertheless, the available evidence indicates that this clade originated in North America, probably

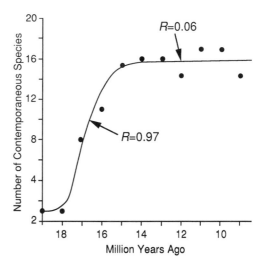

Figure 8.7. *Intrinsic rate of increase (R) for merychippine, equine, and hipparionine horses. Adapted from MacFadden and Hulbert (1988), and Hulbert and MacFadden (1991).*

from an advanced species of *Dinohippus* (MacFadden & Skinner 1979), and then dispersed into South America. On that continent, a small adaptive radiation resulted in about eight named species within about 2 myr. Grasslands had spread throughout South America during the middle Tertiary and had been exploited by the endemic mammals, particularly notoungulates and litopterns. The horses that dispersed from North America apparently were also successful in invading the grazing adaptive zone in South America. The observed diversity of about two to four sympatric equid species (including *Equus*, as discussed later) in South America during the Pliocene and Pleistocene is comparable to that of equid faunas elsewhere in the world at that time.

The adaptive radiation of *Equus* since the middle Pliocene (about 4–3 Ma) serves as an evolutionary microcosm for the success of the grazing Equidae on the whole. As discussed in Chapter 5, the taxonomy of fossil and Recent *Equus* is exceedingly complex. Some 58 *nomina* have been proposed for North American species of *Equus*, of which Winans (1985, 1989) concluded that only five (all extinct) "species groups" were valid for that continent. Eisenmann (1975, 1980, 1981) and Azzaroli (1990) recognized a total of about 10–12 valid species of extinct and living Old World *Equus*. For South America, there probably were about five valid species (Sefve 1912; Hoffstetter 1952; De Porta 1960; MacFadden & Azzaroli 1987), all of which became extinct by the end of the Pleistocene. Depending upon the author, neontologists usually recognize from six to eight valid extant species (e.g., Walker 1975). Therefore, including both extinct and living forms, the genus *Equus* probably consists of

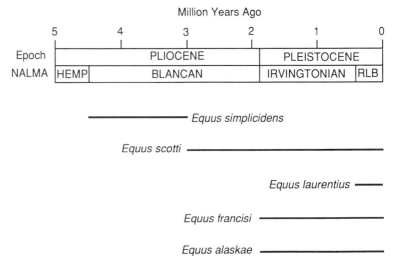

Figure 8.8. *Biostratigraphic ranges of the five currently recognized extinct species of* Equus *from North America. Modified from Winans (1989).*

about 25–30 valid species. Although perhaps an artifact of comparing extant and wholly extinct genera, as well as the relative abundance of the Pliocene–Pleistocene fossil record, the observed diversity for *Equus* is higher than for any other genus of Equidae.

The North American radiation of *Equus* exemplifies the general pattern of diversification for this genus. On that continent the early part of the radiation of *Equus* was characterized by a single numerically dominant and widespread species, *E. simplicidens*, during the middle Pliocene (early Blancan, about 3–4 Ma) (Figure 8.8). During the late Pliocene, *E. simplicidens* was replaced by *E. scotti*, a species that was slightly larger, although otherwise very similar to the former. The mode of speciation may have been anagenesis, because the two are not known to have overlapped in their observed ranges. During the latest Pliocene and into the Pleistocene, three other species occurred, *E. francisi*, *E. alaskae*, and *E. laurentius*, although they usually were not sympatric. The phylogenetic interrelationships are unclear here; these could have resulted from cladogenesis from an ancestral species in North America or from either anagenesis or cladogenesis followed by dispersal from the Old World. Similar evolutionary patterns seem to have occurred in the Old World, with a single dominant and widespread species, *E. stenonis* (and variants), during the Pliocene, and cladogenesis resulting in more than one species afterward (Eisenmann 1975, 1980, 1981; Azzaroli 1990).

The mode and rate of speciation for *Equus* were therefore in contrast to those for the Miocene Equinae described earlier. Rather than experiencing a rapid increase in speciation early on, *Equus* was essentially monospecific in both North America and

Eurasia during the Pliocene, a period of about 2 myr. Thereafter, there was multiplication of species and a greater intrinsic rate of increase. The reasons for that different evolutionary mode and pattern may be as follows: During the Miocene the Equinae were radiating into a newly created adaptive zone to exploit the grazing habitus. In the Pliocene, *Equus* originated in a community that was already inhabited by horses, namely, the hipparions, represented in North America by *Nannippus* and *Cormohipparion* (e.g., MacFadden 1984b; Hulbert 1987). After hipparions became extinct at about 2 Ma, subsequent cladogenesis occurred in *Equus*. Simple explanations for natural systems are rarely all-encompassing in terms of other factors that may have contributed to observed phenomena. Although other changes in community structure and environment undoubtedly occurred, the radiation of *Equus* species after 2 Ma seems at least partially to have been related to the demise and ultimate extinction of other horses throughout Holarctica.

Key characters, adaptive radiations, and clade evolution

A survey of the history of life reveals that certain characters and adaptations have been of great significance in the explosive evolution of many higher taxa. One might even say that the successes of such clades, at least in part, have been related to or have resulted from these key characters or evolutionary innovations. In most clades it has not been only a single key character, but usually a character complex or even seemingly unrelated functional and adaptive complexes that were followed by a significant burst of diversification. The vertebrate fossil record provides many examples of the origin of certain important morphologies and explosive adaptive radiations resulting in significant clade diversification. These have included the origin of key characters, such as the neural crest in vertebrates during the Ordovician, jaws and paired appendages in gnathostomes in the Silurian, the amniotic egg in tetrapods that almost certainly evolved during the Carboniferous, the homocercal tail in teleosts, wings in birds, and endothermy and cusped teeth in mammals.

At a lower taxonomic level, the acquisition of key characters within certain clades of fossil Equidae probably was related to their subsequent evolutionary success (Table 8.2). We have already seen that the transition from phenacodontid condylarths to *Hyracotherium* required no hopeful monsters or large morphoevolutionary gaps. *Hyracotherium* had reduced lateral metapodials and a slightly different dental morphology relative to its plausible primitive outgroup. However, currently it is not clear if these or other characters of feeding or locomotion in early horses were significantly more advanced or beneficial in a com-

Table 8.2. *Key morphological characters recognized within the evolution of the family Equidae*

Taxon	Age	Key character	Source
Hyracotherium	Eocene	Increased E.Q.[a] and expanded neocortex	Radinsky (1976)
Mesohippus and *Miohippus*	Oligocene	Molarized premolars, tridactyl feet	Many (see text)
Equinae	Miocene– Recent	Cement-covered, high-crowned cheek teeth; springing foot	Many (see text)
Equus	Pliocene– Recent	Evolution of the passive stay apparatus	Hermanson & MacFadden (1992)

[a]E.Q. = encephalization quotient; ratio of brain size to body size for extinct species relative to what would be predicted from a living mammal of same body mass (Jerison 1973). Also see discussion in text and Radinsky (1976).

petitive sense. On the other hand, Radinsky (1976; *contra* Edinger 1948) has shown from endocasts that the brain of *Hyracotherium* was significantly more advanced than those of contemporaneous condylarths (see Figure 10.1). He concluded that the features of an expanded neocortex and relatively larger brain were related to increased tactile sensitivity of the lips, expected in mammals specialized for browsing or grazing.

Other Eocene hyracotheres are characterized by increased molarization of the premolars, whereas in *Mesohippus* the P2–M3 are fully molariform (see Figure 11.4B). That, and possibly the acquisition of fully tridactyl manus and pes (in all but some primitive species), may be related to the diversification of species of *Mesohippus* and *Miohippus* during the Oligocene, as recently pointed out by Prothero and Shubin (1989). However, that adaptive radiation had relatively little impact on horse diversity.

A classic and often-cited example of a key character that evolved in the Equidae was the acquisition of cement-covered, high-crowned teeth by the subfamily Equinae during the Miocene. Paleobotanical evidence indicates that during the late Oligocene and early Miocene, grasslands became relatively widespread in North America (Elias 1942; MacGinitie 1962; Retallack 1982), and the evolution of hypsodonty in horses allowed them to invade that newly created adaptive zone to exploit its food resource. Of the 34 currently recognized genera of Equidae throughout the world (Appendix II; see Figure 5.14), 18 are assigned to the Equinae, and all of these apparently were adapted to grazing. In addition to major changes in feeding in the

Equinae, there was a major reorganization in the foot, with a shift toward a more springing style of locomotion in *Merychippus* and later forms (Camp & Smith 1942) (see Chapter 11).

The next major morphological change that occurred in the Equidae was the acquisition of the monodactyl limb, in which the lateral metapodials I and III were reduced to rudimentary or vestigial splints. All monodactyl horses are allocated to the Equinae within the genera *Pliohippus, Astrohippus, Dinohippus, Hippidion, Onohippidium, Parahipparion,* and *Equus* (certain primitive species within the first two genera are tridactyl). The morphocline from tridactyly to monodactyly is frequently cited in textbooks as a familiar example of a long-term evolutionary trend. However, its adaptive significance as a key character is not as clear as, say, that of hypsodonty, because the members of the other late Tertiary equine clade, the tribe Hipparionini (and some primitive Equini, as noted earlier), were tridactyl, and they were relatively diverse. If monodactyly was a key character that conferred an evolutionary advantage on the advanced species of Equini, its effect was relatively subtle.

Recent work on the evolution of the passive stay apparatus (see Chapter 11) suggests a major evolutionary innovation in advanced Equidae. Observations of both hard and soft anatomy show that in extant *Equus* the forelimbs and hindlimbs are "locked" into place during resting posture (while standing), thereby conserving muscular energy (see Figures 11.17–11.19). In the forelimb the biceps brachii tendon has a groove that locks into an intermediate tubercle (INT) of the proximal humerus (Hermanson & Hurley 1900). A recent survey of the origin and taxonomic distribution of the INT shows that its morphology is unique to equids (i.e., it is not found in other ungulates) and first appeared between about 5 and 2 Ma, where it was rudimentary in *Dinohippus* and well developed in Pliocene and all subsequent *Equus* (Hermanson & MacFadden 1992). Similarly, in the hindlimb, the patella (kneecap) locks onto a trochlear crest of the femur. It is not yet known, however, when the hindlimb locking mechanism was acquired in horse evolution, although some current research should resolve this question (J. J. Becker personal communication). In summary, it seems likely that the acquisition of these important character complexes has allowed horses to stand for long periods of time for activities such as resting, feeding, or looking out for potential predators.

Extinction

Many workers have used the fossil record of horses to exemplify patterns of extinction and related causal mechanisms. The most frequently cited examples include the decline of the browsing anchitheres during the Miocene and the hypsodont equines in the Pliocene and Pleistocene (Webb 1983, 1984). This section will focus on those events and their possible causes; rates of extinction will also be discussed in Chapter 9. Extinction is defined here as the extirpation of a lineage or clade. This does not include the concepts of "pseudoextinction," or phyletic transformation, where an ancestral taxon evolves into its descendant.

From the early Eocene until the middle Miocene in North America, originations and extinctions of species within the Equidae were essentially balanced, with relatively low and generally stable diversity and evolutionary replacement of ancestral and descendant forms (Webb 1983, 1984). This seems to have been the "background" amount of extinction within the Equidae, with a diversity usually not exceeding five sympatric species.

Equid diversity increased during the middle Miocene, after about 18 Ma. A close look at the taxonomic composition of that radiation reveals interesting finer-scale evolutionary phenomena. During the Clarendonian, when horses reached their maximum diversity, the browsing anchitherines decreased in diversity, while the hypsodont grazers increased (Webb 1983, 1984) (Figure 8.9A). However, it is almost certainly wrong to conclude that there is a simple competitive explanation for this general pattern of "replacement." It is equally important to realize that during the middle Miocene hypsodont horses were extending their biogeographic ranges and increasing their diversity in response to a corresponding spread of the grazing adaptive zone. Thus, the decrease in diversity and ultimate extinction of browsing horses by about 9 Ma seem to be related as much to a decrease in forested biomes during the Miocene as to direct competition with contemporaneous grazing horses.

If the diversities of artiodactyls and perissodactyls are plotted through time (Figure 8.10), the evolutionary patterns of these two ungulate orders are seen to be fundamentally different. Although the two orders first appeared in the early Tertiary, at about the same time (late Paleocene–early Eocene), the diversification of perissodactyls occurred more rapidly, reaching a peak by the Miocene. On the other hand, the most significant adaptive radiation of artiodactyls occurred later, with the advanced ruminants diversifying greatly from the Miocene onward. Some workers have concluded that given this pattern, the demise of perissodactyl diversity was related to, or caused by, competition with artiodactyls (Simpson 1953a; Stanley 1974a). The competitive edge that is usually ascribed to artiodactyls related to their different feeding and digestive strategies (see Chapter 11), with perissodactyls being hindgut fermenters (in an expanded intestinal cecum (see Figures 5.5 and 11.5), whereas ruminant artiodactyls accomplish much of the digestive breakdown of foodstuffs in the highly specialized stomach complex (Janis 1976). Recent studies have challenged this view and have suggested that the significant changes in climate, and hence

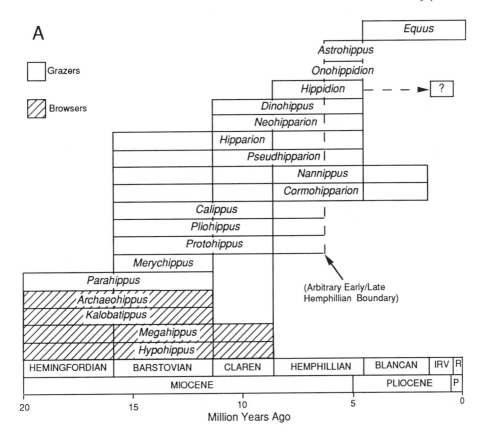

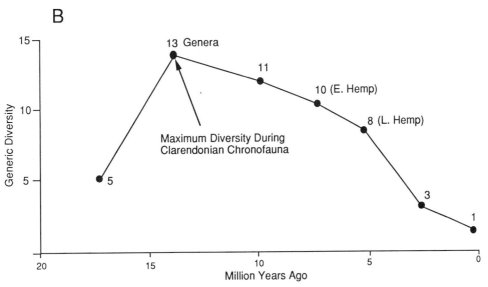

Figure 8.9. *Equid faunal diversity in North America during the Neogene. (A) Stratigraphic ranges of genera and inferred browsing or grazing adapta-* *tion. Range data taken from Figure 9.11 and Appendix II. Note that the range of* Hippidion *is discontinuous, and its Irvingtonian occurrence proba-* *bly represents a "back-migration" from South America. (B) Plot of equid ge-neric diversity for each land-mammal age during the past 18 myr.*

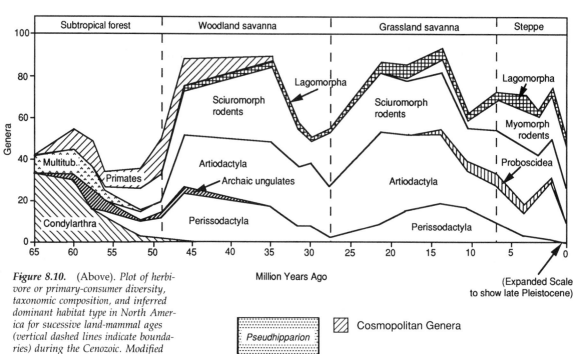

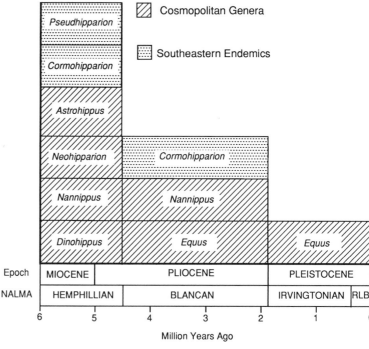

Figure 8.10. (Above). *Plot of herbivore or primary-consumer diversity, taxonomic composition, and inferred dominant habitat type in North America for sucessive land-mammal ages (vertical dashed lines indicate boundaries) during the Cenozoic. Modified from Webb (1987).*

Figure 8.11. *Histogram depicting changes in diversity and regional differences in equid genera in North America during the late Miocene through Pleistocene.*

vegetation, that occurred during the late Tertiary were at least as important in the changes in diversity seen in these ungulate clades (Cifelli 1981; Janis 1989).

The diversity pattern for late Tertiary horses, particularly as exemplified in North America, is similar to that for perissodactyls in general. The height of equid diversity occurred during the middle Miocene, with as many as 11–13 genera (Figure 8.9B) and about 12–15 potentially sympatric species existing between 12 and 10 Ma. Thereafter, diversity decreased, with the extinction of the anchitheres by about 9 Ma. By the end of the Hemphillian, at about 5 Ma, horse diversity at most sites in North America usually was about three genera (and three to five species), with a maximum of six monospecific genera in the more tropical biomes (e.g., in the Hemphillian of Florida, species of *Neohipparion, Nannippus, Pseudhipparion, Cormohipparion, Dinohippus,* and *Astrohippus* co-occur) (Figure 8.11). During the Pliocene, between about 4.5 and 1.8 Ma, there was a further reduction in equid diversity, resulting in a maximum of two to three genera (and sympatric species) at most sites (usually *Equus* and *Nannippus;* also *Cormohipparion* in the southeast) (Figure 8.11).

During the early Pleistocene, at about 1.8 Ma, there was a further drop in generic diversity of horses in North America, with only *Equus* represented (Figure 8.11). Thereafter, the genus *Equus* was widespread, with usually two or sometimes three sympatric species occurring at most sites (Winans 1985, 1989). Thus, although the generic diversity of horses has dropped during the past 5 myr, the number of species has remained essentially constant. The genus *Equus* has been very successful in terms of overall diversity (with about 25–30 valid extinct and living species worldwide). Biogeographically, it also has been the most widespread genus, with a distribution during the Pleistocene that encompassed North and South America, Eurasia, and Africa. That range seems to have existed until the latest Pleistocene, about 10,000 years ago, when, like many other large-bodied mammals, they were part of the mass extinction that wiped out more than 40 mammalian genera (Webb 1984) throughout the New World and parts of Eurasia and Africa. The causes for that "late Pleistocene megafaunal extinction" have been hotly debated, and it probably resulted from a complex series of factors, including dramatic fluctuations in climate and, at least in some regions, the impact of prehistoric humans – the "Pleistocene overkill hypothesis" (e.g., Martin & Wright 1967; Martin & Klein 1984).

9

Rates of morphological and taxonomic evolution

> *How fast, as a matter of fact, do animals evolve in nature? That is the fundamental observational problem of tempo in evolution. It is the first question that the geneticist asks the paleontologist.*
>
> – Simpson (1944:3)

Natural scientists are keenly interested in understanding rates of evolution. This is exemplified by the fact that the foregoing quotation began Simpson's classic book *Tempo and Mode in Evolution*. Thus, in addition to the questions about the processes and mechanisms of evolution discussed in Chapter 8, there is another of fundamental importance: How fast does evolution occur? Evidence to answer this can come from several scales of the biological hierarchy and from a myriad of disciplines, including biochemistry, genetics, neontology, and paleontology.

Rates of evolution can be investigated in several different ways. Simpson (1944, 1949, 1953a) categorized these within three broad groups, namely, genetic, morphologic, and taxonomic. Although he realized that genetic rates of evolution provided the foundation for any observed phenotypic or taxonomic change, he did not concentrate on these in his evolutionary syntheses. This is understandable, because (1) genetic evolution was certainly less quantifiable then, (2) it is difficult to show a one-to-one correspondence between genetic change and evolutionary change, and (3) from a pragmatic point of view Simpson was more concerned with data and interpretations taken from the fossil record. Until biochemical data can be routinely extracted from vertebrate skeletal tissues, the study of "paleogenetics" will remain only a nascent field of inquiry.

Despite their lack of relevant genetic data, paleontologists have unique evidence of long-term evolution, in terms of both morphologic and taxonomic patterns. Until only a few decades ago, a precise geological time scale did not exist, and therefore temporal calibrations of evolutionary patterns were not available. Prior to that time, rates of evolution were studied within a temporal framework that was calibrated by geological processes such as rates of sediment accumulation, increases in oceanic salinity, and even faunal evolution (see Chapter 6). In retrospect, estimates using those processes were of variable accuracy compared with modern geochronological calibrations. Osborn (1934) believed that the evolution of mastodonts, from the Eocene *Moeritherium* to the Miocene *Trilophodon*, had occurred in about 20 myr, an estimate that was too small by a factor of about 2.5. Simpson (1944) believed that the Cenozoic had a duration of about 45 myr, which was about 20 myr shorter than is currently believed. With regard to the subdivisions of the Cenozoic, most earlier workers essentially split up the total duration of that era into equal epochs. Thus, rates of evolution (Figure 9.1) reported in earlier works indicated roughly equal durations for the Tertiary epochs. However, we now know that there is as much as an order-of-magnitude difference between some of these, so that the Miocene has a duration of about 25 myr, whereas the Pleistocene represents the past 2 myr (Berggren et al. 1985) (see Figure 6.10). Similarly, roughly equal amounts of evolutionary change were interpreted to represent similar temporal durations. Surprisingly, the geologist Knopf (1949:8) even remarked that "at present the best estimates of the lengths of the several subdivisions of the Tertiary are based on the evolution of the horses." Until only a few decades ago, biochronology was a universally accepted method of assessing geological time, despite the fact that many workers realized that rates of evolution were not always constant. That situation changed dramatically after World War II as the radioisotopic time scale was developed. Many studies during the 1940s began to discuss rates of evolution in this geochronologically calibrated context (e.g., Jepsen et al. 1949), but there were no uniform means for comparison.

Quantifying rates of morphological evolution: Haldane and darwins

The great British biologist Haldane published a landmark paper (1949) in which he proposed a method for quantifying rates of morphological change. He coined the unit "darwin," and defined this rate of change as

$$r = [\ln(x_2) - \ln(x_1)]/\Delta t,$$

where r is the rate of change (in darwins, d), x_1 is the initial dimension of a character, x_2 is the final dimension of a character, and Δt ($t_1 - t_2$) is the amount of time elapsed between the ages of x_1 and x_2. Haldane (1949) used his equation as a basis for comparing evolutionary rates in dinosaurs, fossil horses, and fossil humans. Although the darwin was a simple and elegant measure for the study of morphological evolution, it was not widely accepted; for a notable early exception, see Kurtén (1960). For example, Simpson (1953a:7) concluded that "at present this seems an unnecessary complication and darwins will not be used as rate units in this book, but the suggestion is repeated here for possible future trial." Within the past two decades, and undoubtedly because of the greatly improved calibration of geological time, the use of the darwin as a common denominator for comparisons of evolutionary rates has increased, particularly for Cenozoic mammals (e.g., Gingerich 1982, 1983, 1987b; MacFadden 1985a, 1988b). There are, however, several limitations and assumptions inherent in its use.

Haldane (1949) established the darwin as a unit of evolutionary change in measured linear dimensions, such as tooth length, femur width, and so forth. Other characters, such as ratios, areas, volumes, and those that are counted (meristic, e.g., the number of toes), cannot be compared with linear dimensions. To say, for example, that a tooth dimension evolved faster than body size is meaningless, because at present there is no valid means for quantitative comparison between these kinds of characters.

In a thought-provoking paper, Gingerich (1983) brought to light another limitation with the method of comparing rates of morphological evolution across widely divergent time scales. He found that laboratory animals with very small Δt values ranging from 1 to 10 years had astonishingly high rates of evolution, between 12,000 and 200,000 d (mean 57,600 d). (As he noted, a rate of 400 d would be sufficient to change a mouse into an elephant in only 10,000 years). At the other end of the temporal spectrum, he found that Paleozoic invertebrates with large Δt's ranging from 0.3 to 350 myr had low rates of evolution, between zero and 3.7 d (mean 0.07 d). Thus, evolutionary rate appears inversely proportional to time interval. If Gingerich is correct,[1] then, as with comparisons of different kinds of characters, it is also currently invalid to compare taxa with widely divergent Δt values.

Calculation of darwins can be used to study both microevolutionary and macroevolutionary changes, for example, within populations of chronoclinal species or between ancestral and descendant species (Figure 9.2). However, in any of these cases, an obvious requirement is that the ages of the populations or species being studied be well calibrated. This method, whether it be within or between species, does not account for evolutionary reversals in individual character trends between measured endpoints; so darwins actually provide an assessment of minimum rates of morphological change.

With regard to the use of darwins as a basis of comparison, several colleagues have presented a challenge that goes something like this: "How can you compare rates of evolution in mice

[1]Gould (1984) challenged Gingerich's observation of the scale-dependent nature of rates of evolution and concluded that this was a "psychological artifact"; see response by Gingerich (1984a). Until this phenomenon is examined further, it does seem that comparisons of evolutionary rates will be of greatest validity when similar time intervals (Δt's) are involved.

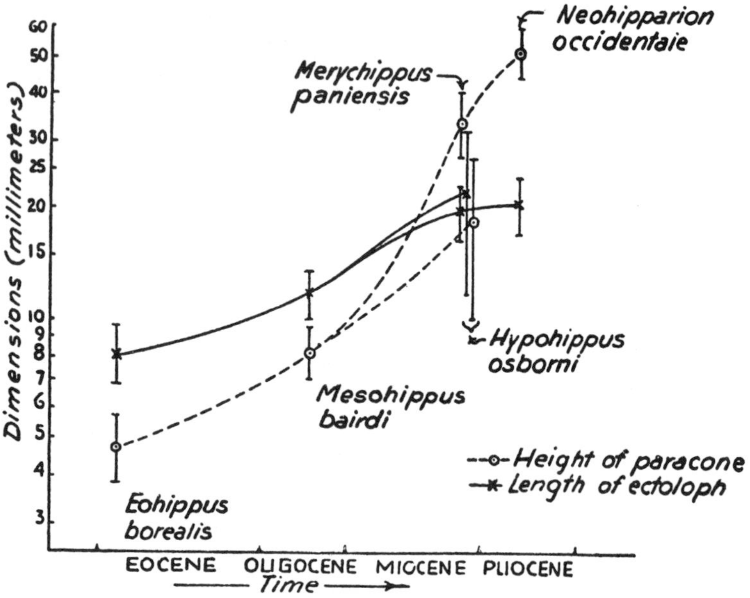

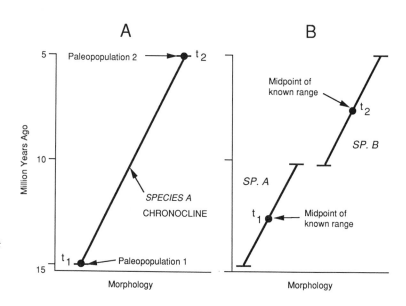

Figure 9.1. *Simpson's plot (1944) of rates of evolution for upper dental characters (i.e., paracone height and ectoloph length) for selected species of fossil horses from North America. Note the roughly equal durations of the Tertiary epochs along the x-axis. Reproduced with permission of Columbia University Press.*

Figure 9.2. *Methods of calculating rates of morphological evolution: (A) between populations within a chronoclinally evolving species; (B) between two presumed ancestral–descendant species pairs. In this case, the midpoint of the known range is used to calculate the time difference separating the two species, Δt $(= t_1 - t_2)$.*

and elephants? They are not the same.'' Haldane's solution was elegant, because as shown in Table 9.1, the use of the original linear dimensions transformed to natural logarithms compensates for wide differences in morphological scale, so long as the amounts (percentages) of change are similar and occur over similar intervals of time. Thus, a doubling of femur size in hypothetical lineages of mice and elephants over 10 myr would result in the same amount of change, 0.069 darwin (Table 9.1).

Table 9.1. *Comparison of rates (in darwins) of morphological evolution of femur in hypothetical lineages of mice and elephants.*

1. Mouse lineage
   Ancestral species that lived at 15 Ma, femur length = 10 mm
   Descendant species that lived at 5 Ma, femur length = 20 mm

$$\text{Rate} = \frac{\ln(20) - \ln(10)}{15 - 5 \text{ Ma}} = \frac{0.693}{10 \text{ myr}} = 0.069 \text{ darwin}$$

2. Elephant lineage
   Ancestral species that lived at 15 Ma, femur length = 1,000 mm
   Descendant species that lived at 5 Ma, femur length = 2,000 mm

$$\text{Rate} = \frac{\ln(2,000) - \ln(1,000)}{15 - 5 \text{ Ma}} = \frac{0.693}{10 \text{ myr}} = 0.069 \text{ darwin}$$

Rates of morphological change in fossil horses

In addition to all the other uses to which their fascinating history has been put, fossil horses have been prominent in discussions of rates of morphological change. The changes in fossil horse skulls, teeth, and limbs during the Cenozoic have provided a familiar evolutionary case study for more than a century, as originally exemplified by Marsh's 1879 illustration (see Figure 3.6). With regard to quantification of evolutionary rates, as discussed earlier, Haldane (1949) calculated that certain dental characters (paracone and ectoloph lengths) evolved in Tertiary horses at a rate of about 0.04 d. Through no fault of his own, Haldane's calibrated time points were understandably few, and his Δt's were relatively large. However, we are now in a position to add many refinements to his pioneering work.

MacFadden (1988b) studied rates of morphological evolution in numerous ancestral–descendant species pairs of Cenozoic horses from North America (Figure 9.3). For each species, four characters of the upper first molar were measured: crown height, greatest length, greatest width, and protocone length (Figure 9.4). In order to factor out ontogenetic variation within samples, for the first three characters only specimens in middle wear were used, and for crown height (which obviously decreases throughout the life of an individual) only unworn or little-worn specimens were measured. In order to calculate Δt, the midpoint of the known range for each species was used (Figure 9.2B). As discussed earlier, because of the possible error in comparing rates of evolution across widely different time scales, all ancestral–descendant species pairs except one (MacFadden 1988b) were limited to Δt's of 1–10 myr.

The results of that study, as well as the findings of MacFadden (1985a) for hipparionines and Webb and Hulbert (1986) for *Pseudhipparion*, highlight some interesting patterns of morphological

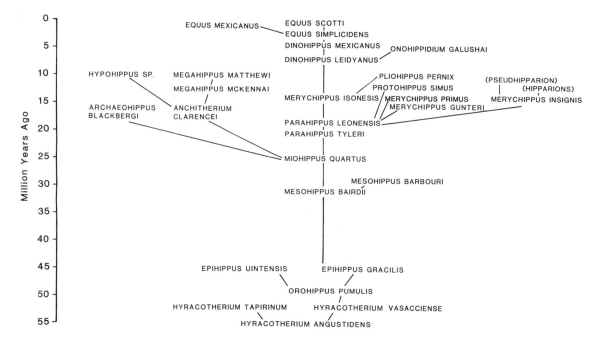

Figure 9.3. *Twenty-six ancestral–descendant species pairs of Cenozoic horses from North America used to calculate rates of dental evolution. Modified from MacFadden (1988b); reproduced with permission of the Linnean Society.*

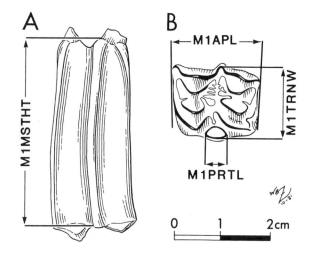

Figure 9.4. *Dental measurements of first upper molar used to calculate rates of evolution: (A) maximum crown height (M1MSTHT); (B) occlusal surface, including greatest length (M1APL), greatest transverse width (M1TRNW), and protocone length (M1PRTL). From MacFadden (1988b); reproduced with permission of the Linnean Society.*

evolution for Cenozoic horses in North America (Table 9.2). For all of the six groups of fossil horses in Table 9.2, the rates of evolution of occlusal dimensions (M1APL, M1TRNW, M1PRTL) were relatively constant, with the values clustering about the means. The only exception was *Pseudhipparion*, a clade of dwarf forms in which these dimensions decreased in size relatively rapidly. On the other hand, the evolution of crown height (M1MSTHT) tells a different story of a wide range of values. Early horses (*Hyracotherium* to *Mesohippus*) showed the lowest

Table 9.2. *Comparisons of mean rates of morphological evolution (in darwins, d) for the same M1 characters for selected groups of North American Equidae*

Group	N (species pairs)	M1 APL (d)	M1 TRNW (d)	M1 PRTL (d)	M1 MSTHT (d)
1. Early horses (*Hyracotherium* to *Mesohippus*, Fig. 9.3)	8	0.049	0.053	0.044	0.037
2. Anchitheres (*Anchitherium*, *Hypohippus*, and *Megahippus*, Fig. 9.3)	4	0.061	0.062	0.071	0.113
3. *Parahippus* and *Merychippus* (Fig. 9.3)	6	0.038	0.039	0.074	0.204
4. Hipparions (MacFadden 1985a)	8	0.040	0.030	0.040	0.080
5. *Pseudhipparion* (Webb & Hulbert 1986)	4	0.070	0.087	0.113	0.242
6. Advanced monodactyl equines (*Dinohippus* and *Equus*, Fig. 9.3)	4	0.027	0.032	0.110	0.086
7. All horses (1–6 above and others in MacFadden 1988b)	43	0.046	0.047	0.068	0.113

Note: See Figure 9.4 for characters and their abbreviations.
Source: From MacFadden (1988b).

rates of change in this character, whereas the various grazing equine groups (entries 3–6) showed high rates of evolution. Surprisingly, the species of *Pseudhipparion*, with a mean rate of 0.242 d, showed the highest rates of crown-height evolution, despite the fact that they composed a clade of dwarf forms. This is in contrast to the possible correlation between rapidly evolving crown height and increasing body size (MacFadden 1987a) observed in the other grazing horses (*Parahippus, Merychippus,* hipparions, and monodactyl equines, Table 9.2). For *Pseudhipparion,* this may be an example of increased fecundity (see Chapter 12). Another interesting group consists of the anchitheres, including *Anchitherium, Hypohippus,* and *Megahippus* (Table 9.2). Traditionally, these horses had been perceived as a low-diversity clade of browsers with relatively slow rates of evolution. The rates reported in Table 9.2, however, belie that notion. The anchitheres have rates for all four dental characters that are near the average, rather than at the low end of the spectrum.

It is as interesting to compare rates of morphological change between taxonomic groups as within monophyletic clades, and yet there have been few papers that have used Haldane's method. Haldane (1949) calculated rates of linear-dimension evolution between 0.002 and 0.060 d for dinosaurs, and much higher rates between 0.3 and 0.6 d for hominids. In a modern context, as Gingerich (1983) noted, these rate differences may be artifacts of the divergent time intervals (Δt's) used for dinosaurs (tens of millions of years) versus hominids (about 1 myr). Kurtén (1960) calculated evolutionary rates for lower molar lengths between 0.4 and 3.2 d for Pleistocene bears (*Ursus*) of Europe. Gingerich has long been interested in quantification of morphological rates of evolution. He has even introduced a "darwinometer," which depicts the rate of stratophenetic evolution (e.g., Figure 8.3), into many of his illustrations. Gingerich (e.g., 1982, 1983, 1987b) found relatively high rates of molar evolution (0.1–1.0 d) for early Tertiary mammals of North America. Relative to fossil horses, Gingerich's results could reflect the artifact of a shorter time duration (Δt) of about 1–2 myr. However, his data could equally as well indicate more rapid mammalian evolution during radiations into new adaptive zones. For all fossil vertebrates, Gingerich (1983) found a geometric mean of 0.08 d. Thus, for fossil horses, the mean evolutionary rates, between 0.113 and 0.046 d (Table 9.2) for the four linear dental dimensions lie within the midrange of those observed for other taxa.

Almost half a century ago Haldane provided a means for quantifying evolution with the introduction of the darwin. Although this currently seems to be the best system for comparing evolutionary rates in such widely divergent taxa as mice and elephants, or invertebrates and vertebrates, Gingerich's important paper (1983) highlights the limitations of this technique. Thus, whether it be different levels within the Linnean hierarchy, or different time scales for fossil taxa, the scale of observation affects our perception of evolutionary phenomena.

As a final note, the results presented here further underscore the principle of mosaic evolution as it was originally proposed by de Beer (1954). All organisms are complex mixtures of primitive and derived characters evolving at different rates. For horses, this can even be seen on a fine scale, where characters within a functionally interrelated complex, such as the dentition, evolve at significantly different rates.

Allometry and scaling of horse evolution: Does ontogeny recapitulate phylogeny?

In addition to examination of rates of evolution of single characters, another popular line of inquiry has been the application of allometry. Allometry is the study of the relative proportions that occur during ontogenetic growth, or size changes within clades. Two kinds of allometry are relevant here. In positive allometry, as size changes, one dimension (usually a linear di-

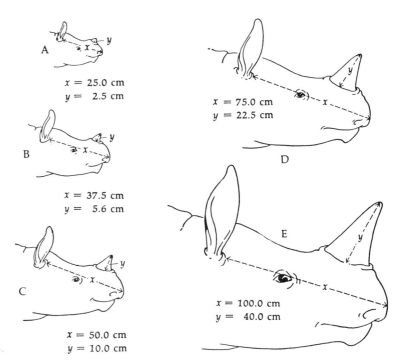

A

x = 25.0 cm
y = 2.5 cm

x = 75.0 cm
y = 22.5 cm

B

D

x = 37.5 cm
y = 5.6 cm

C

E

x = 50.0 cm
y = 10.0 cm

x = 100.0 cm
y = 40.0 cm

Figure 9.5. *Positive allometry of the nasal horn (y) relative to nose-to-ear length (x) for a hypothetical rhinoceros lineage through time. From Moody (1970),* Introduction to Evolution, *original copyright 1953 by Harper & Row, Publishers, Inc. Reprinted by permission of HarperCollins Publishers.*

mension) grows more than another (Figure 9.5). In static allometry there is isometric growth, where all dimensions under study grow by the same proportion. In the hypothetical rhinoceros lineage shown in Figure 9.5, if isometric growth had occurred, then for the youngest species depicted in part E, the horn length (y) would be 10 cm.

Many workers have made the distinction between ontogenetic allometry and evolutionary allometry, that is, respectively, the relative changes in proportions that occur during an individual's lifetime, and those that occur within the phylogenetic history of a clade. This obviously leads to a question that was first asked by Ernst Haeckel during the nineteenth century, as well as by many other natural historians since that time: To what extent does ontogeny recapitulate phylogeny? Related to this, some classic studies of fossil horse cranial proportions and their evolutionary interpretations will be examined here.

Osborn (1902, 1912b) studied the changes in the skull proportions of many fossil mammals, including horses, and coined the term "dolichocephaly" for any type of cranial elongation. He noticed two principal kinds of dolichocephaly, in which, relative to the total skull length, either (1) the preorbital length (i.e., the cheek and muzzle) is enlarged (proopic) or (2) the postorbital length (i.e., the braincase and basicranial region) is enlarged (opisthopic). Osborn (e.g., 1912b) used Tertiary titanotheres as examples of postorbital dolichocephaly (Figure 9.6). Not surprisingly, the Equidae were used as prime examples of preorbital dolichocephaly. Taking the two phylogenetic endpoints, in *Hyracotherium* the preorbital region is relatively short, whereas in *Equus* it is relatively long (Figure 9.7). Osborn's classic study (1912b) of equid craniometry discussed various measuring systems for horse skull proportions and relevant terminology, but there was no evolutionary synthesis or interpretation. The subsequent hypothesis to explain the expanded preorbital region is to accommodate a larger dental battery for mastication of an increased supply of foodstuffs, particularly for the grazing clades that fed on low-digestibility grasses (e.g., Janis 1976; Radinsky 1983, 1984).

Robb (1935) analyzed allometric changes in the cranial proportions of 13 fossil horse species. Like Osborn, he noted positive allometry in which the preorbital length doubled relative to total skull length (preorbital ratios ranged from 0.85 in *Hyracotherium* to 1.70 in large *Equus*). From his data and observations Robb offered several relevant interpretations including the following: (1) Changes in skull proportions were gradual ("insensible gradations"). (2) Elongation of the preorbital region was related primarily to overall body size, so that irrespective of phylogenetic position, all species of similar sizes have similar preorbital lengths. Robb's analysis (1935) and conclusions have been challenged by many subsequent workers, most notably Reeve and Murray (1942); also see Simpson (1953a).

Radinsky (1983, 1984) restudied cranial allometry in fossil horses in light of many new specimens and a better-calibrated time scale. He measured 18 cranial, mandibular, and dental characters from 25 equid species and subjected these to multivariate analysis. In contrast to Robb's conclusions, Radinsky reported that changes in skull proportions had not been gradual. He found that the most profound changes had occurred during the Miocene in the transition from *Miohippus* to *Merychippus*, primarily within *Parahippus*. In addition to a simple lengthening of the preorbital region, the horse skull underwent many (7 of the 18 characters were significantly different) associated biomechanical changes during that time that were interpreted as adaptations to grazing. One of Radinsky's original goals for his studies was to test whether or not ontogeny recapitulates phylogeny. His results were not definitive, and he concluded that (1984:13) "an analysis of changes in cranial morphology in postnatal ontogeny and across a wide range of adult domestic horses shows that some variables scale with similar exponents as in the phylogenetic series, while others scale with significantly lower exponents."

Devillers et al. (1984) studied ontogenetic changes ("trajectories") in modern horse fetuses, adult skulls, and fossil speci-

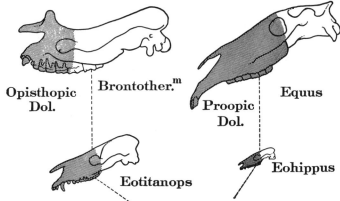

Figure 9.6. *Osborn's view (1912b) of allometric changes in skull proportions showing elongation (dolichocephaly) of the postorbital (opisthopic) region in titanotheres (left) and in the preorbital (proopic) region in horses (right).*

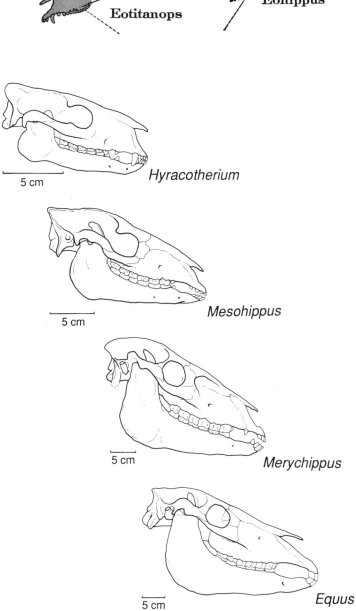

Figure 9.7. *Changes in cranial proportions in* Hyracotherium, Mesohippus, Merychippus, *and* Equus *showing increased preorbital region associated with expansion of the cheek teeth and other characters. Modified from Radinsky (1984). Radinsky (1983, 1984) found that much of this reorganization occurred during the Miocene, with the inferred adaptive shift from browsing to grazing.*

199

mens. Their results, in contrast to those of Robb (1935), indicate that preorbital length is not simply dependent upon body size. For example, they found that in Miocene advanced dwarf forms such as *Calippus*, skull proportions were similar to those of contemporaneous *Pliohippus* (= ?*Dinohippus*), despite the former being the size of Oligocene *Miohippus*.

In summary, studies of skull proportions in fossil and modern horses spanning a wide range of ontogenetic and phylogenetic stages have provided examples of the general evolutionary phenomenon of allometry. A most vexing, unresolved problem, however, is that, like the situation for other groups, the evidence from studies of fossil horses is equivocal for ontogeny recapitulating phylogeny; some data support it, others do not. Thus, Haeckel's "law" may serve as a general approximation in certain instances, but it is by no means a hard and fast rule.

Taxonomic evolution: origination, duration, and extinction

In addition to using morphological rates, Simpson (1953a) stated that another general method for quantifying evolution was to assess taxonomic changes through time within particular groups. There are numerous ways of looking at taxonomic evolution, but they all relate to investigations of patterns of origination, extinction, longevity, and diversity. Of these, here we shall discuss those that have been used in previous studies that are relevant to an understanding of the evolutionary history of the Equidae.

Simpson (1944) coined the terms "bradytely," "horotely," and "tachytely" to qualitatively characterize rates of taxonomic evolution. Bradytelic taxa are those that show slow or arrested evolution, such as pelecypods throughout the Phanerozoic, or the extant brachiopod *Lingula*, a genus that has existed since the Ordovician. Tachytely occurs in rapidly evolving groups, particularly those that are invading new adaptive zones, such as many mammals during the Cenozoic. Horotely describes the "normal" or "average" rates of taxonomic evolution seen in many groups, including horses for much of their history, as we shall see later.

Assumptions and limitations

> The weaknesses of taxonomic data for quantitative studies of evolutionary rates are fairly obvious and do not need belaboring.
> – Simpson (1953a:30)

Some critics have remarked that calculations of taxonomic rates of evolution are so arbitrary and imprecise that their results have little validity, and that view is reflected in the quotation from Simpson. Nevertheless, within the past few decades there has been a great increase in studies assessing this kind of evolution. At the very least, studies of taxonomic rates must take into account the inherent limitations and assumptions of this method. Three of the more important of these are as follows:

THE LINNEAN HIERARCHY. Many studies of taxonomic evolution of fossils rely on analyses of patterns observed for higher taxa, usually families or genera. The reason for this is quite simple and pragmatic. These are the taxonomic levels that are most widely synthesized and compiled for various fossil taxa; although species patterns would be more desirable quantities for comparison, few major clades are sufficiently known at this level to make them useful for analysis of taxonomic rates. The use of higher-level taxa for studies of rates seems valid, because numerous studies have shown that patterns that occur at the family, genus, and species level are similar (Raup & Boyajian 1988). Simpson (1953a) noted similar patterns of diversification for species, genera, and families of fishes. Maas, Krause, and Strait (1988) have shown that for the extinct plesiadapiform primates, diversity patterns through the early Tertiary were similar, whether one looks at the level of genus or species (Figure 9.8).

Simpson (1944, 1953a) believed that the genus was the most useful level within the Linnean hierarchy for studies of taxonomic change. However, the biological species is a less arbitrary taxonomic unit than the genus. How does one compare rates of evolution within a monotypic genus and a genus that has hundreds of species, or between fossil mammals and foraminifera? Can a clade that diversified with many genera of low diversity be said to have evolved more rapidly than another with a few highly speciose genera? These are difficult questions, and they show that no system of comparison provides a panacea. Simpson (1953a) concluded that the more similar the groups being compared, the more valid the comparisons of rates of taxonomic change.

TAXONOMIC DATA BASES. The foundations for most comprehensive studies of taxonomic evolution have been the large, compendium data bases for the various fossil groups, such as the *Treatise on Invertebrate Paleontology* (Moore 1953–86) and *Vertebrate Paleontology* (Romer 1966). From a pragmatic point of view, one doing these kinds of studies must use what is available, and yet there are pitfalls involved in the use of such data bases. Being synthetic in scope, the taxonomy found in these compendia usually results from the research of many different workers and therefore undoubtedly reflects systematic biases. Furthermore, many of the temporal ranges are quite out of date and are poorly constrained, as discussed later. If we are to minimize these limitations for studies of taxonomic evolution, then the preferred data bases are those in which there has been some uniform quality control [e.g., Sepkoski (1982), Smith & Patterson (1988) notwithstanding] for large synthetic studies, or we must concentrate on groups for which there have been recent systematic revisions.

Another recent challenge to the use of compendium data bases is that many of the groups listed are not monophyletic (Cracraft 1981; Novacek & Norell 1982; Smith & Patterson 1988). If we are to better understand patterns within natural, or monophyletic, clades, then those studies that have included polyphyletic or paraphyletic taxa will be of questionable value in understanding evolutionary rates and interpretations of underlying processes and patterns.

TEMPORAL CALIBRATION. The foundation of the data base for studying taxonomic rates in extinct taxa is the stratigraphic range. In Chapter 6 we investigated the various geochronological methods used to calibrate geologic time; however, a few further comments are relevant here.

Statistical methods have recently been devised to test the reliability of a particular biostratigraphic range (Strauss & Sadler 1989; Marshall 1990). However, in virtually all cases the observed ranges have been minimum determinations, because the earliest and latest individuals within that taxon throughout its entire biogeographic distribution usually have not been known. For most of the rate calculations, therefore, this fact tends to increase what we perceive to be the rate of taxonomic evolution (e.g., in cohort analysis, discussed later), because durations artificially appear smaller. Novacek and Norell (1982) advocate using the time of cladistic divergence from the closest sister groups, rather than the first known stratigraphic occurrence (Figure 9.9), to calculate the lower limit of the temporal range of a taxon. In studying 163 fossil primate genera, Novacek and Norell (1982) (Figure 9.10) found that when they used dates of first fossil occurrence, they calculated taxonomic rates of survivorship that were almost three times more rapid (334 μma)[2] than those calculated on the basis of time of phylogenetic divergence from the closest sister group (122 μma).

Even after the temporal range is well established and the analytical errors bracketing the first appearance and last appearance are indicated, there still remains some uncertainty in the range of a taxon. When ranges are compiled into taxonomic compendia, they are traditionally listed as, for example, "early Miocene to late Pliocene." The worker using these data then has

[2] μma = 10^{-6} macarthurs (Van Valen 1973), a rate of taxonomic survivorship.

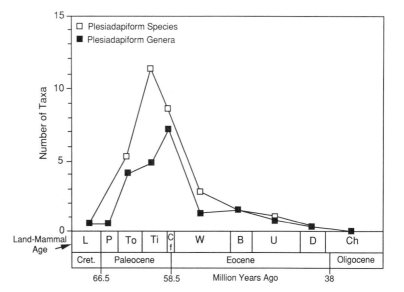

Figure 9.8. Plot of late Cretaceous–early Tertiary plesiadapiform primates showing similar diversity patterns on both the species and generic levels. Land-mammal ages are abbreviated as follows: L, Lancian; P, Puercan; To, Torrejonian; Ti, Tiffanian; Cf, Clarkforkian; W, Wasatchian; B, Bridgerian; U, Uintan; D, Duchesnean; Ch, Chadronian. Modified from Maas et al. (1988).

Figure 9.9. Divergence times based on cladistic interrelationships (A) of three hypothetical taxa a, b, and c showing conflict between cladistic hypothesis and known fossil occurrences (B). Because of an incomplete fossil record ("+" represents fossil occurrences), the biostratigraphic origination times are later than those predicted from the cladistic branching patterns (dashed lines) that in most cases occurred earlier in time. Modified from Novacek and Norell (1982).

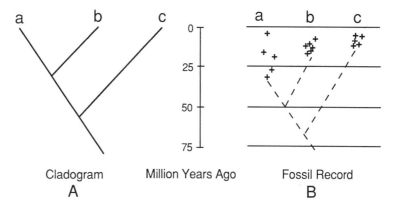

to standardize what this means in terms of when exactly during the early Miocene or late Pliocene (usually impossible to determine in synthetic classifications), and how many million years' duration it represents.

In summary, the assumptions and limitations inherent in the use of biostratigraphic ranges for studies of taxonomic rates of evolution must be kept clearly in mind when assessing the validity of resultant data and interpretations. In general, the further removed the investigator is from the original systematics and temporal calibrations, the more prone will the results be to greater errors. The studies with the narrowest margins for potential error are those with up-to-date systematics calibrated by a precise geological time scale.

Figure 9.10. *Survivorship curves for 163 genera of Cenozoic primates showing differences in extinction rates depending on whether fossil occurrence (A) or phylogenetic divergence (B) times are used. Rates of survivorship are shorter for fossils because the complete biostratigraphic range is rarely known; this also yields the higher extinction rates relative to those calculated using phylogenetic (cladistic) divergence. Modified from Novacek and Norell (1982).*

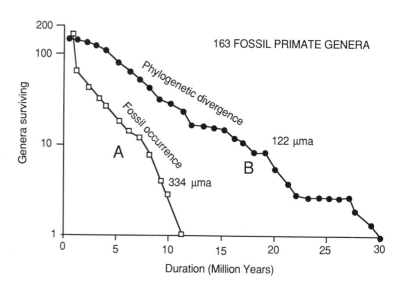

Figure 9.11. *Biostratigraphic ranges for the 27 currently recognized genera of North American horses. Abbreviations are the first three letters in the generic names listed in Appendix II (HIP1 and HIP2 represent* Hippidion *and* Hipparion*); durations are taken from Appendix II.*

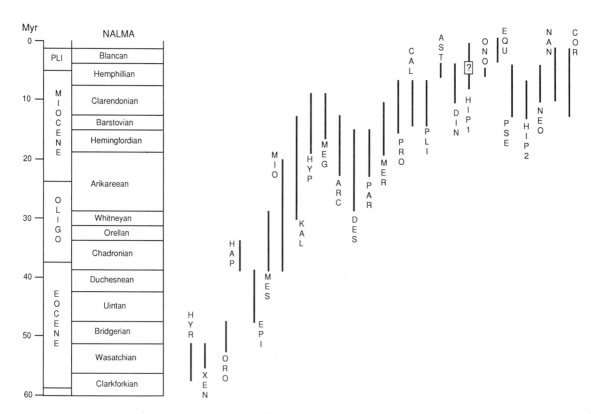

Taxonomic evolution in fossil horses

There are several different ways in which rates of taxonomic evolution can be analyzed and quantified (e.g., Simpson 1944, 1953a; Raup 1978; Stanley 1979). We shall concentrate here on those for which work has already been done for fossil horses, as well as on some new data based on revised biostratigraphic ranges of the North American genera (Figure 9.11).

MEAN DURATION. The observed longevity of a taxon, that is, the time between its first appearance (FAD) and last known appearance (LAD), has been used as a metric for taxonomic rate of evolution (e.g., Simpson 1944, 1953a). Thus, a hypothetical clade whose contained genera had a mean longevity of 5 myr evolved more rapidly than another whose mean longevity was 10 myr. Another related metric is the number of taxa that evolved per million years, or, in Stanley's terminology (1979), "chronospecies/million years." In this section we shall investigate this phenomenon at both the generic and specific levels.

Simpson (1944, 1953a) stated that there were eight genera of North American horses over a period of about 60 myr, giving a mean longevity of 7.5 myr/genus, or the reciprocal 0.13 genus/myr. That was compared with the same rate of 0.13 genus/myr for Tertiary chalicotheres (extinct perissodactyls) and was considered more rapid than the rate for "Triassic and earlier ammonites," a rate of 0.05 genus/myr (Simpson 1953a). The current systematics of North American Equidae indicate that there were 27 genera (Appendix II) that existed over a period of some 57 myr. Using Simpson's system, this yields a mean generic longevity of 2.1 myr/genus or 0.47 genus/myr. This rate is somewhat misleading because it is calculated by simply adding up the number of known taxa, in this case genera, and then dividing by the total group duration. These calculations do not account for the widespread occurrence of taxa that overlap in time, which will increase the average durations. Using the ranges for fossil horses presented in Figure 9.11 (also see Appendix II), the *average of all individual generic durations* (not simply the number of genera divided by the total duration) is 8.4 myr, which is considerably longer than the duration found using Simpson's or Stanley's method. Unfortunately, few higher taxa have been analyzed in such detail, and when they have been, the methods used have not always been the same. This is also the case for extinct species.

Despite these limitations, however, some comparisons of different taxa have yielded interesting results. For the North American Equidae there have been about 150 species over a 58-myr history (B. J. MacFadden unpublished data), yielding a simple mean duration or longevity of 0.38 myr/species or 2.6 species/myr, which because of the method of analysis (Simpson 1953a) is a minimum value. Other studies of fossil groups using actual ranges have reported mean species longevities of 6 myr for echinoderms (Durham 1970), 1.9 myr for Silurian graptolites (Rickards 1977), 1.2–2 myr for Mesozoic ammonoids (Kennedy 1977), 7–16 myr for benthic foraminifera (Buzas & Culver 1984), and less than 1 myr for Pleistocene mammals (Stanley 1979). Raup (1978) reported a mean species longevity of 11.1 myr for all Paleozoic invertebrates, and Simpson (1953a) stated that 2.5 myr was the mean species longevity for all fossil groups. Although not strictly comparable to all of those groups, horses, with a longevity of more than 0.38 species/myr, appear to have had relatively rapid rates of taxonomic evolution.

SURVIVORSHIP CURVES: THE RED QUEEN AND VAN VALEN'S LAW. Simpson (1944, 1953a) plotted survivorships for genera in much the same way as cohort analysis is done in population dynamics (Deevey 1947). Thus, the shorter duration of survivorship in mammalian carnivores implies more rapid taxonomic evolution than in pelecypods (Figure 9.12). Van Valen (1973) constructed taxonomic survivorship curves for a myriad of taxa and found a surprising pattern. The slopes of the curves were linear, indicating that the probability of a taxon becoming extinct was independent of its duration; long-lived taxa were no more prone to extinction than were short-lived taxa. Van Valen (1973) called this the "Red Queen hypothesis" (from *Alice in Wonderland*), and it has come to be called Van Valen's law (Raup 1975).

In addition, Van Valen (1973) proposed the unit "macarthur" (ma) as a unit of taxonomic extinction:

$$\text{ma} = 721.3(\log S_1 - \log S_2)/(\Delta t \log e),$$

where S_1 and S_2 are the numbers of surviving species at the beginning and the end of a time interval Δt $(t_1 - t_2)$.

MacFadden (1985a) constructed a survivorship curve for 16 species of extinct North American hipparion horses (Figure 9.13). Only two species had short durations of less than 2 myr. However, taking into account the portion of the curve in which most of the extinctions occurred, the remaining 14 species had durations of more than 2 myr, whereas only 2 had long durations of more than 5 myr. Using this example, a rate of species extinction can be calculated as

$$\text{ma} = 721.3(\log 14 - \log 2)/(3 \text{ myr} \cdot \log e)$$
$$= 468 \times 10^{-6} \text{ ma} = 468 \ \mu\text{ma}.$$

Few studies have calculated extinction rates for species (most have been published for genera or families). However, Van Valen (1973) calculated rates of 90, 55, and 100 μma for, re-

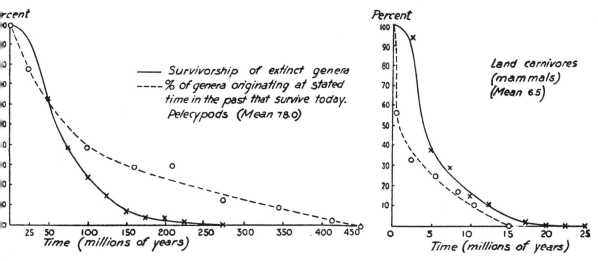

Figure 9.12. *Simpson's original figure (1944) showing generic survivorship of pelecypods (top) and land carnivores (bottom). Relative to pelecypods, the shorter generic duration and survivorship in mammals suggested more rapid rates of taxonomic evolution. Reproduced with permission of Columbia University Press.*

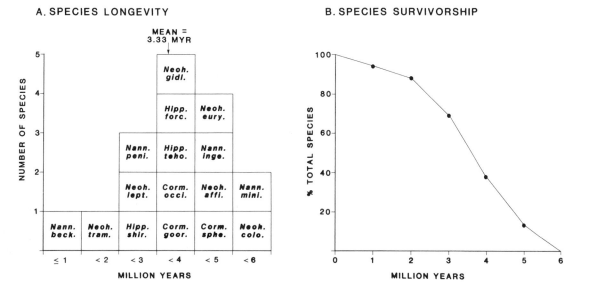

Figure 9.13. *Longevity, or temporal duration (A), and survivorship (B) for 16 species of North American hipparions. From MacFadden (1985a); reproduced with permission of the Paleontological Society.*

spectively, diatoms, dinoflagellates, and planktonic foraminifera. Thus, the rates of species evolution for hipparion horses
were high relative to these other groups.

Using the data for the durations of the 27 genera of North
American Equidae (Figure 9.11), the rate of extinction was 274
μma (Figure 9.14). That was greater than the overall range of 5–
200 μma reported for most groups (Raup 1975), although less
than the 334 μma calculated for fossil primates (using fossil occurrences, not cladistic divergences) (Novacek & Norell 1982).
Comparing with the species of hipparion horses (Figure 9.13),
we see a similar pattern of generic durations for North American
Equidae. Few had short (<2 myr) or long (>14 myr) durations;
the majority (22 genera) were distributed around the mean and
therefore had moderate durations (Figure 9.14A).

RATES OF ORIGINATION AND EXTINCTION. The pace of taxonomic diversification within a clade is dependent upon the rate
of origination (S) balanced by the rate of extinction (E), or $R =
S - E$ (Stanley 1979). When originations greatly outpace ex-
tinctions, rapid clade diversification occurs, as in early phases
of adaptive radiations. When originations and extinctions are
balanced, a steady-state diversity occurs.

Stanley (1979, 1985) used the equation for an exponential-
growth curve to quantify the rate of taxonomic origination, or
diversification, which he referred to as the "intrinsic rate of
increase" (see Chapter 8):

$$R = (\ln N - \ln N_0)/\Delta t,$$

where N_0 is the original number of taxa (usually species or genera), and N is the number of taxa after a duration Δt. For
monophyletic taxa, with the base of the radiation as the starting
point (t_0), $N_0 = 1$, and the equation simplifies to

$$R = \ln N/\Delta t.$$

MacFadden and Hulbert (1988) calculated the intrinsic rate of
increase for Miocene grazing horses in North America. For the
early part of the radiation, between 18 and 17 Ma, they calculated
$R = 0.97$, which then decreased to $R = 0.06$ (a value similar to
those for all bivalves) between 15 and 13 Ma as steady-state
diversity, or taxonomic carrying capacity, was reached (see Figure 8.7). The average R for all 150 fossil horse species over their
58-myr history is 0.09 (see Table 8.1), indicating that the Miocene
was a time of truly exceptional taxonomic evolution for the Equidae. Values for other mammalian clades (see Table 8.1) that are
traditionally thought to have evolved rapidly, such as primitive

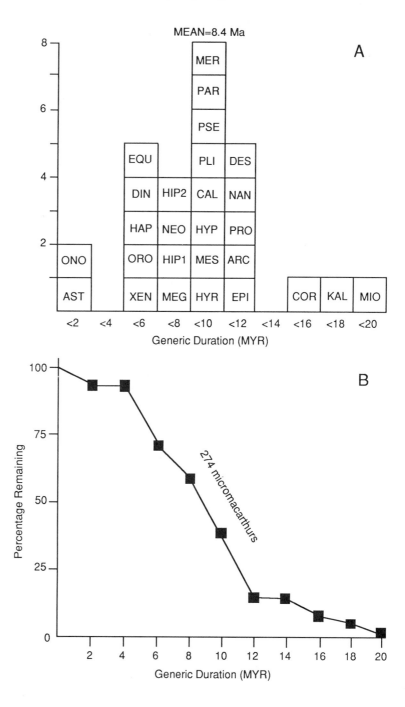

Figure 9.14. *Histogram of temporal duration (A) and survivorship (B) for the 27 genera of North American Equidae. The rate of extinction expressed in B is 274 μma. Abbreviations as in Figure 9.11.*

eutherian mammals (R = 0.12–0.19), the Bovidae (R = 0.15), and the Muridae (R = 0.35), are high relative to all fossil Equidae but are still lower than those seen during the rapid adaptive burst observed for Miocene grazing horses.

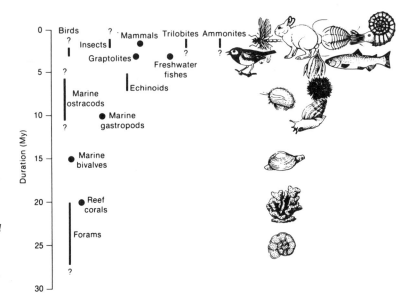

Figure 9.15. A hierarchy of mean species durations for various fossil and extant taxa. From Macroevolution: Pattern and Process *by Steven M. Stanley; copyright © 1979 and reproduced with permission of W. H. Freeman & Company.*

Summary: taxonomic rates

Simpson (1944, 1953a) studied fossil pelecypods and mammals as two contrasting groups with different taxonomic rates of evolution. From his analysis he concluded that the mammals had evolved significantly more rapidly than had the pelecypods. Since Simpson's classic studies, other workers have attempted to identify patterns of taxonomic evolution. Schopf et al. (1975) asserted that the rate of taxonomic evolution is related to genetic and morphological complexity, with more complex forms evolving more rapidly than those of simpler design, a conclusion that tends to confirm that of Simpson. Stanley (1974a) suggested that the more rapid evolution of mammals (and one group of bivalves, the rudists) resulted from more intense competition, resulting in higher rates of extinction. He later (Stanley 1979) proposed that there is a *scala naturae* for rates of evolution, with more complex forms evolving more rapidly (Figure 9.15). Thus, there may be a very general pattern of this nature, but to conclude that an entire group evolved slowly or rapidly for its entire history would certainly be just a generalization.

Investigations of the long-term histories of various fossil groups show that many of these have undergone different rates of taxonomic evolution. Simpson (1944) used the example of the rapidly evolving oyster *Gryphea* that arose from the bradytelic ancestor *Ostrea*. Primitive mammals, whose evolution seems to have been relatively sluggish during the Mesozoic, underwent explosive taxonomic diversification during the early Cenozoic. Horses also exhibited variable rates of taxonomic diversification.

For the first half of their history, rates of taxonomic evolution were relatively slow. That pattern was punctuated by the relatively rapidly evolving clades of Miocene hypsodont horses radiating into the grazing adaptive zone. When horse diversity dropped during the late Miocene, rates of taxonomic evolution correspondingly declined. Thus, taxonomic evolution is a dynamic process that can change, as do the community, ecosystem, and biosphere in which the organisms exist.

Concluding remarks: Are evolutionary rates gradual or punctuated?

An important point to be learned from the fossil record is that morphological evolution and taxonomic evolution do not have to be linked; the two systems can evolve independently. During much of their evolution, particularly the early Tertiary, horses experienced normal, or horotelic, rates of both taxonomic and morphological evolution. However, there are two interesting exceptions to this general pattern in fossil Equidae: (1) The anchitheres, or Miocene browsing taxa, experienced low diversity, but very high rates of morphological change in dental characters (MacFadden 1988b). (2) The Miocene grazers were characterized by very high intrinsic rates of taxonomic evolution, but only average rates of morphological evolution (MacFadden & Hulbert 1988).

Ever since Darwin's work there has been much discussion whether evolutionary rates are constant or sporadic. More recently this has been one of the fundamental questions in the debate about phyletic gradualism versus punctuated equilibria. Although the old-timers usually are said to have been gradualists, some were very perceptive, despite the lack of a well-calibrated fossil record. For example, Stirton (1947) believed that the evolution of hypsodonty in horses and certain other fossil mammals (e.g., beavers) was gradual and progressive, but its rate changed at different times. This also was inherent in Simpson's notions of bradytely, horotely, and tachytely during the 1940s and 1950s. From a modern analysis of fossil horses we can conclude that rates of evolution, whether morphological or taxonomic, were certainly variable. There still remains the lingering question about mode of evolution. Studies of fossil horses do not preclude either possibility – that they are always gradual or always punctuated – or perhaps each of these modes can be valid in different evolutionary situations.

10

Trends, laws, direction, and progress in evolution

A classic example of such a trend is the evolution of the modern horse, whose distant ancestor Hydracotherium [sic] *was a three-toed creature no bigger than a dog. The fossil record shows an apparently steady "progress" through time, with gradual changes in body size and form leading eventually to the familiar* Equus.   – Lewin (1980:884)

From their studies of the results of long-term evolution paleontologists reconstruct phylogenies and interpret general patterns and trends, with the underlying goal of a better understanding of evolutionary processes. Sometimes, if evolutionary patterns and trends are commonplace, they may be called "laws," "principles," or, as exceptions arise, "rules." Thus, from the textbooks we learn about Dollo's law (the irreversibility of evolution), the principle of superposition, and Bergmann's rule (larger-bodied forms in colder climes). The differences among these terms and their usages can be confusing, and as we shall see later, laws in biological sciences are very different from those in physics and chemistry.

The presence of morphological and phylogenetic trends interpreted from the fossil record implies an underlying directionality to evolution. Through the ages this has been emphasized over and over again: from Aristotle's time to the nineteenth-century followers of Lamarckian theory and even up to the present day. It has been commonplace to think of evolution as progress toward more highly adapted forms, perhaps even ultimately striving for perfection. Since Marsh's time, the gradual changes seen in the skulls, teeth, limbs (e.g., Figures 3.6, 3.14, and 3.17) and overall body sizes of fossil horses have been used as examples of general trends, progressing from the less-than-perfect *Hyracotherium* (little improved over its condylarth forebear) to the epitome of the Equidae, the powerful *Equus*. Rightly or wrongly, such thinking is pervasive in the evolutionary litera-

213

ture, namely, hierarchical arrangements of organic diversity on a *scala naturae* from lower to higher forms.

In this, the third and final chapter on evolution, the concepts of trends, laws, directionality, and progress in evolution are investigated using selected examples from the history of the Equidae. This chapter also elaborates on certain themes discussed in Chapter 3 (on orthogenesis). However, the earlier discussion was intended to trace the historical development of orthogenetic concepts, whereas the intent here is to discuss various evolutionary phenomena and thought in a modern context.

Fossil horses and classic trends in evolution

An examination of fossil horse teeth from the American Cenozoic faunas reveals a gradual increase in hypsodonty. – Stirton (1947:32)

The popularity of fossil horses as prime examples of evolutionary trends stems from Marsh's time and the famous illustration described in Chapter 3 (Figure 3.6). Since then, the large majority of discussions about evolutionary trends have begun with the familiar horse story and have included several structural systems, of which we shall discuss here the brain, dentition, and limbs.

Brain

Analysis of fossil brain endocasts requires excellent preservation, often by mud in-filling, followed by the making of latex peels from the inside of the cranial vault. Although previously there had been a dearth of available crania, by the middle of the twentieth century a series of relevant specimens had been assembled that represented many North American fossil horse genera. Edinger (1948) published a classic study in which she depicted the evolution of the horse brain as a gradual increase in size and complexity through time (Figure 10.1). There have been several subsequent criticisms of her work, most notably that the specimens she called *Hyracotherium* probably represented a more primitive condylarth in the same early Eocene (Wasatchian) fauna (Radinsky 1976). Regardless of this and other flaws in her work, the point here is that the evolution of the fossil horse brain is depicted as a clear, unidirectional trend toward increased size and complexity.

Teeth

The evolution of horse teeth frequently is depicted as an orthogenetic (Figures 10.2, 3.6, 3.14, and 3.17) or unidirectional trend, and a further corollary is that this trend represents a gradual progression. Stirton (1947), one of the foremost researchers of

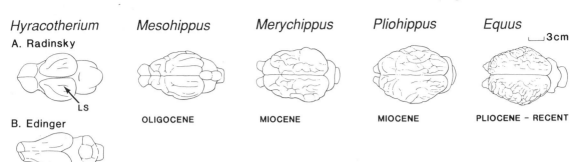

Figure 10.1. Above. *Evolution of the brain in fossil horses, based on the neuroanatomical morphology preserved in fine-grained-sediment internal casts. In the original study, Edinger (1948) believed that* Hyracotherium *had a relatively primitive brain (B). However, Radinsky subsequently asserted that Edinger's specimen may have been of some other contemporaneous mammal, probably a condylarth. In contrast, based on different specimens, Radinsky (1976) asserted that* Hyracotherium *had a more advanced brain (A), including an expanded neocortex and several prominent sulci, for example, the lateral sulcus (LS). Not drawn to scale; bars next to* Hyrachotherium *and* Equus *give size ranges. Modified from Edinger (1948) and Radinsky (1976).*

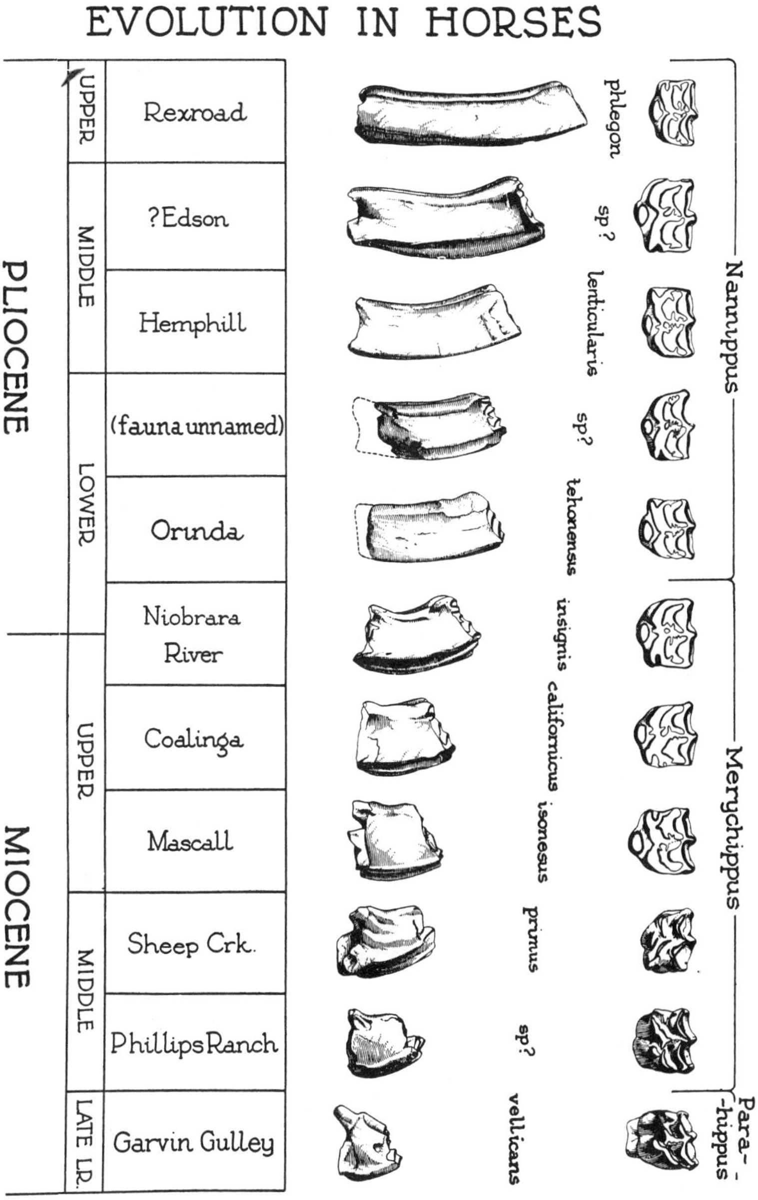

EVOLUTION IN HORSES

Figure 10.2. *Stirton's view (1947) of horse dental evolution depicted as a progressive trend toward increased hypsodonty. Reproduced with permission of the Society for the Study of Evolution.*

PLIOCENE	UPPER	Rexroad	phlegon
	MIDDLE	?Edson	sp ?
		Hemphill	lenticularis
	LOWER	(fauna unnamed)	sp?
		Orinda	tehonensis
		Niobrara River	insignis
MIOCENE	UPPER	Coalinga	californicus
		Mascall	isonesus
	MIDDLE	Sheep Crk.	primus
		Phillips Ranch	sp?
	LATE LR.	Garvin Gulley	vellicans

Nannippus

Merychippus

Para-hippus

fossil horses during this century, was well aware that horse phylogeny was a complexly branching tree; yet for his paper on the evolution of hypsodonty he chose a particular lineage (*Parahippus-Merychippus-Nannippus*) to illustrate gradual evolution and a trend toward increased crown heights. Although Stirton (1947) believed that horses had been essentially gradual in evolution, he also realized that at any given time the rates could have changed depending upon environmental and competitive parameters influencing the evolution of particular lineages.

Limbs

The reduction of the metapodials from 4/3 in *Hyracotherium* to 3/3 in *Mesohippus, Protohippus,* and related genera to 1/1 in "*Pliohippus*" and *Equus* is another classic example of a unidirectional trend in evolution [e.g., Figures 3.6 (using *Orohippus*), 3.14, and 3.17]. Intermediate genera within this sequence, when illustrated, usually are depicted as morphologically transitional, thus conveying the notion of gradual, progressive change consistent with the other functional systems. In the limbs, the reduction and fusion of the ulna to the radius, and the fibula to the tibia, are likewise depicted as gradual trends, although these, too, may be artifacts of simplification.

Since the turn of the century, most of the foremost paleontologists involved in original research on fossil horses (e.g., Matthew, Stirton, and Simpson) have recognized that the gradual, progressive trends depicted for fossil Equidae are at best oversimplifications to illustrate general evolutionary patterns. The problem in the interpretation of these occurs when other scientists and the lay public, themselves far removed from the original data, seize the simplified essence of these general patterns, and consequently many of the details get lost in this process (see Chapter 3).

Cope's law and the evolution of body size in horses

The latest estimate I have seen for the body size of Hyracotherium, *challenging previous reconstructions congenial with the standard simile of fox terriers, cites a weight of some twenty-five kilograms, or fifty-five pounds. . . . Lassie, come home!* — Gould (1988a:24)[1]

[1]On the basis of some recently discovered specimens of a tiny *Hyracotherium,* "Lassie, come home" should now be replaced with "Here kitty, kitty." See Figure 10.7 and discussion in text.

One of the most striking features apparent in the evolutionary sequence of fossil horses is the increase in body size during their 58-myr history from the endpoints of *Hyracotherium* (whose size has been likened to that of a fox terrier) to *Equus* (with a mass around 500 kg, although some large breeds can exceed 1,000 kg). Cope's law states that within a lineage there is a tendency or trend toward increasing body size. This observation has been thought to hold for many fossil groups, ranging from forami-

Figure 10.3. *Lull's illustration (1931) depicting the evolution of body size in fossil horses: 1,* Hyracotherium; *2,* Orohippus; *3,* Mesohippus; *4,* Merychippus; *5,* Pliohippus; *6,* Equus.

niferans to ammonites to dinosaurs to many clades of mammals, including titanotheres (see Figure 3.16) and, of course, horses (Figures 10.3 and 3.14). A corollary to Cope's law is that long-term evolution of body size is "gradual" and "progressive."

In addition to our intrinsic interest in such matters, we find that the body size of a particular species is a character that is highly predictive of a mammal's adaptation to its surroundings (Eisenberg 1981, 1990) (see Chapter 12). However, determining the body size, weight, or mass of a fossil mammal or any other extinct animal (e.g., dinosaurs) is a challenge. With regard to fossil horses, Marsh (1876) and many others who have followed him (although see Gould 1988a) have repeatedly said that *Hy-*

racotherium was the size of a fox terrier. But how much does a fox terrier weigh, and are its proportions similar to those of *Hyracotherium?* In modern mammals of known sizes, different clades have quite different proportions of skin, adipose, muscle, and skeletal masses for similar-sized animals (Grand 1990), which makes it even more difficult to estimate the body mass of a fossil mammal. Lull (1921) stated that *Hyracotherium* stood about 3 hands high at the withers, but with no modern analogues for comparison, that is not much help in determining some uniform metric of body size. Thus, the question of how much a fossil horse weighed is not entirely straightforward.

For our purposes here, body size is synonymous with body mass. One of the better means of estimating the size of an extinct species is to use skeletal characters that scale isometrically with body mass for a modern clade that is closely related to the group under investigation, although even this method has certain pitfalls (Damuth & MacFadden 1990). It is well known that body length and body mass are highly correlated variates in modern mammals and are routinely measured (De Blase & Martin 1981). Radinsky (1978a) used this relationship to estimate a mass of 9.1 kg for *Hyracotherium*, which is indeed within the range of a fox terrier. The problem with that estimate and the others in Radinsky's paper is that the relevant regression equations were based on a heterogeneous assemblage ("ungulate mammals"), a practice that is less satisfactory than determining length/mass equations within the same monophyletic clade in which the body size of the extinct species will be calculated (Damuth & MacFadden 1990).

MacFadden (1987a) reinvestigated the extent to which Cope's law could still be considered valid for the Equidae in light of the modern systematics, phylogeny, and biochronology of this group. Because it is so highly correlated with body size among extant mammals, total head–body length (HBL) (Figure 10.4A) was originally measured in fossil horse skeletons in order to estimate the body masses of extinct equid species. However, there were so few complete horse skeletons available in museum research collections or on exhibit that other characters, highly correlated to HBL, were measured (Figure 10.4B–D).

Another problem encountered in that study was that few body-mass data were available for the large mammals kept in museum collections, and horses were no exception. In the rare instances in which data on body masses are available with osteological specimens, they frequently are for domesticated or zoological-park individuals, rendering such data of questionable significance relative to their wild counterparts. In order to circumvent this lack of important data, body masses and lengths (HBL) were measured (Figure 10.5) for a sample of 34 live individuals of *Equus caballus*, and those data were used to calculate

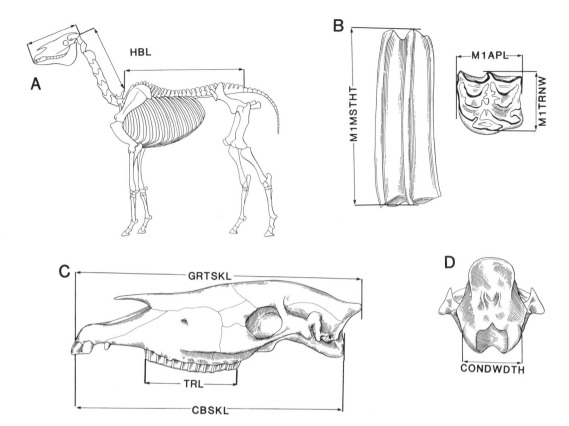

Figure 10.4. *Osteological and dental characters measured by MacFadden (1987a) that are potentially useful in estimating body mass for extinct horses (A) HBL, head–body length; (B) tooth* *measurements (see Figure 9.4); note that occlusal dimensions were taken in middle wear and that M1MSTHT is a poor predictor of body mass; (C) skull (GRTSKL, CBSKL) and cheek-tooth* *row (TRL) lengths; (D) CONDWDTH, with across occipital condyles. Reproduced with permission of the Paleontological Society.*

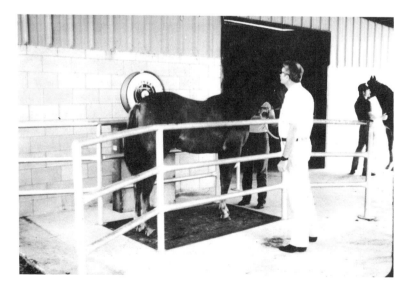

Figure 10.5. *In order to estimate the body masses of fossil horses, living individuals of* Equus caballus *were weighed, as shown here at the University of Florida Horse Farm, and head–body length (HBL) was measured. Photo by author.*

regression equations. Body length was then regressed with the other dimensions used in that study.

Using that method, body masses were estimated for 40 selected species of New World extinct Equidae. Of relevance to the foregoing discussion, masses for three species of *Hyracotherium* ranged from 25 to 35 kg, significantly larger than the traditional estimate of an animal the size of a fox terrier (Figure 10.6). At the other end of the spectrum, estimates of body masses for extinct species of *Equus* ranged from 365 to 498 kg, similar to masses for extant, wild species of this genus (Walker 1975). Those masses had been attained during the late Pliocene and early Pleistocene, at about 1–2 Ma, and therefore any subsequent increases in body size among nondomesticated species of this genus have been negligible.

After that study (MacFadden 1987a) was completed, some more recently described specimens and comparisons with older collections have provided an interesting new look at the body size of *Hyracotherium*. Gingerich (1989) described a new species, *H. sandrae*, from the early Eocene (Wasatchian) of Wyoming. On the basis of an associated partial skeleton, that description represents the smallest known horse, with a size comparable to that of a domestic cat (Figure 10.7), or about 3–5 kg. In his stratophenetic phylogeny of *Hyracotherium* from Wyoming, Gingerich indicated that *H. sandrae* was a side branch from the main sequence beginning with the common *H. grangeri* (see Figure 8.3). Furthermore, he noted that another species, *H. index*, also from the Wasatchian, was comparable to *H. sandrae* in size. These hyracotheres force us to revise our notions of the body sizes of the earliest fossil horses and, along with the size decrease from *H. angustidens* to *H. tapirinum* (MacFadden 1987a), may represent examples of size reduction (dwarfism), a phenomenon that has occurred independently numerous times within the Equidae, as discussed later. In conjunction with larger species of *Hyracotherium* (e.g., *H. tapirinum*), these new discoveries and analyses indicate greater diversity in body size within this genus than had previously been recognized.

Interpretation of pattern

If Cope's law provides a model that is applicable to fossil horses, then two points should emerge from the data: (1) Body sizes increased through time, and (2) the rate of increase was gradual, constant, or progressive. The results presented by MacFadden (1987a) address these points and require a rethinking of the applicability of this law specifically to the Equidae and perhaps also to other groups.

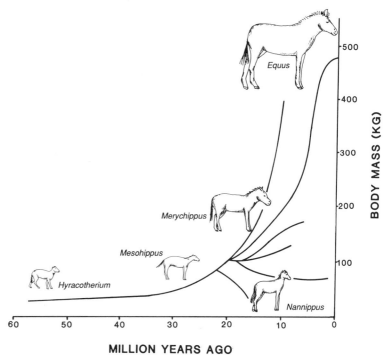

Figure 10.6. *Patterns of body-size evolution in fossil horses from North America (MacFadden 1987a). For about the first 30 myr of their history, body-size evolution was characterized by relative stasis, whereas during the past 25 myr the pattern has consisted of diversification within several clades. Slightly modified from MacFadden (1988a); reproduced with permission of Plenum Press.*

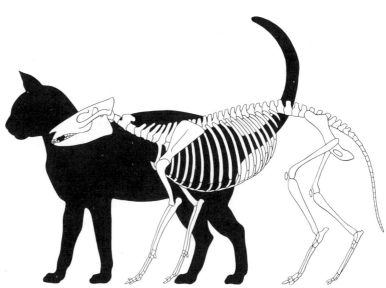

Figure 10.7. *The smallest-known fossil horse; a skeletal reconstruction of* Hyracotherium sandrae *(from the early Eocene of Wyoming) next to a silhouette of an average-size house cat. Modified from Gingerich (1989).*

BODY-SIZE INCREASE.   Of the 24 ancestral–descendant species pairs analyzed by MacFadden (1987a), 19 showed body-size increases; however, the remaining 5 lineages were characterized by dwarfing, where the descendant was smaller than the ancestor. These included species within *Hyracotherium, Miohippus-Archaeohippus, Parahippus, Merychippus,* and *Merychippus-Nannippus*. Other lineages of dwarf forms in North America occurred within *Pseudhipparion, Nannippus,* and *Calippus,* although some species within these genera also showed reversals back to increased size (MacFadden 1984a; Webb & Hulbert 1986; Hulbert 1988a). Dwarf species are also known in *Cremohipparion* and other hipparion clades from the Old World (Qiu et al. 1987; Bernor & Tobien 1989). As the phylogenetic interrelationships of other clades of fossil horses become better understood, more examples of dwarfism undoubtedly will come to light. Thus, although dwarfism is the exception to the rule of phyletic (also termed phylogenetic) size increase, this phenomenon occurred numerous times within the Equidae.

Many examples could also be cited for other fossil groups in which there were exceptions to Cope's law, including the following: Although size increase was a general pattern in Silurian graptolites (Rickards 1977) and Mesozoic ammonites (Kennedy 1977), those taxa demonstrated numerous examples of phyletic size decrease in certain lineages. Jablonski (1987) investigated the extent to which Cope's law applied in Paleozoic bivalves and found that in 58 generic lineages, only 33 broadly confirmed the hypothesis of size increase. As additional taxa are studied in detail, it is probable that we shall encounter new exceptions to Cope's law.

GRADUAL, CONSTANT, OR PROGRESSIVE EVOLUTION. If increases in body size in horses were gradual, then for equal increments of time there should have been corresponding increases in size. The results of MacFadden (1987a) indicate a very different scenario for the Equidae (Figure 10.6). Rather than showing a gradual size increase, for the first half of the history of this family their body sizes were relatively static around 25–50 kg (notwithstanding exceptions such as *Hyracotherium sandrae*), with very slow evolutionary rates. Since the middle Miocene (the past 15–18 myr), body sizes in horses have been characterized by rapid rates of evolution and diversification into several clades, ranging from about 75 to 500 kg. Thus, rather than a unidirectional trend, or progressive increase in body size, an alternate view of the pattern of evolution within the Equidae is that they have demonstrated increased variance and diversification of this character. Gould (1988c) came to a similar conclusion using fossil foraminiferans, namely, increasing diversification of size through time. Thus, diversification, rather than merely unidirectional size increase, probably was a pervasive pattern in the evolution of many fossil taxa, and it undoubtedly extended to other morphological characters as well.

Concluding remarks

In the physical sciences, laws are immutable and absolute as long as the boundary conditions are satisfied (Hull 1987). There are no exceptions to the law of gravity or the second law of thermodynamics. However, in the biological and natural sciences, a "law" is a somewhat different matter. As such, the validity and applicability of these so-called laws should not be confused with the absolute nature of laws in the physical sciences, nor should they be taken as dogma, or the truth. The physical sciences (i.e, physics and chemistry) and natural sciences (i.e., geology, paleontology, and organismic biology) are fundamentally different in certain respects, particularly because of the large historical component of the latter (Kitts 1977; Simpson 1980). The major goals in the physical sciences include observation and prediction of phenomena, whereas natural historians frequently are interested in reconstructing past events and interpreting underlying processes. In reality, few laws in geology and paleontology are absolute; at best they may be considered rules with certain numbers of exceptions, or, in the worst cases, broad generalizations that are correct in some instances. Thus, Cope's law is better considered a "rule," a concept in which there is universal agreement that exceptions occur. To name a "law" after a scientist is a wonderful tribute to the contributions of that scientist, but in the biological and natural sciences any immutability of laws is, in most cases, wishful thinking.

Trends and directionality, though firmly entrenched in the literature and thought to be prime concepts within evolutionary theory, are, at best, broad patterns or generalizations. On one scale they may seem correct, but often at lower levels within the Linnean hierarchy more careful examination will reveal that there are so many exceptions to these patterns or laws that they are of little significance. This conclusion is by no means new, and yet both the general scientific community and the lay public still cling to these concepts as a simple way of understanding patterns and processes in evolution.

Dollo's law, atavisms, and the irreversibility of evolution

That evolution is irreversible is a special case of the fact that history does not repeat itself. The fossil record and the evolutionary sequences that it illustrates are historical in nature, and history is irreversible.
— Simpson (1953a:312)

Dollo's law was originally formulated to describe the observation that once a structure or organ is lost in a lineage, it is never regained. Dollo's law has been broadened further to imply that long-term evolution is irreversible and that as lineages become more specialized, they become increasingly more "canalized" (to use Waddington's term) or constrained in their evolutionary pathway. As such, they never return to the original condition seen in their ancestors (Simpson 1953a).

Atavisms, or morphological "throwbacks" to some previous ancestral state, are partial exceptions to Dollo's law as it pertains to individual character states. The toes of modern horses show such atavisms and have fascinated paleontologists for two centuries. Numerous instances have been described in the literature in which modern horses have had extra toes. Normally, *Equus* is monodactyl, and the lateral metapodials (MP II and IV) are tiny, vestigial side-splints that are homologous to the functioning toes in fossil horses. Marsh (1879, 1892) wrote two articles on "polydactyl" horses in which he described modern "freak" occurrences of *Equus* with elongated MP II (Figure 10.8). More recent studies have indicated that these extra digits can form by two different developmental pathways: either (1) by elaboration of the lateral, normally vestigial metapodial (Figure 10.9) or (2) by duplication of the central metapodial (Gould 1983). The number of extra toes varies in these atavisms; some have one extra on each foot, others two, and still others have different numbers on different feet. Marsh (1879) concluded that the presence of extra toes was more common on the forelimbs, but this has not been confirmed.

The occurrence of atavisms illustrates some important points concerning the evolutionary process. First, although these mutations are rarely incorporated into subsequent generations, they show that in some instances short-term evolution is, so to speak, an imperfect process of two steps forward, one step back. They also show that the direction of evolution, if indeed it has direction, is periodically punctuated by reversals. As discussed earlier for fossil horses, one of the common exceptions to the rule of irreversibility of evolution is body size, which frequently fluctuates and reverses trend (Figure 10.6). However, as a rule, over the long term, reversals of newly acquired or lost characters rarely persist. As Simpson noted, evolution (like history) cannot and does not repeat itself.

Progress and purpose in evolution

The history of the horse family is still one of the clearest and most convincing for showing that organisms really have evolved, for demonstrating that, so to speak, an onion can turn into a lilly.
– Simpson (1951:168)

Progress might have been all right once, but it's gone on too long.
– Ogden Nash, in Provine (1988:49)

Western culture has been heavily indoctrinated with the notion of the value of "progress" in society. It is therefore not surprising

Figure 10.8. *Polydactyl horse (with elongated MP II on each foot),* Equus caballus, *on exhibition in New Orleans in 1878. This horse was known among showmen as the "eight-footed Cuban horse." From Marsh (1879).*

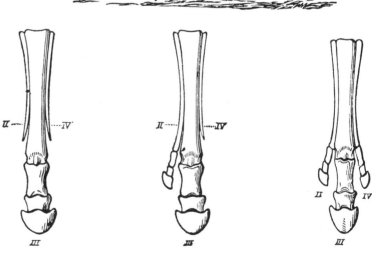

Figure 10.9. *Forefeet of horses showing normal character states in extant monodactyl* Equus *(left) and Miocene tridactyl "Hipparion" (right) and abnormal elongation of MP II with terminal hoof in* Equus *(center). From Marsh (1879).*

that this concept has had a profound impact on how we view evolution. A corollary to the concept of progress is that there is some purpose or goal to evolution.

Hull (1988) believes that the Greek philosophers did not include the notion of progress or directionality in their concept of natural history because they (and later the Romans) viewed the world as a series of eternal cycles. Hull (1988) states that the idea of progress arose from Renaissance thought and was subsequently seized upon by natural historians. That there was a purpose or goal to evolution was strongly embodied in subsequent evolutionary thought. This is exemplified by the concept of *élan vital*, a vital force driving evolution toward more nearly "perfect" forms. Whether or not classic scholars believed in progress, it was implicit in their concepts of directionality and trends along the *scala naturae* from, as Aristotle noted, lowly polyps to humans. Many natural historians over the years have regarded the human species as the culmination of the evolutionary process, a view that today is anathema.

Many dictionary definitions of "progress" equate it with "improvement." However, the biological meaning of "progress" is by no means clear. On a genetic level it might mean increased genetic variability or fitness. On the specific level it might mean increased fecundity, larger body size, increased relative brain size, or being better adapted. Over the long term, it might mean taxa with increased or decreased longevities (both cases could be made, depending upon r or K selection strategies[2]), or a clade whose constituent taxa are less vulnerable to extinction. The list could go on and on (Ayala 1988; Nitecki 1988a). So, to a great extent, arguments centering on whether or not evolution means progress reduce to semantics (Nitecki 1988a). Although this latter point may render the following discussion futile, the idea of progress is so strongly ingrained in Western culture, whether for evolutionary theory or society as a whole, that we cannot ignore it here. For current purposes, I shall stick to the simple view of progress as improvement, however defined.

As discussed earlier, gradual evolution implies progress, and also improvement. Ayala (1988), although not an advocate of the concept of evolutionary progress, cited the example of the progressive molarization of premolars in Eocene and early Oligocene hyracothere horses. He stated that (1988:78) "these trends indeed represent directional change, but it is not obvious that they should be labeled progressive; to do this we would have to agree that the directional change had been for the better in some sense."

The advent of hypsodonty (high-crowned dentitions) in Miocene horses allowed them to invade a new adaptive zone (*sensu* Simpson 1944, 1953a). But does this mean that they were better adapted than the contemporaneous browsing anchitheres? Probably not. It could just as easily be said that the demise of the latter group had little, if anything, to do with their competitive inferiority relative to the grazers, and probably more to do with the decrease in forested biomes during the Miocene.

Rather than look only at the dentitions, let us consider the multitude of evolutionary changes between *Hyracotherium* and *Equus*. We know from empirical evidence that *Equus* is bigger, has a relatively larger brain, and almost certainly is faster than

[2]The r selection strategy favors a rapid rate of population increase. K selection favors a stable population maintained at, or near, carrying capacity (Lincoln, Boxshall, & Clark 1982).

was *Hyracotherium*. But does this really mean that *Equus* was an improved version, better adapted to its environment? Does this imply that extinct horses were less successful than *Equus*? On the contrary, *Hyracotherium* and the other 32 fossil equid genera "did their own thing" for millions of years. Thus, it would seem that they were well adapted to their particular surroundings, despite being unaware of any particular goal that they were working toward on their own "*scala Equidae.*" The former kind of thinking, which often had been either implied or stated in the literature, perhaps was understandable as evolutionary thought was being developed, particularly among those of the strongly adaptationist school. However, the progressive view of long-term evolution is no longer valid in light of modern data and current theory.

The evolutionary process is random, stochastic, with no purposeful design for nature. Nondeleterious mutations may confer upon individuals within populations certain microevolutionary advantages. And, since the time of Darwin, the essence of natural selection has been the "survival of the fittest." However, these concepts were devised within a microevolutionary framework, and over the long term (i.e., macroevolution) the presence of distinctly unidirectional trends resulting from similar phenomena has introduced ambiguity in many instances. It is absurd to think that Paleozoic marine faunas were less "perfect" than their modern analogues or that there were marginally well adapted, or "primitive," land mammals inhabiting the Eocene forests of the world. Phylogenetic position in the *scala naturae* has nothing to do with how good or progressive an animal is.

11

What's the use? Functional morphology of feeding and locomotion[1]

A clear picture of how animals feed and move is essential to understand many of their adaptations and life strategies. Elsewhere in this book we have discussed teeth and, to a lesser extent, limbs of fossil horses in a context of what they tell us about evolutionary patterns and processes. In this chapter we shall explore the functions and biomechanics of these systems in order to gain better insight into the diverse adaptations of various clades within the Equidae.

Feeding: herbivorous ungulates and hypsodonty

With the notable exception of the dinosaurs, mammals are virtually unique among vertebrates in the extent to which they have adapted to herbivory. Of the 34 extant orders of mammals, 24 contain herbivorous species, and 13 are composed almost exclusively of herbivores (Janis & Fortelius 1988). The "cost" of being a herbivore is great and involves extensive modifications to all systems related to feeding, including the structure of the skull and jaws, the teeth, and the gut. The reason is that the amount of digestible foodstuffs in plants that can be converted to usable energy is far less than that in animal protein. Cellulose, a dominant constituent of plants, cannot be digested by mammals (no known multicelluar organism has evolved a cellulase enzyme) (Janis 1976). Many herbivorous mammals have evolved an alternate strategy of breaking down cellulose by the actions of symbiotic microorganisms located in the gut. In addition, other potential nutrients are located within the plant cell, which usually is enclosed in a wall of cellulose. Therefore, much of the function of chewing in herbivores is related to the need to fracture the cell wall to release the nutritious cell contents. The amount of chewing done by herbivores is far greater than that

[1]The first part of this title is adapted from Rainger's (1989) article "What's the Use: William King Gregory and the Functional Morphology of Fossil Vertebrates."

done by mammals with other feeding strategies, such as carnivores. In addition to the fact that herbivores have to chew for a greater portion of their lives than do carnivores, the herbaceous foodstuffs that they eat (particularly grasses) and the contaminant grit from the surrounding environment further increase their rate of tooth wear. Thus, all herbivores have evolved various feeding strategies to cope with these factors.

In recent years several important studies have provided insight into various aspects of developmental constraints, biomechanics, and the structure and function of herbivorous feeding in ungulates, including horses (e.g., Fortelius 1985; Janis 1988; Janis & Ehrhardt 1988; Janis & Fortelius 1988). Herbivorous mammals have solved the problem of increased tooth wear through a variety of strategies (Janis & Fortelius 1988), including (1) increased durability of dental tissues, (2) increased tooth size, (3) additional teeth, and (4) increased tooth height.

Dental durability

In most mammals the occlusal surfaces of the cheek teeth consist of enamel and dentine and, in high-crowned forms, cementum. In general, the chemical compositions of these tissues are remarkably uniform throughout most mammalian orders and consist of the mineral hydroxyapatite and organic compounds (mostly collagen). Of these, the enamel, consisting of more than 95% "inorganic" hydroxyapatite microcrystals, is more resistant to wear than is the dentine, which is about 70% inorganic and 20% collagen (Geneser 1986). Therefore, one way to increase durability is to increase the amount of enamel exposed at the occlusal surface. Some mammals, such as primates, suids, and proboscideans, do this by greatly increasing the thickness of their enamel, whereas others, such as hypsodont ungulates, including late Cenozoic horses, do this by infolding the enamel many times and thereby increasing the occlusal surface area of enamel (Figures 11.1 and 11.2).

In theory, another way to increase the durability of dental tissue would be to change its composition. However, this is not an easy solution because of many potential functional and developmental limitations that might derive from increasing the strength of this tissue (Janis & Fortelius 1988). For example, increasing the strength of a tooth might also make it more brittle and thus more prone to breakage under compressive stresses. Despite these potential obstacles, some clades have modified the general mammalian dental composition. Humans have evolved relatively hard teeth by changing the arrangement of microscopic prisms on the outer layer of thickened enamel. Certain rodents

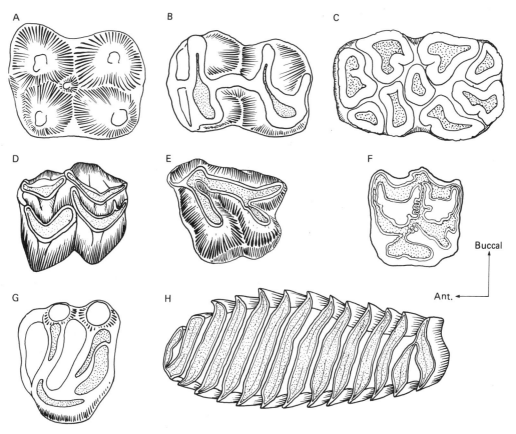

Figure 11.1. *Selected molar (M2 or M3) morphologies and terminologies in herbivorous mammals,* sensu *Janis and Fortelius (1988, Fig. 1): (A) bunodont, peccary (Tayassu sp.); (B) bilophodont, kangaroo (Macropus sp.); (C) columnar, warthog (Phacochoerus aethiop-* icus*); (D) selenodont, deer* (Odocoileus virginianus)*; (E) trilo-phodont, rhinoceros (Subhyracodon sp., Oligocene); (F) plagiolophodont (= hypsodont), horse (Equus caballus); (G) lophodont/bunolophodont, rodent, woodchuck (Marmota sp.); (H) lamel-* lar, rodent, capybara (Hydrochaeris hydrochaeris)*. Shaded regions in B–H represent exposed dentine; buccal equals labial; not to same scales. From Biological Reviews, and reproduced with permission of Cambridge University Press.*

(e.g., lemmings and voles) have also evolved complex, prismatic enamel. Some other mammals have evolved another adaptation for increased durability by depositing other minerals in the enamel, mostly compounds of iron (e.g., in ever-growing rodent incisors), which imparts an orange color to them. Some soricids (e.g., *Blarina* and *Notiosorex*) secrete iron hydroxide in the form of the mineral goethite [FeO(OH)] as a covering over the normal enamel. Pigmented teeth are also known in extinct shrews, extending back into the Oligocene (Repenning 1967; Akersten, Lowenstam, & Walker 1984). Although iron pigmentation is rare, it also may have occurred in some multituberculates (Chow & Qi 1978).

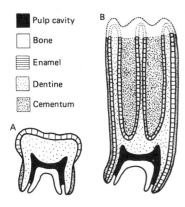

Pulp cavity
Bone
Enamel
Dentine
Cementum

Figure 11.2. Sagittal sections of low-crowned (human, A) and high-crowned (horse, B) molars showing development of various dental tissues. From Janis and Fortelius (1988, Fig. 2); published in Biological Reviews, and reproduced with permission of Cambridge University Press.

Increased tooth size

The paradox of predicted positive allometry but observed isometry of tooth size to body size is resolved by inclusion of the time dimension in the equation. — Fortelius (1985:1)

It would seem advantageous for evolving clades of herbivores to increase the sizes and volumes of their teeth as a strategy for dealing with increased tooth wear. However, extensive developmental and morphological changes are required to accomplish that. Nevertheless, several groups of mammals have been successful in this regard, including certain rodents (Hydrochoeridae), some artiodactyls (Camelidae, Hippopotamidae, and Suidae), and advanced proboscideans (Janis & Fortelius 1988). Despite the potential advantage of relatively larger teeth, as a general pattern most herbivorous mammal clades scale isometrically for tooth surface area versus body size (Fortelius 1985), thus indicating that this has not been a particularly popular evolutionary strategy.

Another way of effectively increasing the occlusal surface area is to "molarize" the premolars. In the primitive condition, many herbivores had a premolar series consisting of tritubercular or triangular-shaped teeth, whereas the molars were more quadrate, resulting from the addition of cusps and associated crests. In the derived condition, cusps were added to the premolars so that each cheek tooth became functionally molariform. That transition was made in many groups, including Eocene horses, certain artiodactyls, and various South American ungulates. It should be noted, however, that in some of these groups premolar molarization was not always complete; for example, it occurred in the P4/p4, was partial in P3/p3, and was absent from P2/p2 in advanced ruminant artiodactyls (see Appendix I for abbreviations). In other ungulates, such as horses, this process was carried to the extreme, where P2/p2–P4/p4 were completely molarized (except in early forms, as discussed later). Premolarization does not, however, take place without the need for other adaptive changes. Therefore, a greater occlusal tooth area is related to many factors, including a greater intake of food, relatively larger masticatory muscles (seen in extant *Equus*), and an increased rate of digestion in the gut (Janis 1976; Janis & Fortelius 1988).

Additional teeth

Another potential strategy for increasing food processing would be to add teeth in order to increase the functioning occlusal area. Although very rare, supernumerary teeth have been reported in the literature as individual occurrences, such as in the Pliocene sloth *Megalonyx* (e.g., McDonald 1978). However, as a rule these are not incorporated into the genome, except in some omnivores (e.g., armadillos), carnivores (e.g., the bat-eared fox *Otocyon*), and some marine mammals (e.g., manatees and cetaceans) (Janis & Fortelius 1988). The fossil record indicates that supernumerary teeth rarely occurred and did not become fixed into the genome in evolving clades of terrestrial herbivorous mammals, including horses.

Increased tooth height

A widespread solution to the problem of greater dental durability is to increase the height of the tooth. This condition is found in a wide variety of modern herbivorous mammals, ranging from certain rodents to proboscideans, as well as numerous extinct families (Figure 11.3). There are several different ways in which mammals have evolved elongated crowns. In some smaller forms, such as certain insectivores and bats, cusps have become relatively elongated (White 1959). However, by far the most common ways of increasing crown heights have been through hypsodonty or hypselodonty, which for our purposes mean, respectively, high-crowned teeth and ever-growing teeth. For more specific definitions of these conditions, see Mones (1982) and Fortelius (1985).

In hypsodont mammals, the formation of the roots and the finite growth of the crown are ontogenetically delayed, resulting in a longer tooth. Late Cenozoic horses certainly were prime examples of hypsodonty, although many other forms evolved this adaptation, and some much earlier in the fossil record. For example, early Tertiary taeniodonts and tillodonts, which were medium-size placentals, were the earliest known hypsodont mammals (Figure 11.3) (Janis & Fortelius 1988). Many clades of early Tertiary South American ungulates were hypsodont earlier than were North American herbivorous mammals (Patterson & Pascual 1972; MacFadden 1985b). In hypselodont taxa, the root never forms, and the tooth is "ever-growing." This condition was first reported for the late Cretaceous (Campanian) *Gondwanatherium*, a eutherian of possible xenarthran affinites from South America (Bonaparte 1986). Thereafter, on that continent several clades of edentates and notoungulates evolved hypselodont dentitions. Hypselodonty is common in many families of lagomorphs and advanced rodents. In Holarctic large mammalian ungulates during the Cenozoic, hypselodonty was relatively rare and was restricted to some rhinoceroses *(Elasmotherium)* and ruminant artiodactyls *(Antilocapra, Ovis)* (Janis & Fortelius 1988). As will be discussed later, no horse clade became hypselodont, though there was an advanced species of North American *Pseudhipparion* that had ever-growing cheek teeth for part of its life span and has been termed "incipiently hypselodont" (Webb & Hulbert 1986).

The standard, simple explanation for the advent of hypsodonty in horses has been that it was an adaptive response to the spread of grasslands on the North American savannas during the Miocene (e.g., Osborn 1910; Simpson 1951, 1953a; Webb 1977). However, careful examination of the timing and geographic distribution of acquisition of high-crowned teeth in a diversity of mammalian clades suggests that other factors were almost certainly involved. One could argue that grasses spread in South America during the early Tertiary (earlier than in North America) and hence explain why there were so many hypsodont notoungulates and litopterns during that time. However, this does not explain the acquisition of hypsodonty in primitive mammals (taeniodonts and tillodonts) that coexisted with Eocene horses on a continent in which grasses did not spread until the middle Tertiary. If Eocene horses did feed on grasses, they may have fed on types that were less abrasive than those

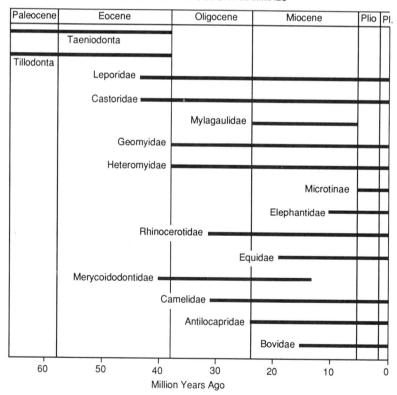

Figure 11.3. Principal clades of hypsodont mammals from Holarctica. Ranges taken from Webb (1977) (for North America), Tedford et al. (1987), and Carroll (1988). Although not depicted here, Webb (1978) also listed 22 families of hypsodont mammals from South America.

in the later Tertiary (Stirton 1947). It is more likely, however, that even though grasses may have been available, hyracotheres probably had a diet quite different from that of later horses, as discussed later.

Workers such as Stirton (1947) have realized the potential influence that the sedimentary substrate can have on terrestrial herbivores. In particular, for those herbivores that feed close to the ground, it has been documented that while they are feeding they ingest gritty dust from their plant food and its surroundings (McNaughton et al. 1985). Extant *Equus* grazing in pastures can ingest significant quantities of sand, which can cause blockage, or colic, in the gut, a common problem encountered by veterinarians. Stirton (1947; White 1959) reviewed the distribution of sediment types in the western interior of North America during the middle Tertiary. Although it is easy to overgeneralize, we can say that during much of the Oligocene, particularly as represented in the classic White River Badlands Group, finer-grained clays were common. In contrast, during the late Oligocene–early Miocene there was a change to a greater proportion of coarser-grained sands. Furthermore, those sediments incorporated a high percentage of siliceous volcanic particles (Hunt 1990). That transition coincided with the time when advanced parahippine and primitive merychippine horses were becoming hypsodont. Thus, the classic notion of hypsodonty in horses as purely an adaptive response to the spread of grasslands is too simple. Although the spread of grasslands should not be dismissed as unimportant to the evolution of ungulate feeding strategies (*contra* White 1959), other factors, such as the change in substrate texture, were almost certainly related to the adaptive responses seen in horses during the Miocene and in other high-crowned mammals at various times during the Cenozoic.

Another noteworthy point is that although hypsodonty may have been correlated to a diet that featured increasingly abrasive material for some mammals (whether it be in the plant foodstuffs or from the surroundings), other hypsodont taxa (e.g., *Antilocapra*, *Gazella*, and *Lama*) (Janis 1984) are not predominantly grazers. Conversely, many other mammals whose diets are known, such as certain baboons and kangaroos, although predominantly grazers, are not strongly hypsodont. Thus, there is not always a close correspondence between crown height and diet. Inferences drawn from the fossil record must always be carefully examined, and wherever possible one should make comparisons with close phylogenetic analogues. Despite these exceptions, modern equids are known to be grazers; they are strongly hypsodont relative to other ungulates (Janis 1988) and thus are good modern analogues for interpreting the dietary adaptations of hypsodont fossil horses.

Horses and herbivory

Early horses, such as Hyracotherium, Mesohippus, *and* Parahippus, *were browsers which fed on leaves from bushes and low trees. As the humid forests retreated and grasslands spread, new horse lineages, such as* Merychippus *and* Pliohippus, *stepped out on the plains and put their reinforced molars to work.* — Benton (1990:280)

Now that we have discussed the general problems, strategies, and adaptations related to herbivory in ungulates, in this section we shall concentrate on the evolution of feeding within the Equidae as interpreted from the fossil record of horses in North America. This discussion will also show that the traditional scenarios presented for these and related adaptive changes, as exemplified by the preceding quotation, require significant revisions in light of modern data.

Hyracotherium: *the equid feeding archetype*

Relative to other horses, the dentition of *Hyracotherium* retained many of the primitive character states of generalized tribosphenic eutherian mammals. In the upper teeth, the premolars were triangular-shaped, whereas the molars were relatively square, due to the presence of the hypocone and metaloph (Figure 11.4A). Although the principal crests, the protoloph and metaloph, in *Hyracotherium* molars were relatively well developed and gave the tooth a characteristic perissodactyl pattern, the cusps remained distinct. In the lower teeth of *Hyracotherium* the primitive trigonid and talonid were apparent in the posterior premolars and molars, although in the anterior parts of these teeth the paraconid was lost, and the paracristid was reduced (see Figure 5.6B).

In primitive horses such as *Hyracotherium* and *Mesohippus*, the principal force between the upper and lower cheek teeth during chewing was compressive and would tend to crush food, rather than grind it by lateral shear. That would have been advantageous for a herbivore feeding on fruits, seeds, or tender leaves (Rensberger et al. 1984; Janis 1990b). This functional hypothesis is perhaps corroborated by fossilized stomach contents containing grape pits from the late Eocene equoid *Propalaeotherium* from Messel, Germany (Franzen 1985) (see Chapter 4). Thus, although the traditional notion had been that primitive horses were browsers, *Hyracotherium* and other early Tertiary equoids may have fed on a variety of vegetation. In addition to fruit and seeds, other foods potentially available to primitive horses included herbaceous dicots, woody shrubs, arid-adapted ferns (Gingerich 1981), and primitive, less abrasive grasses (Stirton 1947; McNaughton et al. 1985).

The varied feeding and digestive strategies in different clades of ungulate mammals (Janis 1976) are excellent examples of the

A. *Hyracotherium*

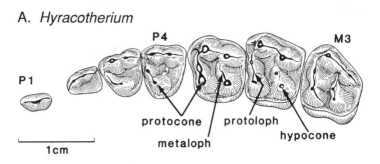

B. *Mesohippus*

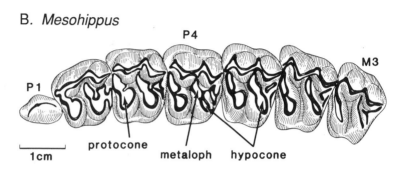

Figure 11.4. *Molarization of premolars in primitive horses. (A)* Hyracotherium *from the early Eocene of Colorado [composite from Stirton (1940) and UF 513] showing the primitive (i.e., triangular) shape of the premolars. (B)* Mesohippus *from the Oligocene of South Dakota (UF 12788) showing derived condition in which the premolars are quadrate, or molarized, by the addition of the hypocone and metaloph. Not to same scale.*

A. Foregut fermenters
Ruminant Artiodactyls

B. Hindgut fermenters
Perissodactyls

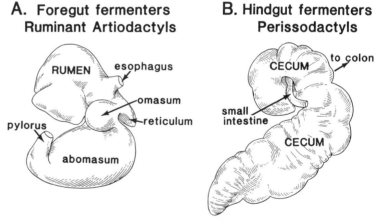

Figure 11.5. *Different ungulate digestive strategies for cellulose fermentation. (A) In ruminant artiodactyls, such as sheep, cows, and deer, the breakdown of cellulose occurs in the rumen of the stomach complex. (B) In perissodactyls, including horses, cellulose breakdown occurs in the cecum. Modified from Getty (1975) and Stevens (1988).*

opportunistic nature of evolution. Modern perissodactyls are "hindgut fermenters," in which much of the digestion occurs in an expanded cecum (extant *Equus* can have a cecum up to 1.25 m long, with a capacity of 25–30 liters) (Getty 1975) located between the small intestine and colon. This adaptation is in contrast to the case of ruminant artiodactyls, in which much of the digestion occurs in the stomach complex (Figure 11.5). Some workers (e.g., Moir 1968) have suggested that the ruminant digestive system was "better" than hindgut fermentation and allowed advanced artiodactyls to "outcompete" coexisting perissodactyls. Although it is true that the rapid diversification of artiodactyls during the later Tertiary coincided with the decline of perissodactyls, many other factors, particularly climate (Cifelli 1981; Janis 1989) also need to be considered. It must be kept in mind that perissodactyls were quite successful during the Tertiary, and their digestive physiology was one of several possible adaptations to herbivory. It is true that there are differences between various ungulate digestive systems, but it is incorrect to think of one kind as better than another.

It is impossible to determine if *Hyracotherium* had an expanded cecum (none has ever been preserved). However, given the fact that this structure is found in modern perissodactyls and closely related outgroups (hyracoids and proboscideans) (Mitchell 1905), it probably was present in *Hyracotherium*. Janis (1976) speculated that hindgut fermentation arose in perissodactyls during the late Paleocene at the base of the adaptive radiation of that order.

During the remainder of the Eocene and Oligocene much of the dental morphology and its presumed function in horses remained essentially similar to those of *Hyracotherium*. Perhaps the most notable change involved increasing molarization of the premolars. In that condition, which was fixed into the genome of *Mesohippus* (Figure 11.4B) and later brachydont horses, the P2–M3 and p2–m3 were all molariform and composed the functional tooth row for the chewing of foodstuffs.

Miocene savannas and the great transformation

The Miocene was a time of great change in mammalian communities throughout the Northern Hemisphere. As mentioned earlier, although fossil grasses are known from the early Tertiary of North America, it was not until the late Oligocene–early Miocene that the grassland biome became widespread. In some regions, such as the High Plains of North America, an influx of tectonically generated sediments resulted in an increased amount of sand in the sedimentary record (Stirton 1947; White 1959; Hunt 1990). Simpson (1951) called that the time of "the great transformation," because of all the changes observed in fossil horse morphology and inferred adaptations.

Fossil specimens show that prior to about 25 Ma, few significant morphological changes had occurred in the skulls of horses. Thereafter, however, judging from a multivariate analysis of 25 relevant characters by Radinsky (1983, 1984), it can be concluded that the early Miocene (between 25 and 15 Ma) was a time of major morphological reorganization of the equid skull. At that time the occlusal surfaces of the cheek-tooth row were shifted anteriorly and also became more ventrally located (see Figure 9.7). That apparently was related to a major change that gave increased power and efficiency to the masseter and pterygoid muscles (Radinsky 1983, 1984). Rensberger et al. (1984) concluded that the early Miocene was a time when chewing changed from a predominantly crushing mode to one in which transverse shearing became more important.

It is well documented that the early Miocene, at about 18 Ma, also was the time when certain clades of horses first became hypsodont (Figure 11.6). Corresponding to that, the depth of the jaw increased to accommodate higher-crowned teeth, and in some later clades, particularly those with deep facial fossae, the teeth became strongly curved (Figure 11.7) to accommodate more crown height within the skull. Whereas all these major morphological changes seem to be related to a general adaptation to feeding on more abrasive grasses and contaminant grit in the diets, other morphological changes suggest that the 10–15 species of horses (depending upon locality) that coexisted during the middle Miocene in North America were more finely partitioning the available resources.

MUZZLE MORPHOLOGY. Several recent studies have stressed the importance of the muzzle and incisor morphology in allowing niche separation between coexisting herbivorous mammals competing for similar food resources (e.g., Owen-Smith 1985). This is understandable, because the front part of the dental battery forms the cropping mechanism, and morphological changes in this region imply specializations in food resources. This is in contrast to the cheek-tooth series in hypsodont horses, which was surprisingly uniform in terms of the general morphology of the shearing surfaces. Janis and Ehrhardt (1988) studied the muzzle and incisor morphologies of 95 species of extant ungulates with known diets within the Perissodactyla, Artiodactyla, and Hyracoidea. They defined grazers as those ungulates that take more than 90% of their year-round diet from grasses. They recognized three different categories of browsers, all of which are similar in taking more than 90% of their diets in dicotyledonous foodstuffs: (1) "Regular browsers" are the most generalized and take a combination of leaves, shrubs, herbs, and succulent plants. (2) "Succulent browsers" take very little leafy material and specialize in fruits and buds. (3) "High-level browsers" feed almost exclusively on tree leaves, usually at or above the levels of their own heads, rather than on lower-level shrubs or herbs nearer the ground.

Several general patterns emerged from their study that can provide a model to interpret diet and niche separation in fossil

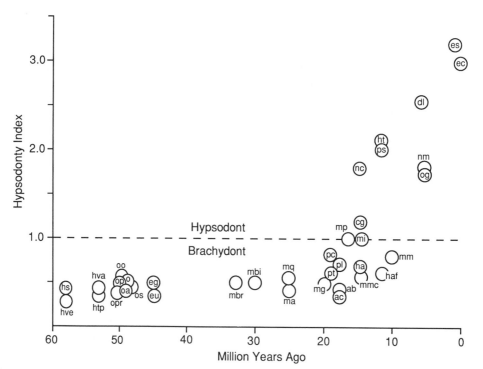

Figure 11.6. Evolution of hypsodonty in fossil horses. Although there are several different definitions in the literature, as used here the hypsodonty index (HI) is the ratio of first-molar unworn or little-worn crown height to occlusal length (M1MSTHT/M1APL, Figure 9.4). Brachydont taxa have HIs less than 1.0; hypsodont taxa are above 1.0 (source: MacFadden, original data). Taxa represented; ab, Archaeohippus blackbergi; ac, "Anchitherium" clar-enci; cg, Cormohipparion goorisi; dl, Dinohippus leidyanus; ec, Equus caballus; eg, Epihippus gracilis; es, Equus scotti; eu, Epihippus uintensis; ha, Hypohippus avus; haf, H. affinis; ht, Hipparion tehonense; htp, Hyracotherium tapirinum; hs, H. angustidens; hva, H. vasacciense; hve, H. "venticolum"; ma, Miohippus annectans; mbi, Mesohippus bairdii; mbr, Miohippus barbouri; mg, M. gemmarose; mi, Merychippus insignis; mm, Megahippus matthewi; mmc, M. mckennai; mp, Merychippus primus; mq, Miohippus quartus; nc, Neohipparion coloradense; nm, Nannippus minor; og, Onohippidium galushai; pc, Parahippus cognatus; pl, P. leonensis; pt, P. tyleri; ps, Protohippus simus; o, Orohippus sp.; oa, O. agilis; oo, O. "osbornianum"; op, O. pumulus; opr, Q. progressus; os, O. sylvaticus.

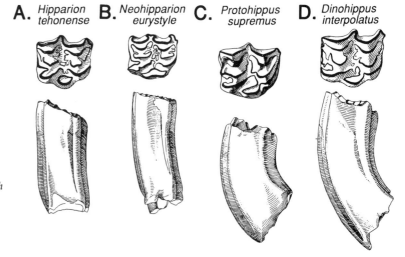

A. Hipparion tehonense **B.** Neohipparion eurystyle **C.** Protohippus supremus **D.** Dinohippus interpolatus

Figure 11.7. Occlusal (top) and lateral (bottom) views of upper cheek teeth of hipparionine (A, B) and equine s.l. (C, D) horses showing differences in curvature, which are particularly strong in the equines. Modified from Stirton (1940).

ungulate mammals. In particular, grazing ungulates have relatively broader muzzles and have incisors all of similar sizes (in contrast to browsers, where the central incisors are larger than the lateral incisors). Furthermore, dietary selectivity is related to muzzle width. More selective feeders have narrower muzzles, and the narrowest muzzles are found in browsers feeding on high-level vegetation or mixed feeders in open habitats. These general patterns seem to provide a valid basis for interpreting and comparing inferred dietary specializations within monophyletic clades, including coexisting horses during the Miocene in North America.

Using the Janis and Ehrhardt (1988) model, the muzzle regions of contemporaneous genera of Miocene horses are illustrated in Figure 11.8. These taxa are all known to have coexisted because they are found in the same Clarendonian chronofauna (e.g., Webb 1969a), a widespread and diverse mammalian community containing 10–15 equid species during the middle Miocene. At one end of the spectrum, the anterior dentition and muzzle of *Megahippus matthewi* (Figure 11.8A) are highly diagnostic. The muzzle was relatively narrow, and the incisor arcade was strongly curved. Using the modern analogue, *Megahippus* was adapted to a browsing habitus, and that is corroborated by the relatively low crowns of its cheek teeth. At the other end of the morphological spectrum, *Calippus* (Figure 11.8C) had a very broad muzzle, a wide symphysis, and a linear arrangement of incisors; it may have been a specialized grazer. Other forms, such as *Pseudhipparion* (Figure 11.8B), were hypsodont and had intermediate curvature of the symphysis. Forms like that were almost certainly grazers (based on their high-crowned cheek teeth), but perhaps fed on a more diverse array of grasses than did the more specialized *Calippus*. Although the other middle Miocene anchithere *Hypohippus* also seems to have been a browser, most other horses in the Clarendonian chronofauna (including *Merychippus, Protohippus, Dinohippus, Pliohippus, Pseudhipparion, Neohipparion, Cormohipparion,* and *Nannippus*) seem to have been grazers, although most probably were less specialized than *Calippus*. Thus, highly specialized forms, either as browsers or grazers, were rare, and relatively more generalized grazers predominated within the middle Miocene equid radiation in North America. That also seems to have been the case in later faunas. Late Miocene (early Hemphillian) species of *Calippus* also featured a relatively wide muzzle and straight symphysis (Hulbert 1988a; B. J. MacFadden personal observation). Advanced *Nannippus*, such as the Pliocene *N. peninsulatus* had very spatulate, procumbent lower incisors, not unlike certain artiodactyls (Figure 11.9). That morphology again allowed a more specialized feeding mode. Within fossil *Equus*, several different incisor and muzzle morphologies evolved (e.g., Eisen-

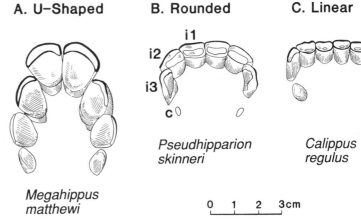

A. U–Shaped   **B. Rounded**   **C. Linear**

Megahippus matthewi   *Pseudhipparion skinneri*   *Calippus regulus*

Figure 11.8. Lower symphysial dentitions (i1–i3, c) of three contemporaneous fossil horse species from the late Miocene Clarendonian chronofauna showing different shapes of the cropping mechanism: U-shaped (A, from Burge Q., Nebraska), rounded (B, from Xmas-Kat Q. zone, Nebraska), and linear (C, from MacAdams Q., Texas). Modified from McGrew (1938), Webb and Hulbert (1986), and Hulbert (1988b).

0 1 2 3cm

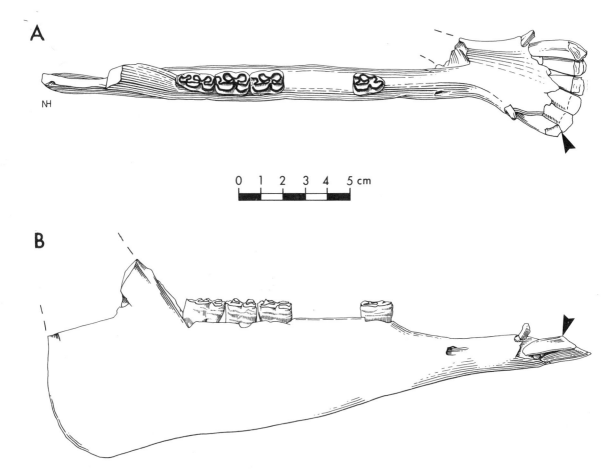

0 1 2 3 4 5 cm

Figure 11.9. Dorsal (A) and lateral (B) views of Nannippus peninsulatus *from the Pliocene (Blancan; specimen UF 22631) of southwestern Florida showing procumbent, spatulate lower incisors (arrows).*

mann 1980), but few equaled the extreme feeding specializations seen in certain Miocene clades.

HYPSODONTY. Prior to the Miocene, all fossil horses had low-crowned teeth (brachydont; Figure 11.6). The earliest hypsodont equids in North America occurred in the advanced parahippine species *Parahippus leonensis* and the primitive equine species *Merychippus gunteri*. On the basis of recent radiometric calibrations of critical sequences, that transition occurred between about 16 and 18 Ma (MacFadden & Hulbert 1988; MacFadden et al. 1991) (see Figure 6.2). As described in Chapter 8, between about 15 and 16 Ma, diversification at the base of the equine radiation occurred very rapidly, resulting in several clades (within the tribes Hipparionini and Equini; see Figure 8.6) of various body sizes (MacFadden 1987a) and degrees of hypsodonty (Figure 11.6). In the Old World the first occurrence of hypsodont horses is represented by the immigration dispersal of hipparions at about 11–12 Ma.

Both total crown height and rate of wear for modern and fossil ungulate teeth are at least partially related to diet (e.g., Janis 1988; Janis & Fortelius 1988). Hulbert (1984) stated that modern browsing ungulates and grazing ungulates have tooth-wear rates of about 1 mm/yr and 3.3–4.2 mm/yr, respectively. He also found that certain early Miocene fossil horses had wear rates of 1.5–2.3 mm/yr (*Parahippus leonensis*) and 2.3 mm/yr (*Merychippus primus*); later hypsodont horses had higher rates. What are the possible causes for these differences in rates of dental wear and acquisition of hypsodonty?

In the traditional view of hypsodont horse evolution, the acquisition of high-crowned teeth was explained as an adaptation to both grazing and increased body size. However, that view is not completely correct, because several different clades were diversifying during the middle Miocene, and even certain lineages of dwarf (*Pseudhipparion*) or small-sized (*Nannippus*) forms became hypsodont, a development that would not be expected if indeed there were a correlation between crown height and body size. Thus, regardless of the particular clade, or size trends within it, during the Miocene and Pliocene hypsodonty evolved very rapidly within the equine horses of North America relative to other dental characters (MacFadden 1988b).

A closer look at the various fossil horse clades reveals that acquisition rates for hypsodonty varied. Surprisingly, some of the most rapid evolution occurred in the late Miocene dwarfing lineage of *Pseudhipparion*, in which crown height decreased at a rate of 0.242 d (Webb & Hulbert 1986; Table 9.2). Furthermore, one advanced species of *Pseudhipparion*, *P. simpsoni*, from the late Miocene and early Pliocene (ca. 4–5 Ma) of Florida, evolved a strategy unique to horses; it was incipiently hypselodont. Root

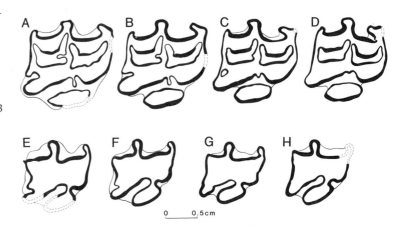

Figure 11.10. *Wear stages and hypselodonty in cross sections of* Pseudhipparion simpsoni *from the late Miocene–early Pliocene of Florida: (A–D), UF 58311, right P4 in early wear showing occlusal surface (A) and cross sections taken at 15 mm (B), 25 mm (C), and 30 mm (D) below occlusal surface; total mesostyle crown height 43 mm; (E–H), UF 57343, left P4 (reversed) in late wear stage characteristically lacking fossettes; original occlusal surface (E) and 5 mm (F), 20 mm (G), and 25 mm (H) below occlusal surface; mesostyle crown height 42 mm; root formation initiated 17 mm below H. From Webb and Hulbert (1986); reproduced with permission of the University of Wyoming.*

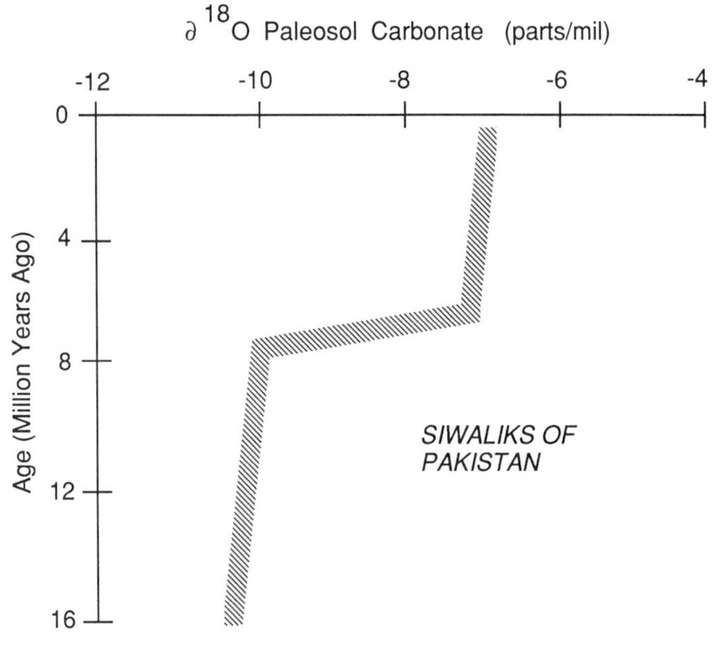

Figure 11.11. *Oxygen isotopic signature from pedogenic carbonates within paleosols in the late Neogene Siwalik sequence of Pakistan, redrawn from Quade et al. (1989b). The isotopic shift after 8 Ma is interpreted as regional climate change, including the onset of monsoon seasonality. Similar patterns in carbon isotopes indicate a change from predominantly C3 to C4 vegetation. These climatic shifts coincided with a major shift in the Siwalik mammalian fauna.*

closure was ontogenetically delayed, thus increasing the potential crown height. Ontogenetic cross sections of cheek teeth show a relatively complex enamel pattern during early wear, followed by a highly simplified wear pattern after the enamel lakes (fossettes) have been completely lost (Figure 11.10). Interestingly, simplification of the enamel pattern is a consistent finding in several other hypselodont mammals (e.g., rodents) and dwarf mammals (e.g., certain primates) (Ford 1980).

Webb and Hulbert (1986) estimated total potential crown

heights of *Pseudhipparion simpsoni*. (Because of the partially ever-growing nature of the teeth, total unworn crown heights could not be measured directly.) Their results indicate crown heights to be 85 mm in the upper premolars and 110 mm in the upper molars, far in excess of values for any other known equid, even including many species of *Equus*. This is even more surprising given the small body size of *P. simpsoni*, which was about 50 kg (it probably was slightly smaller than *Nannippus peninsulatus*, which had an estimated body size of 59.6 kg) (MacFadden 1987a), placing it among the smallest late Miocene–early Pliocene horse species in North America, rivaling the size of the modern prong-horn, *Antilocapra americana* (Walker 1975). Obvious explanations for greatly increased crown heights and smaller body size in *P. simpsoni* might be (1) that this lineage was becoming more selective and was feeding on more abrasive grasses or (2) that the substrate and contaminant grit was more abrasive (the Upper Bone Valley sediments from Florida that yield *P. simpsoni* are relatively coarse). On the other hand, it is also possible that this lineage evolved the innovation of increased individual longevity and therefore greater fecundity, as discussed in Chapter 12.

In the Old World, similar increases in hypsodonty have been reported for evolving horse clades during the Neogene, although they have remained largely unquantified. The traditional interpretation, since Kowalevsky (1873), is that it was a time of increased aridity and the advent of more steppe (extensive arid grasslands) adapted clades of herbivores. Recent work, particularly that involving the deep-sea record, has corroborated the belief that it was a time of major global changes in climate. In the well-calibrated terrestrial Siwalik sequence of Pakistan, Quade et al. (1989b) have shown a major shift in fractionation of carbon and oxygen isotopes preserved in soil carbonates (Figure 11.11). They have interpreted that as indicating the onset of monsoon seasonality, in which most of the rainfall is concentrated in the warm summer months. The consequence of such a climate change for horse evolution would have been that food resources would have been more patchily distributed during the year. Furthermore, carbon isotopic ratios from the Siwaliks indicate a change from predominantly C3 to C4 vegetation, which is interpreted as a shift from principally forested biomes to more open, grassland biomes at about 7 Ma. It was also during that time that there was a major faunal shift among the mammals toward more grazing herbivores (Barry, Lindsay, & Jacobs 1982), including at least two species of hipparion horses (MacFadden & Woodburne 1982).

In summary, rapid acquisition of high-crowned teeth has been documented for numerous clades of Miocene horses throughout the Northern Hemisphere. The advent of this character undoubtedly was related to many of the morphological changes in the skull and jaws observed during that time, resulting in the beginning of horses of modern aspect. Simple explanations for these phenomena, such as the spread of grasslands and increased body size, are not completely satisfactory because other factors, such as increasing contaminant grit in the foodstuffs, are almost certainly involved. Coexisting Miocene horse species experienced major morphological and probable feeding specializations in the cropping mechanism, similar to those of modern ungulates on the African savannas. In addition to the change in relative amounts of browse versus grass, dramatic climate shifts during the Neogene changed the seasonal availability of these food resources.

Locomotion and the evolution of cursoriality

In the horse everything has been sacrificed to speed, making the animal a "cursorial machine." — Scott (1917:46)

Since prehistoric times animal locomotion has fascinated humans for a variety of reasons. Ancient Paleolithic cave paintings most frequently depict extinct species of *Equus* in dynamic motion. In the nineteenth century, "zoopraxography," the science of animal locomotion, arose, although this term is no longer in vogue. Muybridge (e.g., 1899) conducted a classic series of experiments in 1884–5 in which he analyzed the various kinds of gaits and strides of horses (Figure 11.12) and other vertebrates. In addition to the fact that his study was important because of its detailed nature, it was technologically innovative in its use of motion-picture frames taken at intervals, usually between 0.1 and 1 second. The technology developed during those classic experiments contributed to the beginning of the motion-picture industry at the turn of the twentieth century. Thus, the zoopraxiscope was the precursor of the movie camera (Jepsen 1952). In a scientific context, the zoopraxiscope was a harbinger of the x-ray cineradiography of animal locomotion common in modern studies.

Modern horses are endowed with great speed, stamina, and strength. They can sprint by galloping at speeds of up to 70 km/hr, and racehorses complete more than two strides (cycles) per second, which is at the upper range for all mammals. Although slower than some predators (e.g., the cheetah) for short distances, the horse's stamina is almost unparalleled. Equids have been observed to travel long distances at relatively fast, sustained speeds for animals of that size. For example, a wild ass was chased by a car for 26 km at 48 km/hr (Hildebrand 1987). Its tremendous strength in pulling heavy loads has made *Equus* a beast of burden in many cultures for millennia. These attributes have rendered the horse of great value in many human activities,

Figure 11.12. *Animals in motion, original serial photographs, from Muybridge (1899). Top: Rider and thoroughbred horse in transverse gallop at a speed of about 35 miles/hour. Bottom: The mule "Ruth" bucking and kicking. Muybridge (1899:277) noted that "it being difficult to obtain a horse who was sufficiently amenable to discipline as to buck and kick at the word of command, resort was had to a circus mule [Ruth], who had undergone a regular course of instruction in those accomplishments."*

ranging from agriculture to transportation, war, and sport. The functional demands on such an animal, however, are great. The monodactyl limbs of modern *Equus* must together be capable of supporting a body mass frequently exceeding 500 kg and large compressive stresses on the skeletal support system during running. Numerous recent studies (as discussed later) of the osteology and inferred functional anatomy of fossil horses indicate that extant *Equus* is quite different from its forebears in its locomotory systems. In this section we shall review what is known of those differences and discuss possible reasons why the changes occurred as they did.

Hyracotherium

In contrast to later Eocene horses from North America (*Orohippus* and *Epihippus*), whose fossil remains are scattered and fragmentary, we have excellent collections of *Hyracotherium* postcrania from which important morphologies and functions can be described and interpreted. It is not surprising that these remains indicate that *Hyracotherium* was an evolutionary mosaic of primitive phenacodontid condylarth and perissodactyl character states, as well as more derived states that define it as a member of the Equidae.

In the forelimb, the radius and ulna remain distinct throughout their lengths; this represents the primitive eutherian condition. The shape of the elbow joint on the ulna (the trochlear notch) indicates that it was capable of considerable mediolateral rotation relative to the humerus (Kitts 1956). Relative to phenacodontids, the pelvis had a rotated and expanded blade of the ilium, a characteristic of highly cursorial ungulates (but not carnivores), allowing a more dorsal origin of the gluteus medius muscle (Hussain 1975). In the hindlimb, the femur of *Hyracotherium* possessed relatively well developed greater and lesser trochanters, which taken together composed a distinctive perissodactyl condition. The expanded greater trochanter corresponded to changes in the ilium, and those anatomical changes were related to a more forceful action of the gluteus medius muscle. The orientation of the head of the femur with the acetabulum indicates that, as in the forelimb, the range of mediolateral rotation was greater than in later horses. The fibula retained the primitive mammalian condition of being a separate bone distinct from the tibia, although it was slightly reduced relative to that of *Phenacodus* (Kitts 1956; Hussain 1975). In *Hyracotherium* the neck of the astragalus (between the trochlea and navicular facet) was long, and the trochlear crests were more inclined than in later horses (Kitts 1956; Hussain 1975).

Several classic studies have presented detailed descriptions of the metapodials of fossil horses, and from these we generally have an excellent understanding of their anatomy and interpreted function (Camp & Smith 1942; Sondaar 1968; Hussain 1975). *Hyracotherium* possessed four functional digits in the manus and three in the pes, with a greatly reduced MC I and an absence of MT I and V. So far as is known there was a significant change in the arrangement of the wrist bones and phalanges between primitive phenacodonts on the one hand and advanced phenacodonts and *Hyracotherium* on the other. In contrast to the pattern seen in primitive condylarths such as

A. *Hyracotherium*   B. *Mesohippus*

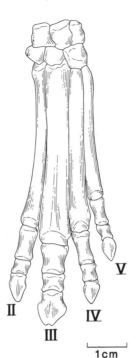
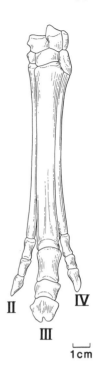

V

II   IV

III

└─┘
1 cm

II   IV

III

└─┘
1 cm

Figure 11.13. Manus in early Tertiary horses; not to scale. (A) Primitive condition in Hyracotherium with MC II–V (AMNH 4832, reversed). Redrawn from Matthew (1926). (B) Derived tridactyl condition in Mesohippus with MC II–IV. Redrawn from Simpson (1951).

Tetraclaenodon, a generally consistent pattern occurs in advanced phenacodonts (e.g., *Phenacodus*), in which each proximal carpal usually articulates with a single distal carpal, as does each distal carpal with the metacarpal (Kitts 19556). This same arrangement is seen in horses, including *Hyracotherium;* for example, the magnum (the central carpal element) articulates with MC III.

In *Hyracotherium,* although there were four metacarpal and three metatarsal digits in the forelimb and hindlimb, MP III was slightly enlarged and slightly longer than the lateral metapodials (Figure 11.13A). The metapodials were closely appressed to each other throughout the lengths of their shafts (Kitts 1956; Hussain 1975). In *Hyracotherium* all functional digits terminated in pads (similar to extant *Tapirus*) and lacked the bony hoof of later horses. Based on muscle and ligament scars on the metapodials, the exquisite reconstructions of Camp and Smith (1942) indicate that *Hyracotherium* probably had a well-developed interosseous muscle in the posterior (caudal) aspect of the manus and pes (Figure 11.14A).

Based on our knowledge of its postcranial anatomy, scenarios that reconstruct the locomotory adaptations of *Hyracotherium* in-

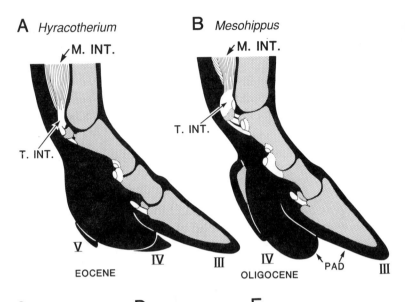

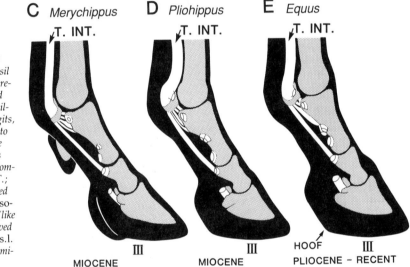

Figure 11.14. *Evolution of the foot (left manus) in selected genera of fossil and Recent Equidae, based on interpretations of restored anatomy. Adapted from Camp and Smith (1942). This illustration shows the reduction of digits, the change from relative digitigrade to unguligrade posture, and the change from the predominantly interosseous muscle (M. INT.; A, B) to the predominantly interosseous tendon (T. INT.; C–E). Camp and Smith reconstructed the more primitive horses (e.g., Mesohippus) with terminal fleshy pads (like extant Tapirus), and the more derived horses (e.g., Pliohippus, probably s.l. in current nomenclature) with a terminal hoof.*

dicate that it was very different from modern *Equus*. The limbs were less restricted in their movements than those of later horses, resulting in what probably was a looser gait. Although the feet were polydactyl, the close appression of the digits to one another indicates that they probably functioned as a unit, rather than as in the primitive mammalian condition, in which movement of individual digits is possible. The distal ends of the limbs indicate that *Hyracotherium* was relatively digitigrade. The digitigrade stance of early horses was significantly advanced relative to the primitive condition in mammals (e.g., primates), called plantigrade (or plantigrady), in which walking is done on the palms and soles. In digitigrady, the palms and soles characteristically are elevated off the ground, and the animal walks on padded toes (Figure 11.14A). In later horses (and certain other ungulates, e.g., artiodactyls) the advanced unguligrade stance evolved, with walking and running being done on hooved toes, as discussed later.

In terms of overall locomotion, a close functional analogue of *Hyracotherium* is the extant *Tapirus*, which has a less springy gait and lower maximum speed than advanced horses (Camp & Smith 1942); a similar mode of locomotion is inferred for other hyracotheres. Unlike *Tapirus*, a predominant forest-dweller, *Hyracotherium* probably was not restricted to dense forests in the modern sense. Thus, hyracotheres may have inhabited somewhat more open-country biomes than was previously believed (Gingerich 1981; although see Janis 1982) and may have been more specialized runners than other contemporaneous mammals in early Eocene faunas (Rose 1990).

Mesohippus *and the advent of tridactyly*

The postcranial remains of other hyracothere horses from North America are of fairly poor quality, but the available fossil specimens suggest that only slight morphological changes occurred relative to those seen in *Hyracotherium*. On the other hand, the Oligocene fossil record of horses is extensive, and from classic studies (e.g., Scott 1941), particularly in the Big Badlands of South Dakota and adjacent regions, there are excellent descriptions of *Mesohippus*, the earliest member of the subfamily Anchitheriinae. *Mesohippus* represents an important point of reference for an analysis of subsequent changes in locomotory anatomy and adaptations of fossil Equidae.

In the lower forelimb, the ulna, although reduced relative to its proportions in *Hyracotherium*, normally remained a distinct bone and extended the length of the radius (Sondaar 1968), although Scott (1941) noted that in some specimens of *Mesohippus* these two bones were fused distally. In the lower hindlimb, the fibula was reduced in proportions relative to *Hyracotherium* and usually was fused to the tibia along the lower third of the shaft. As with the radius and ulna, there was significant individual variation in the degree of fusion of the tibia and fibula (Scott 1941; Kitts 1956; Hussain 1975).

Mesohippus was the first horse with a fully tridactyl manus.[2] Relative to *Hyracotherium*, the central metapodial (III) was enlarged and elongated (Figure 11.13B). Although all three metapodials terminated in digital pads (like *Hyracotherium*), the distal phalanx of MP III had an expanded hoof relative to the more

[2]A possible exception to this is a single specimen of Chadronian *Mesohippus* that retained a rudimentary MC V. However, this seems to have been an individual variant within a fully tridactyl population (R. J. Emry personal communication).

diminutive side toes on MP II and IV. Based on a study of digital scars, Camp and Smith (1942) concluded that *Mesohippus* still retained the primitive interosseous muscle that inserted on the first phalanx of MP III (Figure 11.14B). In *Mesohippus* the first phalanx was about one-seventh the length of MT III; in *Equus* it has increased to about one-third that of MT III (Hussain 1975).

The Miocene: diversification of locomotory adaptations

During the Oligocene and earliest Miocene in North America, the locomotory adaptations of other horses, such as *Miohippus*, were generally similar to those of *Mesohippus*. However, as we have already seen for the dentitions, the early Miocene was a time of great diversification and changes in the locomotory apparatus within several more derived clades of fossil horses.

LATER ANCHITHERES. The advanced anchitheres were represented in North America by the genera *Kalobatippus* (= *Anchitherium* of the Old World), *Hypohippus*, and *Megahippus*. Although their dentitions retained many relatively primitive features, certain characters of the feeding complex, such as occlusal surface area, evolved rapidly (MacFadden 1988b). A striking feature of these anchitheres, particularly *Hypohippus* and *Megahippus*, was their rapid increase in body size (MacFadden 1987a), a character that frequently has been cited as important to understanding the feeding and locomotory changes seen in other clades of Miocene horses. It should be noted, however, that in contrast to those other clades, the larger-bodied anchitheres retained a tridactyl foot that probably terminated with digital pads similar to those of *Mesohippus* and other primitive horses.

MERYCHIPPUS AND OTHER ADVANCED TRIDACTYL HORSES. In *Merychippus* there were many changes in the postcranial skeleton and inferred locomotory adaptations relative to *Mesohippus*. Interestingly, many authors (e.g., Scott 1941; Camp & Smith 1942) have noted that within various merychippine species there was much intrapopulation and interspecific variation in these characters; this suggests evolutionary experimentation.

In the forelimb, the ulna was reduced along the distal part of the shaft of the radius, and these elements were partially fused. In the hindlimb, the fibula was greatly reduced in its length relative to the tibia. Lateral movement of these limbs (away from the midsagittal plane) was restricted relative to that in primitive horses, indicating decreased flexibility. In the metapodials, the central digit was further expanded relative to MP II and IV, and in some advanced forms, such as *Neohipparion* and *Nannippus*,

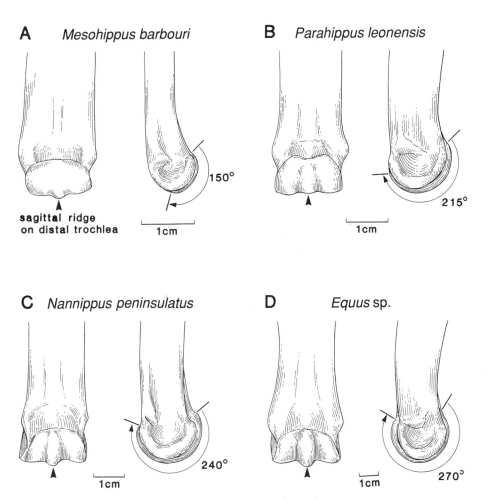

A *Mesohippus barbouri*

sagittal ridge
on distal trochlea

150°

1cm

B *Parahippus leonensis*

215°

1cm

C *Nannippus peninsulatus*

240°

1cm

D *Equus* sp.

270°

1cm

Figure 11.15. *Development of sagittal ridge on distal trochlea of MP III (in this case, right MC III): (A) Mesohippus barbouri (Oligocene of Wyoming,* UF 19632); (B) Parahippus leonensis *(Miocene of Florida, UF 43626); (C)* Nannippus peninsulatus *(Pliocene of Florida, UF 22620, reversed); (D)* Equus sp. *(Pleistocene of Florida, UF 67381). Note the increased angle circumscribed by the sagittal ridge.*

these were greatly elongated, not unlike those of some advanced monodactyl species. The distal articular surfaces of the metapodials had better-developed sagittal ridges (Figure 11.15), enhancing anteroposterior motion and preventing lateral dislocations (Hussain 1975). Camp and Smith (1942) showed that the digital scars of *Merychippus* were greatly expanded. That was interpreted to indicate that the interosseous muscle had become relatively reduced, and correspondingly the interosseous tendon had become elongated (Figure 11.14C), a condition that was further elaborated in later horses. During that transition, the sesamoid bones were elaborated in the lower forelimbs and hindlimbs, giving increased mechanical advantage to the tendons and ligaments (Sondaar 1968). Scars from the lateral sides of the metapodials indicate that in tridactyl horses these bones were closely appressed to one another and probably functioned as a single unit, not unlike that in monodactyl horses. The digital pad in hyracotheres and anchitheres was lost and was replaced with a hoof, the bony core of which was represented by the enlarged distal phalanx of MP III. A related feature during that transition was that the orientation of the terminal phalanx (and hoof) changed relative to the rest of the foot, so that the animal had an unguligrade posture, rather than a digitigrade posture as in primitive horses (Camp & Smith 1942; Thomason 1986). With those changes, the wrist and ankle joints were elevated off the ground, thus giving a longer lever arm during locomotion (Figure 11.14).

Although it is clear that the central metapodial played an important role in weight-bearing and locomotion in tridactyl horses, there has been some speculation about the functions of the lateral toes during different standing and running activities. Particularly for the more advanced tridactyl clades, most workers have stated that the lateral toes contacted the ground only during running. Furthermore, Shotwell (1961) believed that the side toes in *Hipparion* [*s.l.*] increased its traction during dodging movements. Renders (1984; Renders & Sondaar 1987) described a virtually unique situation that addressed this question and added much insight to our understanding of tridactyl locomotion.

Renders (1984) studied footprints made by two adults and one juvenile of the tridactyl *Hipparion* [*?s.l.*] preserved in a soft, wet volcanic mud at Laetoli, Tanzania (Figure 11.16, left). This Pliocene locality, which dates from 3.5 Ma, is otherwise of great interest because in contains fossil humans and their footprints (as well as other associated vertebrates). Studies have suggested that most of the footprints were made within a short period of time. Based on comparisons with modern analogues, Renders concluded that the gait of *Hipparion* was a running walk, with speeds ranging from about 6.5 to 15 km/hr. At that rate of speed, and in contrast to the gallop, all four feet were never off the ground at the same time. The side toes did impact the ground during that type of gait (Figure 11.16, right), although they certainly were of relatively small importance in the weight-bearing function of the foot. Renders concluded that one important function of these toes was to prevent overextension during the springing action of the foot. Extant *Equus* has the option of several types of gaits; for example, at moderate speeds the running walk may be substituted for the trot or pace. Until Renders's study it had not been clear which of these were innate and which had been taught by humans. The evidence from the Laetoli

Hipparion is interesting because it indicates that the running walk was an ancient locomotory adaptation of fossil horses.

MONODACTYL HORSES. It has frequently been stated in the literature that a combination of certain dental characters and monodactyly could be used to define the advanced equines, including *Equus*. However, this notion has more recently been subjected to critical reexamination. In particular, the revised systematics of advanced hypsodont horses indicate that some species of *Pliohippus* and *Dinohippus* previously thought or assumed to have been monodactyl (because they also had distinctive dental characters) were actually tridactyl. Also, just as earlier horses had shown intraspecific character variation within populations, such was the case for the advent of monodactyly. The recent discovery of an exquisitely preserved population of primitive *Dinohippus* [*Pliohippus* of Voorhies (1981)] from Ashfall Fossil Beds in northeastern Nebraska (see Chapter 4) suggests that some individuals were tridactyl, whereas others were monodactyl (Voorhies 1981). Although that also may have been the case for other primitive equine species previously thought to have been exclusively monodactyl, in derived species within this clade this character became fixed into the genome.

Many of the limb characters seen in *Merychippus* were further elaborated in the fully monodactyl representatives of *Dinohippus*, *Astrohippus*, and *Pliohippus*, and in all known fossil and extant species of *Equus*. In the limbs, the ulna and fibula were further reduced and fused to, respectively, the radius and tibia. In most species the metapodials were elongated and frequently thickened, the latter probably resulting from increased body size in many clades. In *Equus* the length of the first phalanx is roughly one-third that of MP III, indicating an elongated limb (in *Mesohippus* it was one-seventh) (Hussain 1975). There are numerous exceptions to this; for example, some South American species of *"Hyperhippidium"* (= ?*Parahipparion*) and *Equus* (*Amerhippus*) were short-legged, apparently as an adaptation to living in mountainous regions (Scott 1913). In *Equus*, the interosseous tendon is further elaborated, and the corresponding interosseous muscle is rudimentary or absent (Camp & Smith 1942; Sondaar 1968) (Figure 11.14E).

Horses spend a large portion of their time standing, sometimes as much as 18 hours per day (Stevens 1988). In order to do this, they have evolved a "stay apparatus" in which the forelimbs and hindlimbs lock into position to reduce the muscular energy required to maintain a standing posture. In the forelimb of *Equus*, the brachial tendon of the proximal part of the biceps brachii has a "dimple" that fits into the bony intermediate tubercle (INT; Figure 11.17) on the proximal articular surface of the humerus, and this forms the locking mechanism of the passive stay ap-

Figure 11.16. Hipparion *footprints preserved at Laetoli, a Pliocene 3.5-Ma site in Tanzania. Left: Trackways of adult (bottom right to top left) and foal (bottom left to top center); also note hominid footprints in upper right. Right: Details showing central MP III hoof and lateral toe prints (arrows). From Renders and Sondaar (1987); reproduced with permission of the Clarendon (Oxford University) Press.*

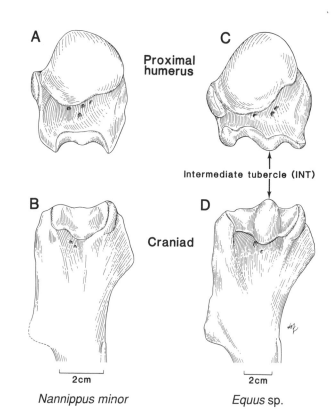

Figure 11.17. *Development of intermediate tubercle (INT) on right humerus. In the primitive condition (A, B), the proximal humerus lacks the INT (e.g.,* Nannippus minor, *from the late Miocene of Florida, UF 69936). In the derived condition (C, D) there is a well-developed INT that serves as a locking mechanism for the brachial tendon of the biceps brachii during resting posture (e.g.,* Equus *sp., from the Pleistocene of Florida, UF 127675).*

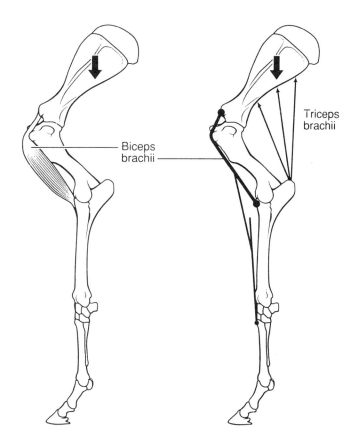

Biceps brachii

Triceps brachii

Figure 11.18. *Forelimb in* Equus *showing the biomechanics of stay apparatus. The action of the biceps brachii (left) flexes the elbow and extends the shoulder, the latter preventing collapse under the force due to gravity (arrow). The action of the triceps brachii (right) also prevents the elbow joint from collapsing. From Hermanson and Hurley (1990),* Acta Anatomica, *137:146–56; reproduced with permission of S. Karger AG, Basel, Switzerland.*

paratus (Figure 11.18). Because this stay apparatus involves a bony structure, its development can be traced in fossil horses. Hermanson and MacFadden (1992) have found that a distinct INT evolved relatively late in fossil horses. Their work indicates that it was rudimentary in some late Miocene species of *Dinohippus* and was better developed in extinct species of *Equus*. In the hindlimb, the patella (kneecap) locks onto the lateral trochlear crest of the femur, and patellar ligaments lock the stifle (knee) joint. In addition, the superficial digital flexor locks the femur with the calcaneal tuber. These associated structures all function to form a parallelogram in the hindlimb stay apparatus that prevents collapse and conserves muscular energy (Figure 11.19).

Thus, there were at least three different adaptive groups of fossil horses with diverse postcranial morphologies that coexisted during the middle and late Miocene. The anchitheres retained the primitive tridactyl foot structure, including a digital pad. The advanced three-toed horses evolved an unguligrade

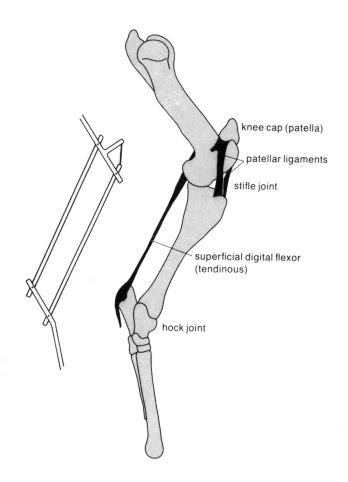

knee cap (patella)

patellar ligaments

stifle joint

superficial digital flexor
(tendinous)

hock joint

Figure 11.19. *Hindlimb of horse in resting position. Modified from Hildebrand (1987). The stifle and hock joints are passively supported within a functional parallelogram (left) by the arrangement of bones, patellar ligaments, and superficial digital flexor. Reprinted by permission of* American Scientist, *journal of Sigma Xi, The Scientific Research Society.*

foot that in many ways was adaptively similar to the monodactyl forms. Finally, the monodactyl horses arose during that time, with a locomotory complex that also occurred later in *Equus*.

Interpretation of locomotory evolution: a summary

Camp and Smith (1942) discussed the reasons for the evolution of the digital ligaments and springing mode of locomotion in fossil horses. They believed that the evolution of this locomotory complex was related to (1) increased body size, (2) the loss of side toes, and (3) the development of the springing mechanism in the lower limbs. In this section we shall investigate these factors in more detail.

BODY SIZE. It cannot be disputed that the increasing body sizes documented for several clades of hypsodont horses placed greater demands on the strength of the limbs. For example, a modern horse weighing 500 kg has 25–50 times more weight to bear than a 10–20-kg *Hyracotherium*. Although horses do not exhibit the graviportal adaptations seen in other large-bodied mammals, such as rhinoceroses and proboscideans, they nevertheless have had to cope with a variety of problems related to bearing the increased weight of a larger body. The mode of locomotion in horses tends to maximize compressive stresses and minimize forces tending to bend the limbs (Camp & Smith 1942; Thomason 1985). Thomason (1985) studied the density and architecture of metacarpal trabeculae, internal structures that function to add strength to the limb bones. He found that relative to *Mesohippus*, later horses, such as *Equus*, had increased their density of trabeculae, which was attributed to the need for more strength. Thomason (1985) concluded that those changes had resulted from a combination of both increased body size and increased compressive stresses incurred in the springing mode of locomotion.

LOSS OF SIDE TOES. Likewise, the loss of side toes placed different constraints on the monodactyl horses. In tridactyl horses the side toes functioned to absorb part of the stresses of impact that fell mainly on the central, weight-bearing digit, as well as to provide more traction in muddy substrates. In addition, Renders (1984) believed that side toes in tridactyl horses prevented overextension of the lower limb during running. With the advent of monodactyly, the digital ligaments functioned to prevent overextension, and, as seen earlier, the increase in the density of trabeculae improved the strength of the limb bones.

EVOLUTION OF SPRINGING LOCOMOTION. The complex of digital ligaments in later horses resulted in a kind of locomotion fundamentally different from that of *Hyracotherium*. In hypsodont tridactyl and monodactyl horses, the lower forelimbs and hindlimbs were elongated, the interosseous muscle was reduced, and the interosseous tendon was elongated (Figure 11.14). When a foot impacted the ground, much potential energy was created within the elastic ligaments. That energy was then released as the foot left the ground, and that cycle created the so-called springing mode of locomotion characteristic of horses. Camp and Smith (1942) likened the elastic foot of the horse to a pogo stick. Along with the evolution of the springing motion came many associated morphological changes, including reorientation of the upper limb bones and musculature for a more efficient, powerful stride and increased restriction (i.e., decreased flexibility) of lateral limb motion.

Paleontologists often have speculated about the reasons for the evolution of the springing mode of locomotion in hypsodont horses. It seems clear that this was a response to the multitude of changes experienced when horses invaded the grazing adaptive zone during the Miocene. On the basis of the foregoing discussion it certainly seems plausible that some of the overall changes seen in Miocene horses resulted from increased body size. However, that is only part of the story. It also seems that later horses ran faster than *Hyracotherium*. It is often stated in the literature that the evolution of faster gaits in horses (and other ungulates) was a coevolutionary response to predator selection (e.g., Benton 1990). It is conceivable that the transition to the more open-country habitat of the Miocene may have made them more readily visible to predators and thus changed the intensity of selection toward faster running. Some more recent workers, however, have challenged predator–prey coevolution as an all-inclusive explanation. Other plausible factors could also have been involved in the evolution of equid cursoriality. For example, McNaughton et al. (1985) presented an interesting hypothesis in which they speculated that the ability to locomote at greater speeds and for longer periods of time would have been a favorable adaptation for animals that may have had to travel long distances during different seasons of the year to exploit patchy food resources. Thus, the many changes in the locomotory systems of fossil horses probably were related as much to increased stamina as to greater speed.

Another point relates to the quotation from Scott at the beginning of this section. Various authors have intimated that modern horses are "perfectly" adapted for fast running and that *Equus* is at the pinnacle of this evolutionary progression. Again, this is evidence of the pervasive orthogenetic paradigm. In contrast, I would argue that during the Eocene, *Hyracotherium*, with its digitigrade posture and tapirlike gait, was well adapted to a set of ecological and environmental circumstances that were quite different from those encountered by horses during the later Cenozoic. There is some evidence to suggest that tridactyl and monodactyl horses coexisting in certain regions did experience subtle ecological differences. For example, in the Miocene of the northern Great Basin, Shotwell (1961) believed that the tridactyl grazers preferred the savanna/forest mosaic habitats that were widespread during that time (see Chapter 7). As those biomes were replaced by more arid, open-country grasslands in the late Miocene, the monodactyl horses became predominant (see Chapter 13).

In summary, it is incorrect to think that the tridactyl horses of the late Tertiary were less well-adapted to the conditions they shared with coexisting monodactyl clades. Both were successful in their own ways, as were the hyracotheres during the Eocene. If clade longevity can be viewed as a measure of evolutionary

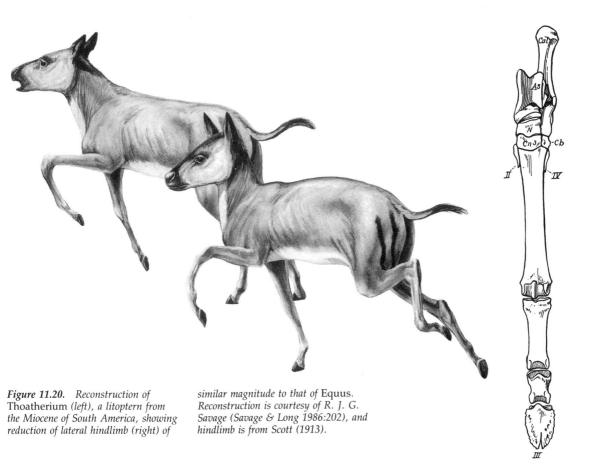

Figure 11.20. *Reconstruction of* Thoatherium *(left), a litoptern from the Miocene of South America, showing reduction of lateral hindlimb (right) of* similar magnitude to that of Equus. *Reconstruction is courtesy of R. J. G. Savage (Savage & Long 1986:202), and hindlimb is from Scott (1913).*

success, then tridactyl equids, which existed in North America for some 35 myr, were the winners in this category.

Concluding comment: horse chauvinism

This chapter smacks of equid chauvinism, and with good reason, because equids are the focus of this book. However, we need to keep in mind that horses were by no means the only group of placental mammals to evolve grazing and monodactyl adaptations. As mentioned earlier, grazing is pervasive in many orders, ranging from rodents to elephants. In their digestive strategy, horses (and other perissodactyls) "did their own thing" in evolving hindgut fermentation to process food of high fiber content and low nutritive value. Artiodactyls evolved similar diets, but with a fundamentally different evolutionary strategy that included a highly complex stomach in ruminants.

Several groups of ungulate mammals evolved reduced limbs, although few did so to the extreme seen in monodactyl equids. Notable exceptions were the litopterns (Figure 11.20), a group of extinct South American herbivores that became monodactyl during the early Miocene, about 8 myr earlier than did horses in North America. Simpson (1951) referred to these as "false-horses," and although that is perhaps a misleading analogy, it does convey the notion of an example of extreme parallel evolution of the locomotory complex in two distantly related eutherian mammals. Other groups also evolved reduced limbs, but in different ways. For example, loss of the lateral metapodials and fusion of the central metapodials (forming a cannon bone) in ruminant artiodactyls were functionally similar to adaptations in monodactyl horses and allowed sustained running. In summary, from study of the fossil record of feeding and locomotion among horses, we have a better understanding of the long-term functional changes that occurred to these important morphological and adaptive systems. From comparisons with other groups of herbivorous mammals we also can see the opportunistic nature of evolution, because there are many ways to evolve adaptations for feeding and locomotion.

12

Population dynamics, behavioral ecology, and "paleoethology"

Horses, like all other mammals, have relatively highly evolved behavioral ecologies and social systems. A better understanding of these important characteristics has come from a large body of research involving countless hours of observation of wild equid populations, such as those that naturally occur in Africa and western North America, as well as domesticated populations of ponies in Great Britain, as discussed later. Thus, we are fortunate to have modern analogues that can be used to extrapolate back in time and infer the behavioral ecology and social organization of extinct horses. One might argue that it is difficult, if not impossible, to reconstruct the behaviors and social systems of extinct mammals, because obviously such traits do not fossilize. However, as will be discussed later, if several important parameters of fossil ungulates are known, most notably body size, feeding strategy (i.e., browser versus grazer), habitat type, and climate, then strong predictions can be made regarding the population and social biology of a particular extinct species. Furthermore, if the ages and sexes of individuals within a population can be determined, then several additional inferences can be made about evolutionary and ecological phenomena, such as the intensity of long-term natural and sexual selection (see Chapter 8). There are limitations to such interpretations, but with thorough knowledge of the inherent assumptions, a better understanding of the population ecology and paleoethology can be presented for fossil horses, which is the purpose here. These final two chapters are devoted to a discussion of the ecology of extinct horses to the extent that it can be interpreted from the fossil record. In this chapter we shall investigate their population dynamics, behavioral ecology, and social organization. In Chapter 13 we shall consider extinct horses as parts of their ancient terrestrial communities and ecosystems and how they all evolved through the Cenozoic.

Behavioral ecology and social organization of modern equids

Extant equids fall into two groups with distinct types of social organization, and although these patterns are relatively widespread and similar in diverse regions, there also are intergradations (Klingel 1975; Eisenberg 1981).

Type I bands

The first kind of equid social organization, which Klingel (1975) termed Type I [= category D1 of Janis (1982)], is represented by feral *Equus caballus, E. przewalskii*, the mountain zebra *E. zebra*, and the plains zebra *E. burchelli* (Table 12.1). Within Type I there are two distinct kinds of bands. The first, which represents the population reproductive unit, is the polygynous harem, with a dominant stallion, one to several reproductively mature mares, and usually several juveniles. Dominance within the band is recognized by various acts, including visual display, communication, and maintenance of fecal piles (Feist & McCullough 1976) (Figure 12.1). The sizes of these equid harems vary greatly, with the greatest frequency of 3–6, but usually less than 10 (Leuthold & Leuthold 1975; Feist & McCullough 1975; Berger 1977, 1986) (Figure 12.2). The harem usually is a very stable unit, with few changes in composition, except for the voluntary departure of juveniles, particularly males, usually at an age of about 2 years (Berger 1977). This band stability has the selective advantage of higher fecundity. In the rare instances when the composition of a harem band changes, or is disrupted by intruding males, reproductive success declines. For example, in the Granite Range population of feral *Equus caballus* in the Great Basin of the western United States, Berger (1983b) found a significantly higher foal production by females within stable harems, in contrast to those in unstable harems (82% and 38%, respectively, of reproductively mature mares produced foals in a given year).

In addition to the harem bands, Type I equid social organization is characterized by celibate bachelors, usually younger members of the population "on their own" after having voluntarily left the original bands in which they were foaled. Bachelor males either spend time as solitary individuals or form bands, usually consisting of two or three individuals, seldom with more than eight individuals (Feist & McCullough 1975; Berger 1977). Klingel (1975) found that in some cases older or sick males joined the younger bachelors. Like harems, bachelor bands are characterized by a hierarchical dominance of individuals. In contrast to the stability of the harems, the individual composition of a bachelor band is relatively dynamic from year to year (Feist & McCullough 1975; Berger 1986). Berger (1981) believes that these bands are adaptively very important in improving the fighting skills of young males, which will become important later in their lives, particularly during aggressive (agonistic) behavior related to competition for reproductively available mares.

Type I equid bands characteristically live in dry regions with high seasonality, resulting in significant changes in temperature and rainfall over the year. For feral bands of *E. caballus* in the western United States, births usually occur within a two-to-three-month period and are concentrated within a six-week period during late spring and early summer. With these kinds of variable resources, home ranges of bands vary seasonally, but usually are around 10–50 km^2 (Feist & McCullough 1976; Berger 1986).

Type II bands

This kind of equid social structure is characterized by male-defended territories, and it is seen in Grevy's zebra, *E. grevyi*, wild burros, and African and Asiatic wild asses, *E. africanus* and *E. hemionus* (Klingel 1975, 1977) [= category D2 of Janis (1982)]. The defending males have exclusive mating rights within these territories, which are frequently marked by fecal piles. In contrast to Type I harems, there are no permanent bonds, and the compositions of these bands can change within hours. In Type II bands, resource availability is more constant, and breeding can occur throughout the year (Klingel 1975). For fossil equids, Klingel (1975) speculated that Type II (territorial) social organization evolved first during the Eocene, followed later by Type I during the Miocene. Thus, the savanna equids of modern-day Africa are the closest analogues of the Miocene adaptive radiation in North America, and Type I social organization probably is the appropriate model for interpreting most of the extinct grazing equids throughout Holarctica.

All ungulate mammals have evolved highly characteristic intraspecific display and agonistic behaviors, whether for territorial defense or competition for mates, particularly with rival males competing for females. Many of these ungulates have evolved cranial appendages in the form of horns and antlers, and yet, with the curious and anecdotal exception of rudimentary, bony "frontal protuberances" in domesticated *Equus caballus* (Chubb 1934), these structures are conspicuously lacking in all fossil and extant members of the Equidae. Janis (1982, 1986) attributes the absence of cranial appendages in equids to the need to range widely for forage, and thus the relatively diminished need to defend resource territories. If she is correct, then this pertains more to Type I than to Type II equids. In the absence of horns, modern horses have evolved an elaborate competitive behavior that involves a ritualized sequence of threat display,

Table 12.1. *Categories of ungulate life-history characteristics and selected examples of fossil and living taxa using the scheme proposed by Janis (1982)*

Category	Habitat	Diet	Morphology	Body weight, range (and mean)	Social and reproductive behavior	Dimorphism	Possible living ungulates included	Possible fossil ungulates included
A1	Forest-seasonal woodland; small home range	Folivorous, but take nonfibrous food when available; diet seasonally variable	Molars brachydont; limbs short–medium	15–200 kg (25 kg) (artiodactyls); 150–1,000 kg (perissodactyls)	Solitary or ♂/♀ pairs defending territory; monogamous–solitary	Monomorphic (slightly dimorphic); no horns in either sex (or small horns in ♂♂)	Chinese water deer (*Hydropotes*); ?okapi (*Okapia johnstoni*); tapirs (*Tapirus*); small rhinos (*Rhinoceros sondaicus* and *Dicerorhinus sumatrensis*)	Some traguloids (e.g., large species of *Dorcatherium*) and moschoids (e.g., *Dremotherium*) ?Some primitive extinct equids
A2	Savanna-seasonal grassland; medium home range	Highly selective for nonfibrous parts of open habitat plants (e.g., grass seeds)	Molars mesodont–hypsodont; limbs long	2–40 kg (6 kg)	Solitary, or ♂/♀ pair bonding; monogamous	Monomorphic–slight dimorphism (no horns in either sex, or small horns in ♂♂ only)	Open-country antelope in category A; e.g., steinbok (*Raphiceros campestris*); orebi (*Ourebia ourebi*); musk deer (*Moschus*)	North American traguloids and moschoids (e.g., *Hypertragulus*, *Leptomeryx*, *Blastomeryx*); ?stenomyline camels
B1	Open woodland; large home range	Browse, fruit; some seasonal variation in diet	Molars brachydont–mesodont; limbs short	10–200 kg (50 kg)	Mixed-sex feeding herds; males maintain dominance hierarchy; polygynous	Primitively monomorphic (no horns in either sex)	Peccaries (although more folivorous than oreodonts); many cervids in post-Pleistocene habitats, e.g., North American odocoilenes, and Asian cervids, e.g., chital (*Axis axis*)	Oredonts (families Agriochoeridae and Merycoidodontidae); browsing anchitherine equids

Category	Habitat	Feeding	Morphology	Body weight	Social organization	Horns	Modern examples	Fossil examples
C1	Woodland–savanna–grassland	Selective for plant parts, and grass, and browse	Molars mesodont–hypsodont; limbs long	15–1,000 kg (80 kg) or (300 kg)	Some mixed-sex feeding; males hold small breeding territories and/or harems; polygynous	Monomorphic (no horns in either sex), or slight dimorphism (small horns in ♀♀, larger horns in ♂♂)	Gazelles (*Gazella*); pronghorn antelope; (*Antilocapra americana*) South American camelids (*Lama* and *Vicuna*)	Most fossil camelids; some dromomerycids (*Aletomeryx* and *Pediomeryx*); antilocaprine antilocaprids; ?Some Miocene equids, e.g., *Parahippus*
C2	Woodland–savanna	Predominantly high-level tree browsing, or take both browse and shrubs	Molars mesodont; limbs medium length, or highly elongate	40–2,000 kg (1,000 kg)	Loose aggregation mixed-sex feeding herds; some male territorial–hierarchical dispute; no fixed territories; polygynous–serially monogamous	Monomorphic (no horns in either sex, or small horns in both sexes)	Giraffe (*Giraffa cameleopardis*); Indian rhino (*Rhinoceros unicornis*); black rhino (*Diceros bicornis*)	Most fossil giraffes; aepycameline camels; chalicotheres; aceratherine rhinos
D1 (= Type I)	Open woodland–savanna–grassland	Predominantly grass, or grass plus low-level browse	Molars mesodont–hypsodont, or highly hypsodont; limbs long	80–400 kg (200 kg)	Males hold harems with fixed female membership; polygynous	Monomorphic (no horns in either sex, or similar-size horns in both sexes)	Most equids (e.g., Burchell's zebra, *Equus burchelli*)	Most fossil grazing equids; ?some protolabine and cameline camels; ?brachypotherine rhinos
D2 (= Type II)	Arid savanna–grassland	Predominantly grass	Molars hypsodont; limbs long	250–2,500 kg	Males hold exclusive feeding and breeding territories; polygynous	Monomorphic (no horns in either sex, or similar-size horns in both sexes)	Some equids (e.g., Grevy's zebra *Equus grevyi*); white rhino (*Ceratotherium simum*)	?Some grazing equids; some rhinos (e.g., *Elasmotherium*)

Note: For categories D1 and D2, respectively, Type I and Type II pertain to designation of band composition in modern equids (Klingel 1975).
Source: Modified from Janis (1982); reproduced with permission of Cambridge University Press.

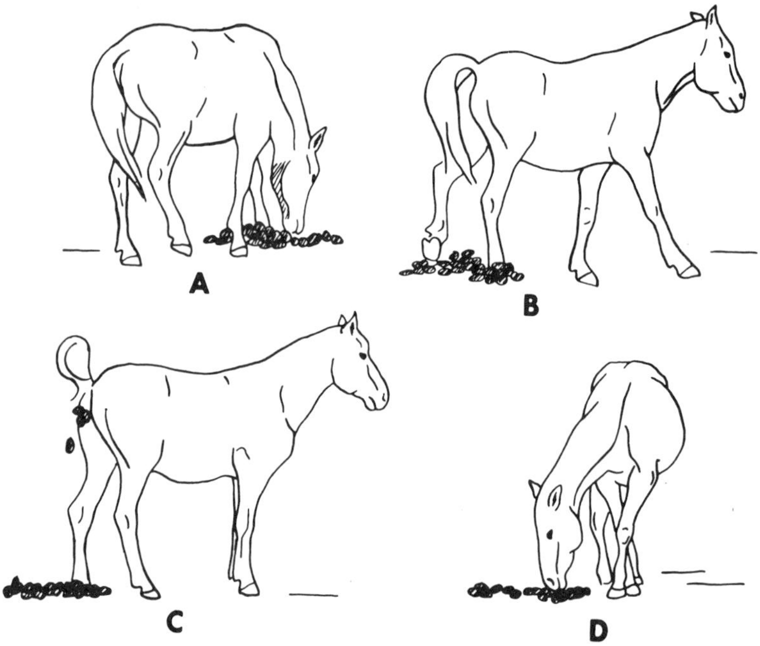

Figure 12.1. *Behavior of feral* Equus caballus *stallion at fecal piles: (A) smelling the pile; (B) stepping over the pile; (C) defecation; (D) smelling his own feces. From Feist and McCullough (1976); reproduced with permission of Verlag Paul Parey, Berlin.*

Figure 12.2. *Histograms showing sizes of harem (A) and bachelor (B) groups (bands) for populations of feral* Equus caballus *from the Pryor Mountain Wild Horse Range in the western United States. Modified from Feist and McCullough (1975).*

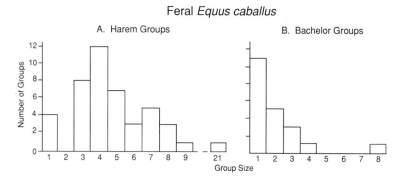

biting, and kicking (Figure 12.3). The frequency of combat among wild equid males varies from population to population, and Geist (1974) attributes such variation, at least in part, to the presence or absence of predators within the mammalian community. Thus, he believes that in areas with lower predation, rival male equids can afford a greater risk of being wounded during combat, because they will not also have to fend off predators. In addition to these observations of modern behavior, certain characters of extinct horses allow interpretations of ancient agonistic behavior, as discussed later.

In summary, an understanding of the social structure of extant *Equus* suggests a framework for interpreting similar systems

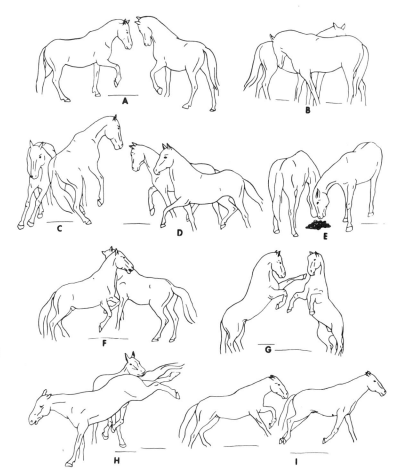

Figure 12.3. *Typical behavior during aggressive encounter between two stallions of* Equus caballus: *(A) approach and smell noses; (B) smelling flanks; (C) pushing and threatening with small hindfoot kick; (D) parallel prance; (E) smelling fecal pile together; (F) shoulder-pushing and biting; (G) rearing and forefoot kicking; (H) hindfoot kicking; (I) defeated stallion chased away by the winner. From Feist and McCullough (1976); reproduced with permission of Verlag Paul Parey, Berlin.*

among extinct horses. In addition to carrying out neontological studies, we have other parameters preserved in the fossil record whose study can increase our knowledge of the paleoethology and population biology of extinct horses. These include body size, diet, habitat, climate, and sex-related characters, all of which are discussed in the sections that follow.

Behavioral ecology and social organization of fossil horses: juggling the parameters

Body size

If I can have only one parameter to interpret the life-history strategy of a fossil mammal, give me body size. It is a powerful predictor of so many things. On the other hand a molar tooth, if it can be placed within an evolutionary lineage, could allow you to infer size and to fairly well infer dietary specialization.

– John Eisenberg [personal communication (paraphrased), ca. 1985]

Within the past few decades, biologists have become increasingly aware that many characters scale allometrically with body size. This stems from the classic paper by Kleiber (1961) demonstrating that metabolic rate scales to body size (rate $= W\delta^{.75}$, where W is body mass). In another important paper, McNab (1963) noted the correlation between body size and home-range area; thus, larger mammals have larger home ranges. Subsequently, many papers and books have elaborated on what sorts of variables scale to body size (e.g., Eisenberg 1981; Peters 1983; Calder 1984; Schmidt-Neilsen 1984). Western (1979) and Owen-Smith (1988), for example, demonstrated that body weights of mammals scale with such life-history parameters as gestation time (Kiltie 1984), growth rate, age at first reproduction, litter size, life span (Prothero & Jürgens 1987), intrinsic rate of natural increase, birthrate, net reproductive rate, and litter weight. Other authors have noted that body size correlates with other parameters, such as feeding and digestive strategies (Geist 1974; Janis 1976, 1986) and social organization (e.g., Janis 1982). Thus, body size is important, but until recently its estimation in fossil mammals had been poorly constrained. Recently, because of the great interest in this character among neontologists and its potential in paleoecological reconstructions, many papers have been published estimating body masses for fossil mammals (Damuth & MacFadden 1990).

Diet

As discussed in Chapter 11, on the basis of dental morphology and known diets of extant mammals, the general dietary preferences (i.e., carnivore, omnivore, frugivore, herbivore) of fossil mammals can be identified with confidence in most cases. In addition, for the herbivorous ungulates, unworn crown heights usually indicate whether a particular species was predominantly a browser (low-crowned or brachydont; >90% of diet from dicots) or a grazer (high-crowned or hypsodont; >90% of diet from grasses) (Janis 1988). There are some exceptions to this pattern. However, of relevance here, Janis (1988) found that African equids fit the grazing dietary habit that would be predicted from their dentitions. The assumption that relative crown height is an indicator of browsing versus grazing adaptation, particularly in horses, has been inferred by paleontologists for more than a century. Recent research by workers such as Janis (1988) has confirmed this dietary interpretation based on morphological evidence from the fossil record. In the future, in addition to traditional morphological studies of dentitions, techniques such as dental microwear and geochemical studies of fossil mammals (see Chapter 2) will routinely provide further evidence of ancient dietary perferences.

Habitat and climate

There are numerous geological and paleontological techniques available to interpret habitats of ancient ecosystems. Studies of sedimentary environments can provide evidence regarding whether a fossil locality was part of, for example, a floodplain or an over-bank or stream-channel deposit. Evidence from fossil plant materials can indicate whether a region was predominantly forested (e.g., Bown et al. 1982) or predominantly grassland (e.g, Retallack 1982). Related to that and also to climatic interpretations, within the past few decades an increased ability to recognize various types of fossil soil horizons (paleosols) has been of great utility in reconstructing the ancient habitats and paleoclimates of mammal-bearing terrestrial deposits during the Cenozoic (e.g., Retallack 1981, 1983a,b).

There also are numerous techniques currently available to discern ancient climates. Traditionally, paleobotanical evidence, mostly from plant macrofossils, has been used to interpret the climatic context (and habitat) of extinct terrestrial mammalian communities (e.g., Elias 1942; MacGinitie 1962). More recently, workers such as Axelrod and Bailey (1969) and Wolfe (1978) (see Chapter 7) have done much to improve our understanding of long-term climatic fluctuations during the Cenozoic; their research has also been based on paleobotanical evidence. In addition to the recent use of paleosols mentioned earlier, modern methods such as stable-isotope geochemistry allow a high degree of paleoclimatic resolution previously unavailable to mammalian paleontologists. In particular, oxygen isotopic ratios from deep-sea sediments are extremely valuable in discerning long-term (i.e., 0.1 to 1-myr) global temperature cycles throughout the Cenozoic (e.g., Opdyke 1990) (see Figure 7.6). In terrestrial sequences, carbon isotopic ratios are extremely valuable in understanding regional climatic and vegetation patterns, as well as long-term change. A noteworthy example of this is the interpretation that rapid shifts in oxygen and carbon isotopes at about 7–8 Ma in the Siwaliks indicate the onset of monsoon seasonality in the Indo-Pakistan subcontintent (Quade et al. 1989b) (see Figure 11.11).

The distinction made in this section involving separate interpretations of habitat, vegetation, and climate is indeed arbitrary, because an understanding of each of these reinforces knowledge of the others. One's understanding of these physical and biotic parameters must be as comprehensive as possible. With the varied interdisciplinary techniques and data now available, the gap is being narrowed between the limits of environmental parameters that the paleontologist can determine in the fossil record versus what the neontologist can observe in modern ecosystems. Coupled with a knowledge of body size and diet, these parameters are powerful predictors of the behavioral ecology and ethology of extinct horses.

Sex in fossil horses

Fortunately, present-day mammals exhibit such a wide range of dimorphism patterns that most instances in fossils may probably be interpreted with the help of recent analogies. The necessity of using neozoological data as a frame of reference in paleozoology must be emphasized.   – Kurtén (1969:233)

In addition to understanding the behavioral ecology and social organization of extinct mammals via knowledge of body size, diet, habitat, and climate, it is highly desirable to know the sexes of individuals within an ancient population. To this end the paleontologist is hampered by the lack of soft anatomical structures. However, in many mammals there are preservable hard parts that, by comparison to modern analogues, can allow the determination of sex. Although extant equids are monomorphic relative to other medium-sized grazing ungulates (Janis 1982), there are some morphological structures that can be used to determine sex. The value of determining the sex of an individual or the socionomic ratio within a fossil quarry population cannot be overemphasized, because once that information is available, then many accurate interpretations can be made regarding the population biology of that extinct species.

Kurtén (1969) described two kinds of sexual dimorphism in fossil mammals: qualitative and quantitative. In the first, structures such as horns and antlers frequently can be used to determine the sex of an individual. In modern horses, the presence of canines in males and their smaller size or absence in females is another example of qualitative dimorphism. Quantitative sexual dimorphism occurs when there is a statistically determinable size difference between the males and females of a species (Figure 12.4). Size dimorphism is quite common, although not universal, in many groups of mammals, and in most cases the males are larger than the females. Among certain cetaceans, the males can be twice as long and about eight times as heavy as the females. For example, in the sperm whale, *Physeter macrocephalus*, the maximum length is 25.5 m for males and 12.3 m for females. Cetaceans also exhibit the exception to the rule, with females of the river porpoise *Platanista gangetica*, for example, being, on average, larger than the males (Kurtén 1969). Some other mammals also exhibit the exception, with larger females, but in the great majority of cases males are statistically larger.

Because horses are relatively monomorphic, extraordinary specimens or fossil samples are needed before sexual dimorphism can be recognized. Some early workers believed that even with well-preserved material it was impossible to determine the sexes of individual fossil equid specimens (Gaudry 1867). Although sexual dimorphism in fossil horses was noted early in this century (e.g., Osborn 1918), little was concluded in terms of assessing these secondary sexual characters for a better understanding of behavioral ecology. Grazing horses are quantitatively monomorphic (MacFadden 1984a), with no discernible size difference in measured characters of the cheek teeth. Thus, in order to recognize sexual dimorphism, it is first necessary to have large quarry samples in which relative sizes of canines can be seen. Pirlot (1952) analyzed the lower canines of Miocene *Hipparion* from Pikermi and found that in the males these teeth were about twice as large (in anteroposterior length and transverse width) as in the females, and the cross-sectional shapes were different. MacFadden (1984a) also showed canine dimorphism for a Miocene *Hipparion* quarry population from Texas (Figure 12.5). With sufficiently large and well-preserved samples, canine dimorphism is seen to have been common in virtually all extinct horses. Trumler (1959) noted the large canine in a presumed male of the extinct *Miohippus acutidens* and, on the basis of extant *Equus*, speculated that it might have functioned as a display structure when the lips were drawn upward. Thus his *Rossigkeitsgesicht*[1] display behavior has an antiquity of at least 20 myr. Other evidence to be presented later suggests a much greater longevity for this display structure and possible associated behavior.

In addition to aiding studies of sexual dimorphism and notions of sexual selection, the determination of sex in fossil horses allows interpretations of the socionomic sex ratio, which is the ratio of males to females in a breeding group (Berger 1983a). These ratios have important consequences for the paleobiology of extinct mammals, particularly when analyzing population dynamics and age structure.

Age structure and population dynamics

If individuals within a sample can be aged, then the proportions of the entire population at different age classes can be determined. If the sexes can also be determined, then, taken together, these data provide powerful indicators of the population biology of a species, whether it be fossil or extant. Voorhies (1969:24) stated that "population dynamics refers to the balance between natality (births) and mortality as well as the distribution of ages within a natural population of animals."

In a classic paper, Deevey (1947) was the first to present the population dynamics, including life tables and survivorship curves, for a wide variety of animals ranging from the sessile rotifer *Floscularia conifera* to the extant Dall mountain sheep, *Ovis d. dalli*. In addition to Deevey's pioneering work (1947) on extant animals, many studies have analyzed the population dynamics

[1] *Rossigkeitsgesicht* can be roughly translated as "the stallion's facial expression when desiring a mare."

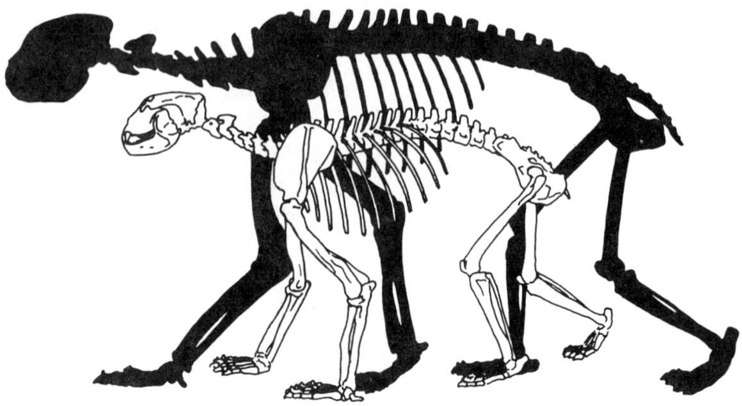

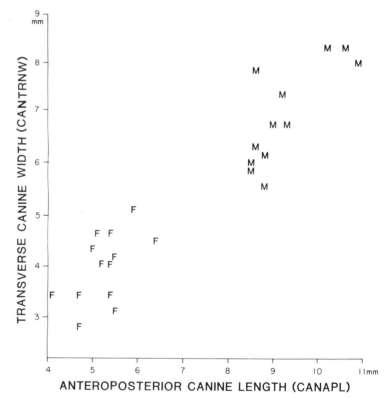

Figure 12.4. *Reconstructions of male (silhouette) and female (outline) skeletons of the spectacled bear,* Tremarctos floridanus, *showing sexual dimorphism in body size. Specimens from the late Pleistocene (Rancholabrean) Devil's Den in Florida. From Kurtén (1966, 1969).*

Figure 12.5. *Bivariate plot of canine lengths and widths showing size dimorphism of presumed males (M) and females (F) for a sample of* Hipparion tehonense *from the late Miocene (Clarendonian) MacAdams quarry in the Texas panhandle. From MacFadden (1984a).*

of extant ungulates, such as the plains zebra, *Equus burchelli boehmi* (Spinage 1972). Kurtén's classic paper (1953) was the first to analyze the population dynamics of fossil mammals. That was followed by many interesting studies, several of which included fossil grazing horses from North America (Van Valen 1963, 1964; Voorhies 1969; Hulbert 1982, 1984).

A critical prerequisite for studies of population dynamics is

Figure 12.6. *Selected age-dependent wear stages in the horse showing loss of the infundibulum (cup) and development of the dental star. Modified from Getty (1975).*

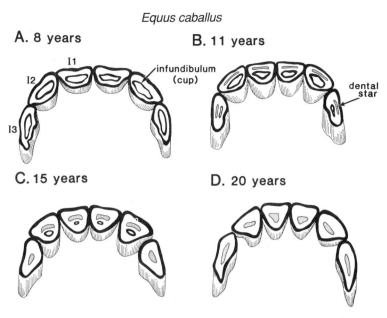

Equus caballus

A. 8 years

B. 11 years

I1

I2

infundibulum (cup)

dental star

I3

C. 15 years

D. 20 years

that the age of the individual can be determined. Neontologists have a larger array of potential characters to assess age, although some of these are the same as those available to the paleontologist. This is especially true for horses, whether they be extant or fossil. As every veterinarian or horse-lover knows, the amounts of wear on the incisors and cheek teeth, as well as the stage in the dental-eruption sequence, are powerful indicators of age. In most instances (except in very old individuals) they can be used to determine the age of an individual to within one or two years, and sometimes within six months (Sack & Habel 1977; W. O. Sack personal communication) (Figure 12.6). It is therefore fortunate that teeth are among the most durable remains of fossil horses. Thus, we can assess the ages of individuals within extinct populations by the amount of wear on the teeth and by reference to modern analogues. However, this is not a foolproof method. The further back in time one goes within a clade, the more changes in developmental patterns (heterochrony) there could have been, resulting in different timing of dental wear. However, these changes can be controlled for, as discussed later.

There are two kinds of "paleopopulations" that are normally encountered in fossil assemblages, and usually these can be discriminated on the basis of an analysis of the age structure of the individuals in the sample (see Chapter 4). In an attritional assemblage, fossil specimens accumulate over time as a result of such factors as carnivore predation or localized "mini" mass mortalities resulting from natural phenomena such as floods, quagmires, droughts, or snowstorms (Kurtén 1953; Berger 1983c)

Figure 12.7. *Foals of wild* Equus caballus *that died in a mud quagmire in the Owyhee Desert, Nevada. Photograph courtesy of the Bureau of Land Management, U.S. Department of the Interior (Winnemucca, Nevada); also see Berger (1983c).*

Figure 12.8. *Comparison of typical samples of fossil mammal quarry populations representing catastrophic and attritional mortality, assuming little or no taphonomic biases. (A) In an attritional assemblage, such as* Pseudhipparion gratum *from the Miocene Burge quarry, Nebraska, the younger and older age classes are relatively overrepresented, implying a higher predator-induced mortality. Modified from Van Valen (1964). (B) In a catastrophic assemblage, such as* Protohippus cf. Perditus *from the Miocene Verdigre quarry, Nebraska, the distribution of individuals is representative of the living herd. Modified from Voorhees (1969).*

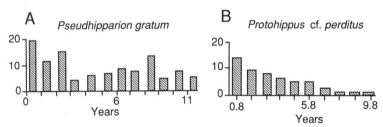

(Figure 12.7). In attritional fossil assemblages, particularly those resulting from natural selection (such as predation), the age structure of the population usually is biased toward increased numbers of juveniles and older individuals (Figure 12.8A). In practice, however, juveniles are almost always underrepresented in fossil assemblages because their bones are less likely to be preserved (e.g., Kurtén 1953; Voorhies 1969; Spinage 1972), and therefore correction factors need to be used based on modern population age structures (e.g., Hulbert 1982, 1984).

The second type of fossil assemblage is catastrophic, which results from a significant event, often a natural phenomenon such as a large flood, a blizzard, or even a "Pompeii-style" ash eruption, as at Ashfall Fossil Beds, Nebraska (e.g., Voorhies 1981). In this kind of assemblage the entire terrestrial community is annihilated, and thus one has a picture frozen in time of the population dynamics of the constituent species. The age-class representation typically is quite different from that found in an attritional assemblage (Figure 12.8B). If sampling biases can be accounted for (principally involving the underrepresented juveniles), then catastrophic fossil assemblages usually are similar in age distributions to population censuses of modern mammals. In contrast to the biases encountered with attritional samples, catastrophic assemblages provide a better understanding of important population parameters such as birthrate, age-related mortality, longevity, and survivorship, and, if sex can be determined, socionomic ratio. In summary, assuming that one can control for the various pitfalls and assumptions inherent in examining the fossil record, studies of the population dynamics of extinct mammals, including horses, allow the determination or interpretation of characteristics of particular species that are vital to an understanding of their paleobiology.

Evolution of body size, vegetation, habitat, and climate: implications for the population biology of fossil Equidae

In this section we shall review the 58-myr history of fossil Equidae in North America from a perspective of what can be learned about the evolution of behavioral ecology and social organization. Most of this discussion will focus on the area that encompasses the northern Great Plains and adjacent Rocky Mountain regions of western North America. This is done because many of the important samples of fossil horses come from that region, and many studies have been published on the paleoenvironments in which those extinct horses lived.

The early Eocene: Hyracotherium *and the equid archetype*

The classic interpretation of North American *Hyracotherium* is that it was the size of a fox terrier, fed on browse, and dwelled in forests. Given these parameters, it would be reasonable to interpret the behavioral and population ecology of the oldest North American horse as similar to that of small, extant, forest-dwelling artiodactyls that lack cranial appendages such as chevrotains (*Tragulus* and *Hyemoschus*) (Janis 1982). Klingel (1975:11) speculated that equids "evolved from small, forest-dwelling Eocene species of the genus *Hyracotherium*, who were browsers like the present-day duikers, and who may well have had the same type of social organization, i.e., territoriality of males and females." As we shall see later, however, recent work has shown that there is another possible scenario for primitive browsing horses during the early Tertiary. We therefore may need to reassess the archetypical behavioral ecology and social organization of primitive horses.

Although *Hyracotherium* is a commonly encountered fossil in early Eocene deposits in western North America, until recently we had no well-preserved quarry samples from which the population biology of this pivotal genus could be interpreted. That changed with the discovery of the large Castillo Pocket quarry sample of *Hyracotherium* from Colorado and its subsequent de-

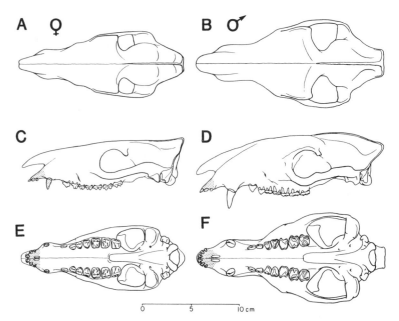

Figure 12.9. Dorsal, lateral, and ventral views of reconstructed crania of presumed female (A, C, E) and male (B, D, F) specimens of Hyracotherium tapirinum *from the early Eocene Castillo Pocket sample from Huerfano, Colorado, showing distinctive sexual dimorphism in overall dimensions and size. Stippling indicates areas not represented in the available sample. Modified from Gingerich (1981); reproduced with permission of the Paleontological Society.*

scription by Gingerich (1981) (see Chapter 4). From this sample, 24 individuals are attributable to the more common species *H. tapirinum* (although a smaller species *H. vasacciense*, also occurs there, its smaller sample size precludes analysis of its population biology).

Gingerich (1981) found marked sexual dimorphism in *H. tapirinum*. The canines were 40% larger, and cranial dimensions 15% larger, in the males than in the females (Figure 12.9). Although the population is represented by different wear (age) classes (P. D. Gingerich personal communication) and is believed to represent a catastrophic assemblage (Gingerich 1981), the population dynamics were not studied in this sample. Therefore, empirical data are not currently available to assess parameters such as longevity, mortality, and reproductive strategy (e.g., continuous or seasonal breeding). We can only speculate about them on the basis of data for other available parameters, as discussed later. Judging from the numbers of males and females in the quarry sample, Gingerich calculated a ratio of 1 male to 1.5–2.0 females. Although this socionomic ratio is much higher than those for entire populations of modern equids (nearer 1:1, although there is some variation) (e.g., Feist & McCullough 1975; Berger 1986), the Castillo Pocket sample may indicate something about smaller population subsets, such as band composition [also see Berger (1983a) for a cautionary note about interpreting fossil assemblages]. Perhaps it represents a catastrophic accumulation of harem(s) rather than solitary bachelors or bachelor bands.

Although earlier size estimates for *Hyracotherium* were about 10 kg (e.g., Radinsky 1978b), MacFadden (1987a) gives a mean estimate of 34.9 kg for *H. tapirinum*. Paleobotanical evidence for western North America indicates a time of change from widespread dense forested woodlands in a relatively wet climate during the Paleocene to a mosaic of more open forests and savannas during the Eocene. Although evidence indicates that grasslands as we know them today did not exist in the early Eocene in North America, the open savanna habitat probably consisted of a vegetative cover of herbaceous dicots, dry-adapted ferns, and shrubs. This latter biome spread during the early Eocene (Wasatchian), and there was a corresponding decline in the forested habitats (Hickey 1980; Wing 1980; Gingerich 1981).

Given the results from Castillo Pocket, the new interpretations of the behavioral ecology and social organization of *Hyracotherium* (and hence other primitive horses as well) are quite different from the traditional notions. Crown heights for mammals feeding on relatively nonabrasive vegetation (Hulbert 1984) as well as estimated body size (MacFadden 1987a), would suggest a potential longevity of about three or four years. As a rule, litter size for extant mammals larger than 10 kg is usually one offspring, with only rare occurrences of twins (Eisenberg 1981) [the Messel palaeothere seems to have had only one fetus (Franzen 1985)]. With drier, perhaps seasonal and patchier food resources, breeding probably was synchronized; this could be corroborated in the future if discrete wear classes are found in catastrophic quarry samples of *Hyracotherium*.

Gingerich (1981) speculated that the body-size dimorphism observed in *H. tapirinum* was related to intraspecific (sexual) selection that would increase the size of males, or else indicated different food-resource utilizations in males and females. Enlarged canines have been found in many early Tertiary fossil mammals, and given their small body sizes, these may have functioned as weapons during combat [in larger-bodied forms, other structures, e.g., horns, were also involved (Janis 1982)], although, like body size, these canines also could reflect defense or feeding specializations.

Gingerich (1981) speculated that the social structure of *H. tapirinum* was polygynous, with year-round harem bands. Given its body size, its daily home range undoubtedly was smaller than those of extant equids (McNab 1963), probably on the order of 1 km^2. With the possibility of patchy food-resource availability, in addition to the probable synchronous births, the foregoing evidence suggests that *H. tapirinum* underwent seasonal migrations.

After Gingerich's study (1981) of the Castillo Pocket *Hyracotherium*, Janis (1982) published an important synthesis on the evolution of social structure in ungulates, including horses. In that paper she maintained the traditional view that although there may have been more open country during the early Eocene, *Hyracotherium* inhabited the more forested biomes. Thus, the social structure of the earliest equids would have been different from that of the polygynous harems postulated by Gingerich and would have been more like that of small, forest-dwelling artiodactyls of today, which are solitary and maintain nonexclusive home ranges (Klingel 1975). Before we choose between these two hypotheses, we might consider that both could be correct: Perhaps the observed diversification of body size within contemporaneous species of *Hyracotherium* (see Chapter 10) indicated habitat preferences and different population social structures. If this interpretation is correct, then smaller forms such as *H. sandrae* may have been forest-adapted, whereas larger forms such as *H. tapirinum* may have inhabited more open forests or savannas. If forced to choose, currently it would be difficult for us to say which of these hypotheses has more support. At the very least we have learned that in view of plausible alternatives, the traditional view of *Hyracotherium* as a forest-dweller may be too simplistic.

The available paleoenvironmental evidence suggests that for the remainder of the Eocene, much of western North America had a climate similar to that of the early Eocene described earlier. This observation, along with the fact that body masses for the other Eocene genera *Orohippus* and *Epihippus* remained relatively static (MacFadden 1987a), suggests that the behavioral ecology and social organization also remained similar for primitive equids during that time. However, that would be in marked contrast to the greatly increased body sizes observed in Eocene palaeotheres from the Old World. Whereas it would be difficult to invoke a regional climate change in Europe during that time (i.e., different from that of North America), a possible explanation may be related to competition for food resources from other coexisting perissodactyls (Radinsky 1969). Another explanation might be that palaeotheres represent an example of the Jarman/Bell principle (Geist 1974), in which body size increases as a result of a change to a lower-digestibility, higher-fiber diet.

The Oligocene: a time of change

The late Eocene–early Oligocene transition was characterized by a major shift in global climate toward generally colder mean annual temperatures (by about 10°C), increased aridity, greater seasonal temperature fluctuations, and contraction of the tropical/temperature belts across the earth (see Chapter 7). Evidence of that transition is suggested by distinct changes in land-mammal faunas, including numerous Holarctic immigrations, extinctions, and origination events (Opdyke 1990). In North America, the two dominant Oligocene equid genera were *Mesohippus* and *Miohippus* (Prothero & Shubin 1989). Like Eocene forms, those genera had teeth with very low crowns and therefore usually had been interpreted as forest-dwelling browsers. However, that notion now seems oversimplified, because recent paleoecological reconstructions of the western interior of North America (e.g., Clark et al. 1967; Retallack 1983a,b) suggest a more complex environmental mosaic than has traditionally been thought to have existed at that time.

With body masses estimated between 42.2 and 53.8 kg (MacFadden 1987a), *Mesohippus* and *Miohippus* species were about 50% larger than most Eocene horses (although they were an order of magnitude smaller than extant equids). However, that change by itself probably was not enough to drastically alter their behavioral ecology. On the basis of data from extant mammals (e.g., Eisenberg 1981; Prothero & Jürgens 1987) it would appear that their important life-history strategies, such as potential longevity and reproductive parameters, probably remained similar to those of Eocene horses. Recent environmental reconstructions of the classic sequence in the Big Badlands of South Dakota indicate that during the Oligocene there was a complex habitat mosaic ranging from riparian forests to open-country savannas, as well as many intermediate types of biomes (Clark et al. 1967; Retallack 1983a,b). In these recent interpretations, *Mesohippus* usually is seen as having lived in gallery forests near rivers (Figure 12.10) and in widespread open woodlands, whereas *Miohippus* (not depicted in Figure 12.10) may have been adapted to more arid conditions, inhabiting areas ranging from savanna woodlands to grasslands (Retallack 1983a). It is possible that this habitat displacement provides at least a partial explanation for the recent interpretation that these two morphologically similar horses coexisted for about 5 myr (Prothero & Shubin 1989). With these possible environmental preferences, *Mesohippus* and *Miohippus* also may have had different social structures, with the former being relatively monomorphic (B. J. MacFadden unpublished data), monogamous, and pair-bonded, like extant small, forest-dwelling, browsing artiodactyls (Janis 1990a; category A1 in Table 12.1), and the latter being polygynous and migratory, like some hyracotheres (Gingerich 1981). If this scenario is correct, then it represents an early instance of diversification of behavioral ecology and social systems within coexisting equid genera; this pattern appears to have continued into the Miocene during the height of the adaptive radiation of horses.

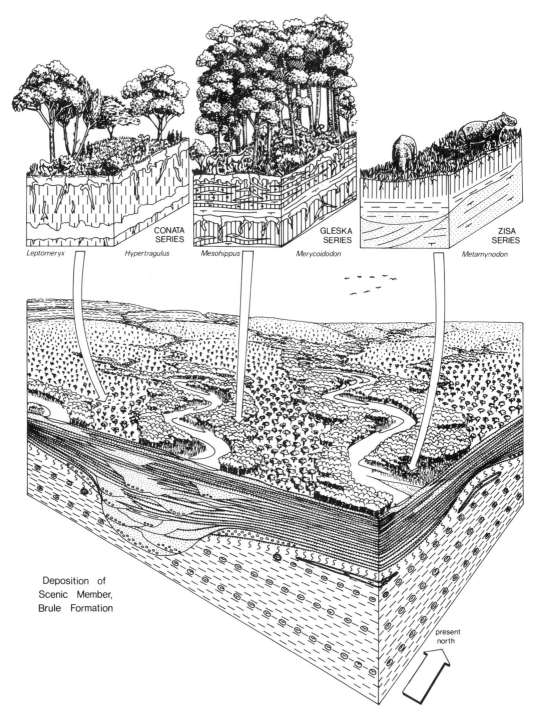

Figure 12.10. *Geological cross-section and environmental reconstruction of Oligocene time based on paleosols (represented by ticked lines) from Big Badlands National Park, South Dakota. Of* the three different paleosol types (Conata, Gleska, and Zisa series), the horse Mesohippus predominates in the one interpreted to represent a riparian, forested habitat. Leptomeryx, Hyper- *tragulus, and* Merycoidodon *are artiodactyls, and* Metamynodon *is a rhinocerotoid. From Retallack (1983a); reproduced with permission of the Geological Society of America and author.*

An interesting corollary to the foregoing scenario is related to the empirical observation that horses never developed cranial appendages (i.e., horns and antlers) as did other ungulates that surpassed a critical size (about 20 kg) (Janis 1982), most notably artiodactyls. Janis (1986) attributed this lack to dietary specializations in which, so far as can be interpreted, equids have characteristically specialized in high-fiber foodstuffs of low nutritive value. Thus, relative to artiodactyls, equids have occupied larger ranges (for their body size) within woodlands, or savanna-woodland mosaics, to acquire a larger volume of foodstuffs. Consequently, the practice of defending smaller, resource-rich territories, which is characteristic of many horned artiodactyls, did not develop in horses. Some modern equids are territorial (Type II), but they select discrete, defended territories in open areas (e.g., Klingel 1975; Janis 1982, 1990a). Janis (1986:17) concluded that "horses, like camels, probably never passed through an evolutionary stage where the right combination of diet, body size, and size of foraging area provided the threshold conditions for the development of territorial behavior, and the subsequent evolution of horns."

Miocene horses in the forests and savannas of North America

As we have already seen in other contexts, the Miocene was the time of the principal diversification of grazing horses in North America. Before that important evolutionary phase is discussed, however, the Miocene browsing horses will be examined. This latter group is often neglected, perhaps because of their relative rarity and because they are frequently overshadowed by the grazing clades that existed in many of the same communities during that time.

MIOCENE BROWSERS. There were at least two distinct clades of browsing horses during the Miocene in North America. The first is represented by the dwarf genus *Archaeohippus* and possibly by primitive, low-crowned parahippines (*Desmatippus*); the second is represented by the more advanced anchitheres, including *Kalobatippus* (= *Anchitherium*), *Hypohippus*, and *Megahippus*.

Archaeohippus is widely distributed in early Miocene deposits throughout much of North America, including California, Nebraska, and Florida. Although the exact habitat preference of this tiny horse has not been analyzed, its molar crown heights suggest that it either was a forest-dweller (perhaps like *Mesohippus*) or inhabited an open woodland/savanna mosaic (like *Miohippus*) (*fide* Retallack 1983a); it almost certainly was not adapted to the drier, open savannas. Given these parameters, and with a body size estimated at 43.9 kg (for *A. blackbergi* from Florida) (MacFadden 1987a), *Archaeohippus* was very similar to the ungulates in category A1 of Janis (1982; Jarman 1974) (Table 12.1). From modern analogues we can infer that *Archaeohippus* had a seasonally variable diet and was a folivore, but also ate other nonfibrous foods when available. It probably was monomorphic, or slightly dimorphic. Individuals either were solitary or formed monogamous pairs, with defended territories, and had small home ranges. Given its size and crown height, *Archaeohippus* probably had a potential longevity of four or five years. Although this scenario is indeed speculative, available samples (e.g., from Thomas Farm in Florida) should allow testing of these hypotheses in the future.

In North America the browsing anchitheres included the genera *Kalobatippus, Hypohippus,* and *Megahippus.* Despite the traditional notion that this clade represented evolutionary stasis, recent analysis suggests that for some characters, such as body size, they evolved rapidly and were highly specialized in their dentitions (MacFadden 1988b). There are numerous morphological characteristics that allow an interpretation of the behavioral ecology and social organization of these interesting horses. For the most part, anchitheres had low-crowned dentitions and relatively short limbs. Body masses are estimated to have ranged from 131.7 kg for *Kalobatippus* ("*Anchitherium*" *clarenci*) to 266.2 kg for *Megahippus* (*M. matthewi*) to 403.0 kg for *Hypohippus* ("large sp."), all values from MacFadden (1987a). Given those sizes, as well as their low-crowned dentitions and relatively short limbs, anchitheres probably inhabited forested or open woodland/savanna mosaic habitats. They probably occupied relatively large home ranges and had seasonally variable diets, including fruit and browse. They were relatively monomorphic and possibly lived in polygynous bands, with a male-dominance hierarchy. As such, these horses were similar to those in category B1 of Janis (1982) (Table 12.1). As in the case of *Archaeohippus*, a rigorous analysis of the population dynamics has never been presented for any North American anchithere sample. Unfortunately, it will be difficult to test these predictions, because anchithere horses are universally rare, and consequently the available sample sizes are small.

MIOCENE GRAZERS. *Parahippus* was a pivotal genus in horse evolution. Although primitive parahippines (*Desmatippus*) had low-crowned teeth, the advanced species of *Parahippus*, including *P. leonensis*, were higher-crowned and represented the closest outgroup for the explosive adaptive radiation of grazers that occurred during the Miocene, at about 17–18 Ma (MacFadden & Hulbert 1988; Hulbert & MacFadden 1991; MacFadden et al. 1991). Hulbert's study (1984) of Florida *P. leonensis* has given us a better understanding of the population dynamics of this important species.

Hulbert's sample (1984) was based on 89 individuals of *P. leonensis* from the middle Miocene Thomas Farm quarry located in northern Florida. *P. leonensis* is the dominant mammal occurring at that site and represents 70% of all of the macrovertebrate remains collected there. Hulbert believed this to be an attritional sample, possibly resulting from a steady accumulation of individuals over time (although not necessarily due to predators). Other evidence indicates a sinkhole/fluvial deposit, possibly with seasonal influx of streams. The climate is interpreted to have been subtropical to tropical, and the vegetation predominantly woodland, although savannas or woodland/savanna mosaics could have existed nearby (Pratt 1990).

Hulbert's multivariate analysis (1984) failed to reveal discrete cohorts, but statistical peaks indicated nine age classes. These were corroborated by qualitative differences in eruption and wear of the cheek teeth (Figure 12.11). The lack of statistically discrete cohorts suggests that either births or mortalities were not seasonal. In addition to canine dimorphism, bimodal distributions of some of the cheek-tooth dimensions suggested sexual dimorphism, although it was not pronounced. Survivorship analysis indicated that the average life span was three to four years, and the potential longevity was about nine years (Hulbert 1984). Using the same sample from Thomas Farm, MacFadden (1987a) estimated a mean body mass of 76.6 kg for *P. leonensis*.

As mentioned earlier, *P. leonensis* was transitional between low-crowned browsers and high-crowned grazers. Klingel (1975) speculated that these mixed-feeding *Parahippus* also were transitional in behavioral ecology and social structure. Using the modern analogue of *Equus grevyi*, Klingel (1975) stated that the ancestral condition in these Miocene horses was to have territorial males defending local resources (Type II = D2 in Table 12.1). The problem with that interpretation is that Type II social organization occurs in highly arid-adapted equids today (e.g., *Equus grevyi*); yet *Parahippus* and other transitional fossil horses inhabited biomes with considerably more rainfall. It would be difficult to tell whether *Parahippus* had one type of behavior rather than the other; the work of Janis (1982) indicates a transitional behavioral ecology. It is possible that *P. leonensis* and similar fossil horses fell into category C1 (Janis 1982) (Table 12.1), whose members fed selectively on plant parts and grass, were monomorphic, fed in mixed-sex herds, were polygynous, and lived either in small breeding territories or in harems (Figure 12.12). This scenario thus combines the essence of the extremes suggested by Klingel (1975) and also seen in extant equids.

Like *P. leonensis*, *Merychippus primus* was at the base of the adaptive radiation of hypsodont grazing horses (Hulbert &

Miocene *Parahippus leonensis*

A. Wear-class 2

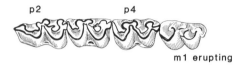

p2   p4

m1 erupting

B. Wear-class 5

C. Wear-class 7

1cm

Figure 12.11. Tooth eruption and wear stages in a fossil population, Parahippus leonensis, *from the middle Miocene Thomas Farm site in northern Florida. The wear classes 2, 5, 7, and 9 represented in A–D correspond to those identified by Hulbert (1984, Figs. 2 and 3). Specimens: A, UF 47780, reversed; B, UF 20008; C, UF-V 6451, reversed; D, UF-V 6437, reversed.*

D. Wear-class 9

p2   m3

Figure 12.12. Reconstruction of *Parahippus leonensis* based on the sample from the middle Miocene Thomas Farm site in northern Florida. This represents a harem band consisting of one stallion, three mares, and two juveniles that were foaled at different times within a long, nonsynchronized breeding season. The open habitat in the foreground is a relatively xeric pine flatwood with saw palmetto and wiregrass. This merges in the background with a slightly more mesic hammock, including deciduous trees such as live oaks (habitat types based on modern analogues) (Abrahamson & Hartnett 1990).

MacFadden 1991). We are thus fortunate to have the classic work on the population dynamics of this species from the large catastrophic assemblage preserved from the middle Miocene Thomson Quarry of western Nebraska (Van Valen 1964). Although this sample was discussed in Chapter 4 (on exceptional discoveries) and Chapter 8 (for its evolutionary significance), it is briefly summarized here within the context of what can be learned about the population ecology of this important species.

Van Valen analyzed a sample of about 150 individuals and from their lower dentitions was able to recognize discrete wear classes representing 11 ages from 0.1, 1.1, 2.1 up to >10.1 years. He estimated mortality to have been highest (ca. 30–40%) during the first year, dropping thereafter for intermediate age classes (ca. 10%), followed by a slight increase during old age. Median age expectancy was about eight years for those individuals that survived the first year of life. In order for them to have achieved a stable population age structure, given a hypothetical birthrate (and survivorship past infancy) of 0.5 offspring per year, Van Valen speculated that there was an average fecundity of about three or more progeny per female during her lifetime. Potential

Figure 12.12. Caption on facing page.

longevity was estimated to have been about 11 years. There was no discernible sexual dimorphism in tooth dimensions, except in canine size (B. J. MacFadden unpublished data). The mean body mass for *M. primus* was estimated to have been 71.0 kg (MacFadden 1987a).

Van Valen (1964) also analyzed a sample of *Pseudhipparion* (= *Griphippus*) *gratum* from Burge quarry, Nebraska (Figure 12.8A). *Pseudhipparion* is important because it was a representative of one of the major clades of hypsodont horses within the tribe Hipparionini (Webb & Hulbert 1986; Hulbert & MacFadden 1991) (see Figure 5.15). Relative to *M. primus*, wear classes were less discrete in *P. gratum* [for a different interpretation, see Webb (1969a)]. Thus, he interpreted this sample to represent an attritional assemblage in which there had been no seasonal peaks in mortality. Another possibility, however, is that breeding likewise may not have been seasonal. Mortality was estimated at 30% for the first years, followed by 15–20% in subsequent years; these values are roughly similar to those for *M. primus*.

Voorhies (1969) studied a sample of *Protohippus* cf. *perditus* from the late Miocene Verdigre quarry of eastern Nebraska. This species is important because it is a primitive representative of the tribe Equini (Hulbert 1988a; Hulbert & MacFadden 1991) (see Figure 5.15), the other major clade of hypsodont horses during the Miocene, and it is closely related (relative to the other species discussed here) to extant *Equus*. Voorhies's study was based on a minimum of 57 individuals of *P.* cf. *perditus* that were interpreted as a nonselective catastrophic assemblage, and therefore, assuming that sampling biases are known, it represents an accurate picture of the living herd prior to death. Five discrete age classes were identified and interpreted to represent individuals of −0.2 year [i.e., unborn foals; as discussed by Voorhies (1969), because of poor fossilization this cohort is not shown in Figure 12.8B] to 3.8 years. Older age classes were less distinct, probably because of a combination of wear variation and smaller sample sizes, but seemed to represent six cohorts from 4.8 to 9.8 years (Figure 12.8B). Given these data, plus other evidence from the site, Voorhies believed that this assemblage accumulated rapidly during a winter, possibly as a result of a storm, about two months before normal breeding season for *P.* cf. *perditus*. The relative proportions of age classes of the herd were similar to those for *M. primus* (Van Valen 1964); there had been high mortality among the juvenile and "senior citizen" (to use Voorhies's term) age cohorts, and lower mortality for the intermediate age classes. Based on unworn crown heights, and assuming wear rates of about 3 mm/year, the potential longevity of *P.* cf. *perditus* was estimated to have been 12–15 years. This represents an increase over *Parahippus* and *Merychippus*, but is significantly less than that of *Equus*.

Although body mass has not been estimated for *P. perditus*, on the basis of similar forms listed by MacFadden (1987a) (e.g., *P. simus*) it would seem to have been about 150 kg. Voorhies (1969) reconstructed the habitat of the Verdigre locality as a woodland/grassland mosaic. The relatively high-crowned dentition of *P. perditus* suggests that it was specialized for grazing on a high-fiber diet consisting predominantly of grass. Paleobotanical evidence and the presence of cold-intolerant vertebrates, such as crocodiles and land tortoises, indicate a generally frost-free climate.

Judging from these examples it would seem that by the middle Miocene, at 15–16 Ma, hypsodont grazing horses in North America had evolved a behavioral ecology and social organization similar to those seen in modern equids. It would be difficult to say, however, whether particular clades of extinct grazing horses would fit into one or the other behavioral category (Table 12.1), such as D1 (= Type I), forming harems, or D2 (= Type II), defending territories. There have been suggestions that, for example, Miocene hipparions of the Great Basin were better adapted to moist, forested climes, whereas equines were adapted to drier, more open habitats (Shotwell 1961) (see Chapter 7 and Figure 7.2). However, those may have been localized patterns not necessarily applicable to all members of these two clades. If hypsodonty is an indicator of feeding specialization, and given the observation that contemporaneous hipparions and equines had roughly equivalent crown heights, then so far as can be told from the fossil record there was no discernible character displacement in resource utilization.

It therefore seems that the development of widespread grasslands during the Miocene was a principal causal factor in the evolution of behavioral ecology and social organization of advanced horses. According to the Jarman/Bell principle (Geist 1974), the rapid diversification of body size during the Miocene (MacFadden 1987a) was related to exploitation of high-fiber, low-digestibility foodstuffs (i.e., predominantly grasses). As body size increased, so did the home range, because equid bands had to range widely to exploit locally patchy resources, or else make long migrations to exploit seasonally variable food resources. In addition to facilitating predator evasion, a high degree of cursoriality allowed for highly mobile herds. This adaptive strategy would have favored the polygynous-harem/bachelor-band type of social organization, although not to the exclusion of some species exhibiting the pattern of male-defended territories, as seen in some extent equids. Under this scenario, Klingel (1975) may have been correct that Type II organization was primitive, and Type I evolved secondarily, although this may be too simplistic a scenario. With as many as 12–15 potentially coexisting equid species during the Miocene, intraspecific recognition was important, and in addition to other sensory mechanisms, ritualized display and vocalization probably were very important in communication. As discussed in Chapter 5, the complex array of facial fossae observed during the Miocene probably was related to different facial displays and possibly vocalizations. Therefore, when equid diversity dropped during the late Miocene, the facial fossae became reduced or lost.

Evolution of selected life-history strategies from *Hyracotherium* to *Equus*: a summary

Potential longevity

Prothero and Jürgens (1987) analyzed maximum life spans for a variety of extant mammals with different adaptations and inhabiting different environments. They found that life span was highly correlated to characters such as brain size and, of relevance here, body mass. Thus, in general, larger-bodied mammals live longer. There are exceptions to this; for example, bats live longer than would be predicted from their body size. However, ungulates [Prothero & Jürgens (1987) analyzed artiodactyls] and other medium-sized mammals have a more predictable scaling relationship between body size and longevity. To my knowledge, no previous study has analyzed the evolution of longevity for a fossil group; this will be done here for the Equidae.

Van Valen (1964) and Hulbert (1984) described potential longevity as the mean age at which individuals within a population will die because their teeth have worn down to the roots, rendering them incapable of feeding. Among horses, the rates of wear on browse versus grass can be determined, and therefore if maximum unworn crown heights can be measured, then potential life span (under optimal circumstances) can be estimated. As mentioned earlier, this can be corroborated by analyzing the population dynamics of fossil horse species to determine how many cohorts are represented in a given quarry sample.

Figure 12.13 summarizes the data on estimated potential longevity for selected fossil horses. This plot shows that over the 58-myr history of the Equidae there has been a five-fold increase in longevity, from about 3–4 years in *Hyracotherium* to about 20–25 years in *Equus*. However, these extremes do not tell the interesting part of the story. Rather than having experienced a gradual increase in longevity, 75% of the increase has occurred over the past 15–20 myr. This is not at all surprising, because longevity scales with body size; larger equid species should therefore have the potential to live longer. The general form of this graph is nearly identical with that of body-mass evolution (see Figure 10.6) (MacFadden 1987a). Thus, the interpretation here is that for the first 38 myr there was stasis in potential longevity, followed by a rapid change after 20 Ma, mostly in the direction of an increase.

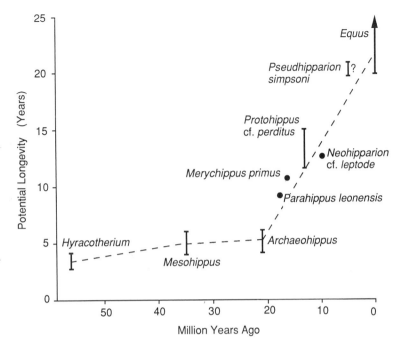

Figure 12.13. *Evolution of potential longevity in fossil horses using selected taxa. Sources: Hulbert (1984) for Parahippus leonensis; Van Valen (1963, 1964) for* Merychippus primus; *Voorhies (1969) for* Protohippus cf. perditus; *Hulbert (1982) for* Neohipparion cf. leptode; *for other taxa, see discussion in text.*

What happened to potential life span in dwarfing lineages such as *Pseudhipparion?* Is potential life span genetically fixed within clades, such that once it increases it is not likely to decrease? Or is it plastic enough to revert back to a former state? These are interesting questions yet to be answered, although shortly I shall suggest that later, tiny species of *Pseudhipparion* increased their reproductive potential by maintaining the potential longevity of their larger-bodied ancestors.

Reproductive potential/fecundity

One aspect of reproductive strategy and population age structure in animals is the socionomic ratio (i.e., the ratio of males to females). Charles Darwin was interested in this subject, and after studying 25,560 domestic foals he concluded that parity was the norm (Berger 1986). In general, however, it is thought that equids produce slightly, but not significantly, more male offspring than females (Berger 1986). Despite equality, or near equality, at birth (this is still controversial), in many mammalian species male mortality increases with age, usually because of factors such as intraspecific competition or greater nutritional stress (Berger 1986). Thus, by maturity there usually are more females than males in a given population. Berger (1983a, 1986) found a ratio of 1.0/1.5 adult males/females in his study of wild

Equus caballus from the Granite Mountains of Nevada; similar values have been found for other North American feral horses (Berger 1986) (Figure 12.14). On the other hand, Spinage (1972) found a 1.0/1.0 sex ratio in extant African *Equus burchelli*, as did Feist and McCullough (1975) for feral *Equus caballus* from Montana and Wyoming.

In the study of *Hyracotherium tapirinum* from Colorado, Gingerich (1981) found a sex ratio of 1 male to 1.5–2.0 females. This seems slightly higher than for an entire population of polygynous ungulates and, with a sample size of 24, may represent an accumulation of several harems and probably no bachelor bands.

Pirlot (1952) was one of the first to seriously consider socionomic ratios in fossil horses. He found a ratio of nine males to seven females in a sample from Pikermi, Greece. This slight difference may represent small-sample error and suggests a value near unity. In a review of North American hipparions, MacFadden (1984a) presented some data on fossil quarry populations that are relevant to the present discussion. One of the samples consisted of *Hipparion tehonense* from the middle Miocene (Clarendonian) MacAdams quarry in Texas. Although the population dynamics were not analyzed, general tooth-wear classes (Figure 12.15) and, when possible, sexes of 40 individuals were determined. Of relevance here, the roughly equal numbers of males and females in all four wear classes suggest that the socionomic ratio was about 1.0. In another sample of *Cormohipparion occidentale* from the Clarendonian Hans Johnson quarry in north-central Nebraska, MacFadden (1984a) also found a socionomic ratio of 1.0 (5 males/5 females) (Figure 5.17) for the individuals within the quarry for which sex could be determined, although admittedly this is a small sample size.

With the exception of Gingerich's findings (1981) for *Hyracotherium* (which may represent harem bands), the other data, although admittedly scanty, suggest socionomic ratios near unity for entire populations of fossil horses. If there was higher male mortality, which would have tended to skew the sex ratios, it could not be discerned from the available fossil samples. Because of the problems of determining the sexes of individuals, particularly in relatively monomorphic species such as horses, studies of socionomic ratios will continue to be difficult to accomplish with fossil populations.

"Fecundity" refers to female reproductive success. Although there are various definitions that could be offered, for our purposes this refers to how many offspring an individual female produces during her lifetime. Many studies have shown that smaller mammals have more litters, with greater numbers of littermates, and along with this there usually is less parental care and higher infant mortality. However, there are so many other

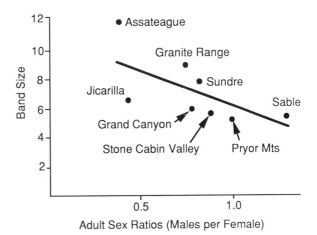

Figure 12.14. *Adult sex ratios and mean band sizes for eight populations of North American wild horses. Modified from Berger (1986).*

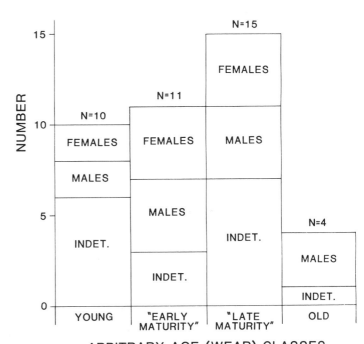

Figure 12.15. *Histogram of males and females (determined from relative canine size) for a sample of* Hipparion tehonense *from the late Miocene (Clarendonian) MacAdams quarry in the Texas panhandle. The four arbitrary age classes are separated on the basis of relative amount of cheek-tooth wear. From MacFadden (1984a); reproduced with permission of AMNH.*

factors that affect fecundity that it is very difficult to juggle all of the variables.

For fossil horses, some factors that are known to affect fecundity include body size and developmental timing of reproductive maturity, and these factors are interrelated. If reproduction was not synchronized into breeding peaks, then some primitive, smaller-bodied forms (with shorter gestation times), such as *Hyracotherium*, would have had the potential to produce more than one offspring per year under optimal circumstances. [In modern populations, annual foal production varies widely, between about 50% and 95% of all reproductively capable females in the population (Van Valen 1964; Berger 1983b, 1986).] We can never know for sure how many foals an individual female *Hyracotherium* produced in her lifetime. For Miocene *Merychippus primus*, Van Valen (1964) speculated that in order to maintain a stable population age structure, each female would have had to give birth to about three foals during her reproductive years, a value less than half that for extant *Equus* (Berger 1986).

As mentioned earlier, fecundity in dwarfing lineages poses an interesting question. Did these phyletic dwarfs experience evolutionary reversals in longevity, thus decreasing fecundity (within a reproductive regime of seasonally synchronized, annual breeding)? Although each of the many clades of extinct dwarf horses had its individual differences, an interesting observation can be made for *Pseudhipparion*. During the late Miocene the advanced species *Pseudhipparion simpsoni* became very small, with a body mass estimated at about 50 kg (MacFadden 1987a) (see Chapter 10), or the size of the extant pronghorn *Antilocapra americana* (which ranges between 36 and 60 kg) (Walker 1975). Surprisingly, and unpredictably, at the same time the cheek teeth of *P. simpsoni* became incipiently hypselodont (partially ever-growing), with effective upper-molar crown heights of about 100 mm (Webb & Hulbert 1986). Even if *P. simpsoni* was highly specialized to feed on the most abrasive grass, it is unlikely that its rate of cheek-tooth wear was much greater than 5 mm/yr, because most grazing equids typically have annual cheek-tooth wear of 3.3 to 4.2 mm/yr (Hulbert 1984). Even with high wear rates of 5 mm, a crown height of 100 mm would yield a minimum estimate of about 20 years for the potential longevity of *P. simpsoni* (Figure 12.13). This estimate is undoubtedly greater than that for any other fossil horse, and it approaches that for extant *Equus*, despite the fact that the former had an order of magnitude less body mass. That increased longevity could have resulted in greater reproductive success for any given female during her lifetime. If adequate quarry samples ever become available for *P. simpsoni* or related forms, this hypothesis can be tested by cohort analysis of the ancient population.

Evolution of sexual dimorphism and sexual selection

And this leads me to say a few words on what I call Sexual Selection.
This depends, not on a struggle for existence, but on a struggle between
the males for possession of the females; the result is not death to the

unsuccessful competitor, but few or no offspring. . . . Generally, the most vigorous males, those which are best fitted for their places in nature, will leave the most progeny. But in many cases, victory will depend not on general vigour, but on having special weapons, confined to the male sex.   – Darwin (1859:88)

Although many types of polygynous mammals exhibit body-size dimorphism, extant equids are notably monomorphic in this character. In contrast, males of early Eocene *Hyracotherium* were significantly (about 15%) larger in cranial dimensions than were the females (Gingerich 1981). Similarly, body-size dimorphism (represented by dental dimensions) appears to have existed, albeit slightly, in middle Miocene *Parahippus leonensis* (Hulbert 1984). Thereafter, body-size dimorphism is not discernible for hypsodont grazing horses (MacFadden 1984a); this condition has continued for the past 15 myr and is characteristic of extant *Equus*.

Gingerich (1981) speculated that certain sexually dimorphic characters found in *Hyracotherium tapirinum* were related to different patterns of food-resource utilization by males and females within that primitive equid species. Although this has also been observed in extant mammals, such as the red deer, *Cervus elaphus* (Clutton-Brock, Guinness, & Albon 1982), sexual differentiation in use of food resources has otherwise not been inferred for most fossil mammals, including horses. An exception to this may come from a population of *Cormohipparion occidentale* from the late Miocene of Nebraska (MacFadden 1984a) (Figure 12.16). In this fossil sample the males tend to have broader muzzles and relatively more complex enamel plications (at similar ontogenetic stages) than the females. On the basis of functional analogues in extant ungulates (Janis & Ehrhardt 1988), this observation would suggest different feeding patterns in which the males were more generalized (broader muzzle) and ate more abrasive foodstuffs (more complex enamel is more durable) than the females. A similar pattern has been observed in extant ungulates and is related to the need for more specialized, higher-quality forage for pregnant and lactating females (Clutton-Brock et al. 1982; Van Soest 1982).

As noted earlier, Darwin was one of the first to observe that in species with intense male competition for mates, there is a tendency and selective advantage over time to evolve relatively larger structures for display or combat. The reasoning is that a particular individual with larger or more effective display/fighting structures will increase his possibilities of mating. This concept has been termed "sexual selection" and has received widespread attention in the literature (e.g., Stanley 1979).

I shall discuss here a possible example of sexual selection for

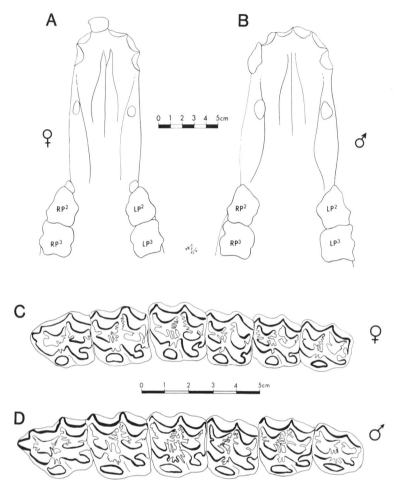

Figure 12.16. Diagrammatic ventral views of muzzles, anterior parts of maxilla, and dentitions of female (A, C) and male (B, D) of Cormohipparion occidentale *showing sexual dimorphism (as determined by relative canine size) and possible feeding specializations between the sexes. Specimens are from Hans Johnson quarry, late Miocene (Clarendonian) of northern Nebraska.*

larger canines in the males among extinct horses. The prediction is that if body size can be factored out, then the males will have a greater positive canine allometry than the females. Samples of 13 species of fossil and extant horses were analyzed, with sufficient numbers of both males and females. From these, canine and molar dimensions were measured, the latter of which are known to be highly correlated to body size (MacFadden 1987a). Over time, the ratio canine area/molar area decreased in females (Figure 12.17A). In contrast, there was a significant increase in the ratio canine area/molar area in the males (Figure 12.17B). These results suggest that the canine has been important in male intraspecific competition for display and biting, as also previously noted by Trumler (1959) and Feist and McCullough (1976). Thus, larger relative canine size in male horses has been selected for through time.

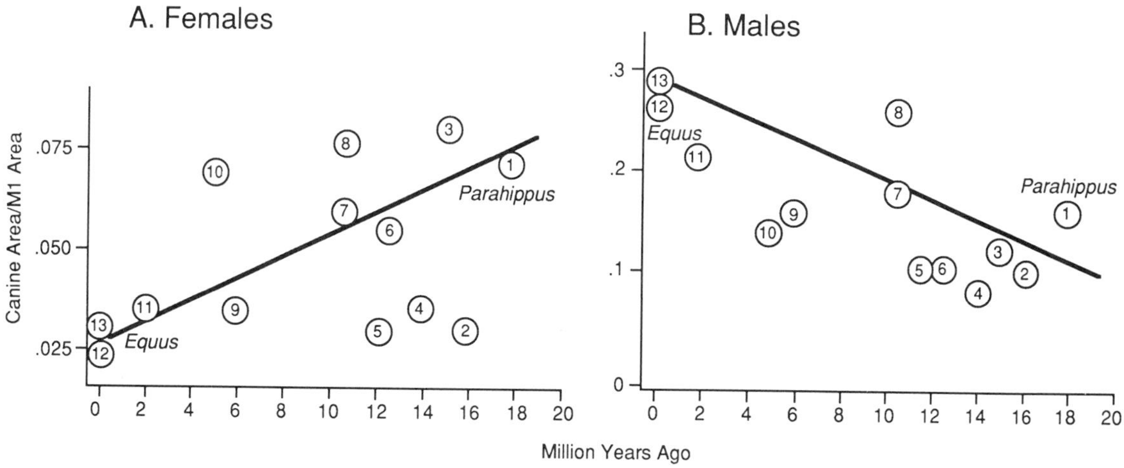

Figure 12.17. Example of sexual selection in Neogene horses. Upper canine and M1 dimensions were measured in 11 quarry samples of fossil horses from North America and two modern populations of Equus from Africa. The M1 area was used to factor out body-size differences, because previous work had shown that this character is highly correlated to body size (MacFadden 1987a). The ratios of the upper canine area to M1 area were plotted for females (A) and males (B). Relative to the females, the slope for the males indicates that canine size has become larger, and a plausible explanation for this is sexual selection. Numbers refer to species and locality (and collection, acronyms in Appendix I) studied: 1, Parahippus leonensis *from Florida* (UF); 2, Merychippus primus *from Nebraska (AMNH, F:AM); 3,* Merychippus insignis *from Nebraska (F:AM); 4,* Megahippus mckennai *from California (F:AM); 5,* Protohippus supremus *from Nebraska* (F:AM); 6, Pliohippus pernix *from Nebraska (F:AM); 7,* Hipparion tehonense *from Texas (F:AM); 8,* Cormohipparion occidentale *from Nebraska (F:AM); 9,* Dinohippus leidyanus *from Oklahoma (F:AM); 10,* Onohippidium galushai *from Arizona (F:AM); 11,* Equus simplicidens *from Idaho (AMNH, USNM); 12,* Equus burchelli *from Kenya (USNM-M); 13,* Equus grevyi *from Kenya (USNM-M). Source: MacFadden (original data).*

Concluding remarks: Does behavior fossilize?

At first glance one might lament that it is indeed difficult to reconstruct the autecology of fossil species, and it is impossible to then infer aspects of their behavioral ecology and social organization, because obviously these characters do not fossilize. On the contrary, by studying certain morphological characters (such as body size and sexual dimorphism), as well as the associated vegetation, habitat, and climate of the ancient terrestrial communities, and by referring to modern analogues, the "paleoethology" of fossil mammal species can be fairly well constrained. Furthermore, these kinds of studies should prove to be even more interesting as other important, well-preserved quarry samples of ancient equid populations are collected and studied in the future.

13

Fifty-eight million years of community evolution

Modern ecologists are keenly interested in understanding the structures and patterns within modern ecosystems and communities. In Chapter 12 we saw how the population biology of extinct equid species ("paleoautecology") can be interpreted from the fossil record. In this chapter we attempt to reconstruct the communities ("paleosynecology") in which extinct horses lived and trace the evolution of those biotic systems through time. Studies such as these make the contributions of paleontology unique, because although modern ecologists have the potential to study short-term (i.e., usually over days to years) ecological change within communities, their goals and data bases do not lead to an understanding of synecological patterns that may have occurred over millions of years. That is the domain of the paleobiologist.

Chronofaunas and paleoguilds

Chronofaunas

In a pioneering study of the paleoecology of the terrestrial/aquatic faunal succession from the classic Permian redbed sequence of north-central Texas, Olson (1952, 1966) set the stage for subsequent research in community paleoecology. He coined the now well-known term "chronofauna" (1952:181): "a geographically restricted, natural assemblage of interacting animal populations that has maintained its basic structure over a geologically significant period of time."

Olson (1966) identified three types of continental communities that are differentiated by the relative mix of aquatic versus terrestrial plants and invertebrate versus vertebrate consumers (Figure 13.1). Of relevance here, his dominantly terrestrial Type II (Figure 13.1B) is similar to the Cenozoic communities in which horses lived.

All of Olson's communities are characterized by great antiquity; for example, Type II first occurs during the Permian, with

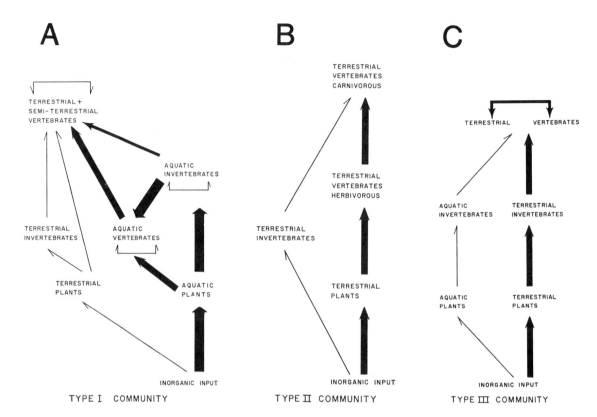

Figure 13.1. *Olson's classification of three kinds of terrestrial communities, based on examples from the classic Permian localities throughout the world. (A) Type I, with principal component of food chain derived from aquatic plants and animals consumed principally by terrestrial vertebrates and, to a lesser extent, terrestrial invertebrates. (B) Type II, dominantly terrestrial producers and consumers, with relative independence from aquatic food ties. Cenozoic terrestrial communities containing fossil horses generally fit into this category. (C) Type III, terrestrial, but with a larger component of invertebrate consumers than in Types I and II. In Type III communities, terrestrial vertebrate herbivores are relatively rare. Modified from Olson (1966); reproduced with permission of the Ecological Society of America.*

therapsids, and continues until the present day, where it is widespread and is dominated by mammals (Olson 1966). Thus, the general community structure in which horses evolved has a longevity of over 200 myr. Another characteristic of the earliest terrestrial communities is that they arose several times independently of one another in different regions, including the Permian in North America, Eurasia, and South Africa. An important point here is that even though the taxonomic compositions of the families, genera, and species of these synapsid/therapsid communities are quite different, there are many ecological similarities, most notably the trophic specializations and the relative numbers of herbivores versus carnivores seen in them.

A widespread characteristic of chronofaunas is their remarkable stability over long periods of time. Ecological turnover usu-

ally is represented by lower-level (species, genera) phyletic replacements of adaptively similar types. Profound changes in
community structure usually follow major climatic fluctuations,
which often result in major vegetative changes at the base of the
food chain, and the primary consumers adapt accordingly. Although the "big picture" of large-scale community changes is
interesting, equally interesting are the patterns of lower-level
replacements characteristic of all chronofaunas. Olson predominantly discussed reptilian community evolution in his formulation of the chronofaunal concept. As will be discussed later,
Webb (1969a,b, 1974, 1977, 1978, 1983, 1984, 1987) has done
much to enhance our understanding of the widespread Cenozoic
terrestrial (Type II) chronofaunas dominated by placental mammals. Of relevance here, Webb (1969a) based his original example of a mammalian chronofauna on the rich middle Miocene
sequence from the midcontinental United States (Figure 13.2)
that contains the maximum diversity of extinct horses. He
termed that interval, which extended from the Barstovian to
early Hemphillian, the "Clarendonian chronofauna."

Guilds and paleoguilds

Recent work has allowed a better understanding of the structures
and patterns at various levels within the food chain, for example,
the study of carnivore guilds in Cenozoic mammalian chronofaunas. "Guild" is a term used in modern ecology to describe a
group of species that exploit a given class of resources in similar
ways (Root 1967). The adaptive scope and taxonomic composition of a guild can range from being as specialized as nectar-
feeding sunbirds to being as broad as large terrestrial predators,
the latter being defined as any mammal within a community
that weighs over 7 kg and preys on other vertebrates (Van Valkenburgh 1982).

There are certain characteristics, or patterns, common to
guilds regardless of their antiquity. For example, all guilds of
large terrestrial predators have had a variety of dietary types,
ranging from hypercarnivorous catlike mammals, specializing
almost exclusively on meat, to omnivores, with a wider variety
of dietary preferences ranging from meat to high-fiber foods.
The relative proportions of the different dietary adaptive types
may vary (Figure 13.3) depending on the history of the predators
and climatic regime. In addition to feeding preferences, specializations within the predatory guild can include locomotory
adaptations, where individual species may be fossorial (e.g.,
badger), arboreal (e.g., binturong), or highly cursorial (e.g., fe-
lids and canids). Thus, feeding and locomotory adaptations are
of paramount importance in understanding the positions of particular species within guilds. Fortunately for the paleobiologist,

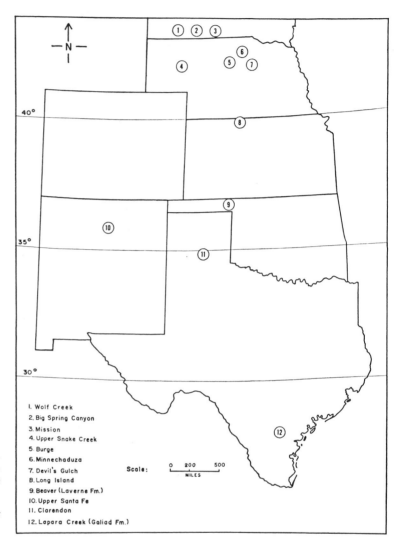

Figure 13.2. *The major localities producing local faunas in the Great Plains and Gulf Coast regions on which Webb based his concept of the Miocene Clarendonian chronofauna. From Webb (1969a); reproduced with permission of The University of California Press.*

1. Wolf Creek
2. Big Spring Canyon
3. Mission
4. Upper Snake Creek
5. Burge
6. Minnechaduza
7. Devil's Gulch
8. Long Island
9. Beaver (Laverne Fm.)
10. Upper Santa Fe
11. Clarendon
12. Lapara Creek (Goliad Fm.)

Scale: 0 200 500
MILES

these functional systems frequently are preserved in fossil species, thus allowing accurate reconstructions of "paleoguilds" (Van Valkenburgh 1982, 1985, 1988).

From a study of paleoguilds we can better understand the long-term relative stability and changes in taxonomic composition through extinction and, when it occurs, replacement. For example, Van Valkenburgh (1982, 1985) found that during the Oligocene in North America the predator guild had a stable taxonomic composition for at least 3 myr. Later, the "archaic" predators (creodonts) were ecologically replaced by functionally analogous "true" carnivores, and during the late Miocene the

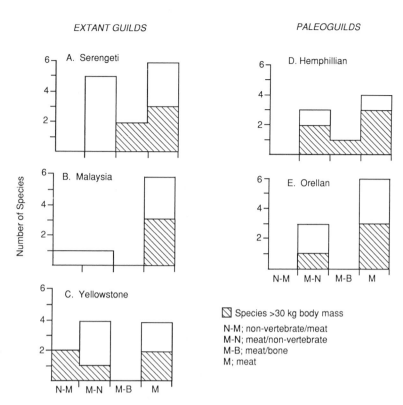

EXTANT GUILDS

A. Serengeti

B. Malaysia

C. Yellowstone

PALEOGUILDS

D. Hemphillian

E. Orellan

Number of Species

N-M   M-N   M-B   M

Species >30 kg body mass
N-M; non-vertebrate/meat
M-N; meat/non-vertebrate
M-B; meat/bone
M; meat

Figure 13.3. *Compositions of extant (A, B, C) and extinct (D, E) guilds of large, predatory mammals. Modified from Van Valkenburgh (1988). Categories are as follows: N-M, predominantly fruit and/or insects, with less than 50% meat; M-N, 50–70% meat, with remainder fruit and/or insects; M-B, more than 70% meat, with remainder bones; M, more than 70% meat in diet.*

diversity within this guild was destroyed by extinctions, without replacements.

Furthermore, from studies of guilds of large terrestrial predators we gain a better understanding of the associated prey species, notably the ungulates, and for the Cenozoic of Holarctica, horses. Van Valkenburgh (1988, 1989) noted that predator diversity was higher in communities with increased herbivore biomass and diversity. Although studies of ungulate paleoguilds perhaps have not been as refined as Van Valkenburgh's pioneering research on carnivores, several recent studies have provided important insights in this general field of community paleoecology. Most notable in this regard is the work on incisor and muzzle morphology and feeding specializations in ungulates (Owen-Smith 1985; Janis & Ehrhardt 1988). In conjunction with knowledge of locomotory adaptations and body-size distribution, studies of ungulate paleoguilds have the potential to yield significant insight into ungulate paleocommunity dynamics and associated chronofaunal evolution.

Although some aspects of guilds can be studied from both ancient and modern examples, some of their characteristics can best be understood only by turning to the fossil record. For example, one of the most profound mass extinctions in the history of terrestrial vertebrates occurred during the late Pleistocene, about 10,000–12,000 years ago, when large-mammal diversity was devastated on most continents. As a result of that event, the low diversity in some modern guilds and communities can be understood only with the aid of the historical component derived from the fossil record.

Cenozoic terrestrial communities in North America and fossil horses

The Cenozoic record of terrestrial vertebrates, dominated by therian mammals, is more extensively documented than any other part of vertebrate history.   – Webb (1987:445)

The evolution and adaptations of mammalian clades can best be understood in a context of corresponding knowledge of the physical (climatic) and biotic factors that existed in the ecosystem and how those changed through geological time. As Webb noted, there is an extensive record of Cenozoic terrestrial communities from which the paleoecology of extinct horses can be elucidated.

Horses existed continuously in North America for about 58 myr, from the early Eocene to the late Pleistocene. Yet the biotic factors that affected the terrestrial communities in which horses originated extended back into the Paleocene, a time when the relative proportions of different biomes were quite different from those seen today. Most notably, the global climate was significantly warmer and more equable, with expanded tropical and subtropical belts. In North America the dominant biomes included subtropical cypress swamps and multistoried, broadleaved forests. Studies of functional morphology indicate that there were many clades of Paleocene mammals with arboreal or forest-dwelling adaptations. During that time the large-bodied herbivores were relatively rare, were primarily browsers (although perhaps some groups may have been grazers), and belonged to the orders Condylarthra, Taeniodonta, Pantodonta, and Dinocerata (Webb 1987). The closest ecological/adaptive analogues of primitive horses probably were the advanced phenacodontid condylarths, such as *Phenacodus*, a clade that also is considered, in one view, to be the closest outgroup of perissodactyls (see Chapter 5).

The earliest horse, *Hyracotherium*, first appeared during the early Eocene in North America and Europe. That time was marked by a significant climate change toward more arid conditions and a corresponding decrease in the extensive forested biomes of the Paleocene (Hickey 1980; Gingerich 1981). Although the first record of grasses (Gramineae) in North America occurs

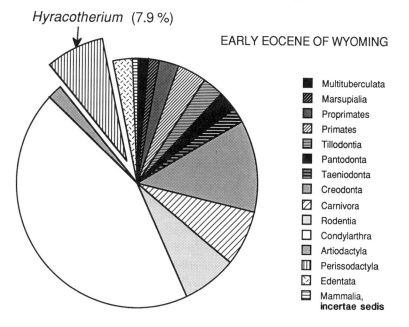

Hyracotherium (7.9 %)

EARLY EOCENE OF WYOMING

■ Multituberculata
▨ Marsupialia
▦ Proprimates
▨ Primates
▤ Tillodontia
■ Pantodonta
▤ Taeniodonta
▨ Creodonta
▨ Carnivora
□ Rodentia
□ Condylarthra
▨ Artiodactyla
▥ Perissodactyla
▨ Edentata
▤ Mammalia, **incertae sedis**

Figure 13.4. *Pie diagram showing relative abundances of all mammalian species in the early Eocene Wasatchian (Wa$_0$) fauna of Wyoming. Data from Gingerich (1989). As shown here, Hyracotherium, which is the only perissodactyl in this fauna, represents a relatively small (7.9%) portion of the assemblage, in contrast to, for example, Condylartha (44.1%).*

in Eocene deposits (MacGinitie 1969), the grassland biome was not extensively developed there until the middle Cenozoic. Paleobotanical evidence indicates that the more open regions in North America probably consisted of a ground cover of arid-adapted ferns, herbaceous dicots, and low shrubs, forming a woodland/savanna mosaic (Hickey 1980; Gingerich 1981). Coexisting with archaic clades during the early Eocene, several important orders of modern mammals originated, including both the Perissodactyla and Artiodactyla. Of the medium-sized herbivores during the early Eocene (Wasatchian), condylarths such as *Ectocion* and the tapiroid perissodactyl *Homogalax* probably were forest-dwellers, whereas *Hyracotherium* may have had habitat preferences ranging from forests to the more open woodland/savanna mosaic (Gingerich 1981; Janis 1982). As the latter biome increased (relative to the Paleocene), *Hyracotherium* increased as an important member of the Eocene terrestrial communities (Gingerich 1981). In most cases *Hyracotherium* was of average abundance relative to most of the other coexisting mammals (Figure 13.4). However, that changed, so that by the middle Cenozoic the relative abundance of horses in North American mammalian faunas increased dramatically.

Although the equids *Orohippus* and *Epihippus* are found in later Eocene (post-Wasatchian) deposits in North America, they also are characteristically small components of their constituent faunas. Furthermore, with the exception of increasing molarization of the premolars, they seem to have been adaptively similar to *Hyracotherium*.

The "terminal Eocene event," or "Grand Coupure," resulted in many changes in the climate, floras, and mammalian faunas of the Northern Hemisphere (see Chapter 7). In North America, significant faunal turnover occurred in the terrestrial communities, which resulted in decreased diversity or extinctions. These events affected several archaic orders of mammals, including condylarths, multituberculates, primitive primates, and uintatheres that had been dominant members of earlier Tertiary terrestrial communities. Correspondingly, during the later Eocene and into the Oligocene there was a relatively rapid diversification of several modern orders, including Rodentia, Perissodactyla, and Artiodactyla (see Figure 8.10). Within the perissodactyls, horses retained a relatively stable diversity, usually consisting of two to six contemporaneous species at any given time.

The classic Big Badlands sequence of present-day South Dakota provides one of the richest records of Oligocene land-mammal communities, including horses. Although classically thought to represent an orthogenetic, ancestral–descendant, generic sequence, Prothero and Shubin (1989) suggest that *Mesohippus* and *Miohippus* coexisted during the Oligocene for about 5 myr (see Figure 8.5). Within this sequence, habitat and resource partitioning seems to have occurred with these two morphologically similar horses. *Mesohippus* apparently was adapted to a diversity of habitats, ranging from gallery forests and swamps to savanna/woodland mosaics. *Miohippus* apparently was adapted to the drier spectrum of biomes ranging from savanna/woodlands to grasslands (Clark et al. 1967; Retallack 1983a,b). In the forested and swampy habitats, *Mesohippus* was among the most common mammals. Along with the ubiquitous oreodont *Merycoidodon*, these two ungulates represented 55% of the fauna in the near-stream, gallery-forest habitat. In the drier mosaic and more open habitats, however, *Mesohippus* was a relatively rare (usually about 5%) component of the mammalian fauna, whereas highly cursorial, hypsodont rhinocerotoids (*Hyracodon*) and small artiodactyls (e.g., *Leptomeryx*) were very common. It was in those more open habitats that Retallack (1983a,b) believed that *Miohippus* predominated, although there may have been numerous examples of overlap with *Mesohippus* in ecotones and habitat mosaics. Oligocene horses were widespread in North America, and in addition to the rich sites in the midcontinent, they are known from the Cypress Hills of Saskatchewan, the John Day Beds of the Columbia River region in Oregon, the Vieja Group in western Texas, and northern Florida (Emry et al. 1987) (see Figure 4.4). Thus, they probably were well adapted to a variety of habitats. However, until large, paleoecologically useful collections are made elsewhere, a detailed analysis of habitat partitioning will be available only for the classic deposits of the White River Group in the midcontinent.

The Miocene and the Clarendonian chronofauna

The Miocene epoch was a time in which major changes occurred in the terrestrial floras and faunas in North America. Wolfe (1978) stated that it was a time of continued cooling relative to the earlier Tertiary. The Miocene also seems to have been the time of the spread of the savanna/grassland biome (Elias 1942; MacGinitie 1962; Retallack 1982). As mentioned earlier, that latter event had profound influences on the evolutionary and ecological diversifications of the Equidae and their places in the Clarendonian chronofauna.

In North America the early Miocene (Arikareean–Hemingfordian) was characterized by a relatively low generic diversity of horses, on a level equal to or slightly greater than that of the preceding Oligocene, including *Miohippus* (extinct by the late Arikareean), *Desmatippus*, *Kalobatippus*, *Archaeohippus*, and *Parahippus*. *Desmatippus*, *Kalobatippus*, and *Archaeohippus* represent the beginnings of the diversification of body sizes and adaptations relative to late Oligocene forms (MacFadden 1987a). *Parahippus* was a pivotal genus, because advanced members within it, including *P. leonensis* and close relatives, evolved the key evolutionary innovation of hypsodonty during the middle Miocene, at about 17–18 Ma. From that monophyletic origin there was an explosive adaptive radiation of two principal clades of grazing horses, the Hipparionini and Equini (MacFadden & Hulbert 1988; Hulbert & MacFadden 1991), so that by 15 Ma (during the Barstovian) equid diversity had increased dramatically in North America. That was the beginning of the acme of equid diversity (Figure 8.9), resulting in as many as 13 contemporaneous genera (and 15 species) in North America at the height of the Clarendonian chronofauna. That chronofauna had the greatest diversity of land mammals known for the entire Tertiary, rivaling that seen in the rich terrestrial communities of modern-day East African savannas (Webb 1977, 1983, 1987). The general taxonomic pattern and integrity of that chronofauna persisted for about 10 myr, until the late Miocene, although, as will be discussed later, it experienced several notable changes during that time.

Since the diversification of grassland savannas, the large majority of the mammalian herbivores in terrestrial communities have been grazers. Thus, 59–79% of the biomass in extant East African communities is made up of grazing mammals, as were 55–87% of Miocene communities in North America (Webb 1983)

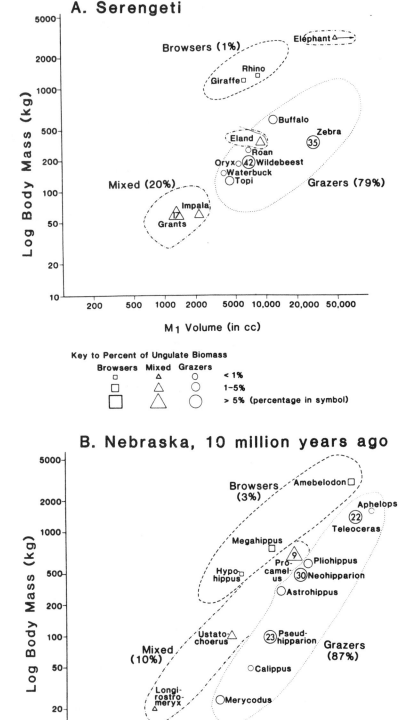

Figure 13.5. Percentages of browsers, mixed feeders, and grazers in modern (A) and extinct (B) mammalian faunas. Modified from Webb (1983).

(Figure 13.5). Webb (1983) stated that, in general, the primary-consumer biomass in the Clarendonian chronofauna consisted of about 10% or less browsers (in the Serengeti, it is about 1%), 10–30% mixed feeders, and 60–80% grazers.

At the height of the Clarendonian chronofauna, at about 10 Ma, the predominant browsers consisted of artiodactyls, gomphothere proboscideans, and horses. The latter included the taxa *Hypohippus* and *Megahippus*, which had relatively curved incisor arcades, like modern-day browsers (Janis 1988; Janis & Ehrhardt 1988), and rapidly attained relatively large body sizes, between about 266 and 403 kg (MacFadden 1987a). Other minor elements of the medium-large browsing guild included tapirs (and chalicotheres during the early Miocene and also in the Old World), although these are characteristically rare in the fossil record.

A minor component (10%) of the total estimated biomass during the Clarendonian chronofauna was composed of mixed feeders, mostly peccaries, camels, and, to a lesser extent, oreodonts. However, some species of *Merychippus* (= *Protohippus*) and primitive hipparions, which were only moderately hypsodont relative to contemporaneous horses in North America, may also have been mixed feeders (Webb 1983).

By far the most diverse component of the Clarendonian chronofauna was made up of grazers, high proportions of which were perissodactyls. The dominant rhinocerotid grazers during that time included *Aphelops* and *Teleoceras*, the latter apparently an ecomorph of modern-day *Hippopotamus* of East Africa. Although rhinoceroses (and proboscideans) usually are not numerically as abundant as horses in fossil samples [density is inversely proportional to body size in mammals (Damuth 1981)], because of their large body size they contributed greatly to the total mammalian biomass in Miocene terrestrial communities. Taking an example from the Love Bone Bed in Florida, with estimated body masses of 635 and 889 kg, respectively, for *Teleoceras proterum* and *Aphelops malacorhinus* (MacFadden & Hulbert 1990), the total preserved biomass (Table 13.1) represented by these two rhinocerotid species (46,228 kg) is similar to that of the nine contemporaneous species of horses (51,415 kg) in that ancient local fauna. For the late Miocene Ashfall Fossil Beds in northeastern Nebraska, Voorhies (1981) found that the biomass of the rhinoceros *Teleoceras* was greater than that for any other group of grazing mammals within the community, including horses. Therefore, although horses were both diverse and numerically dominant (Figure 13.6) during the Clarendonian chronofauna, they were not always the greatest contributors to the ungulate biomass.

The maximum diversity of ungulates, including horses, continued with only modest declines (e.g., the demise of browsing anchitheres) until the middle Hemphillian (see Figure 8.9). By

Table 13.1. *Minimum numbers of individuals (MNI), body mass, and total preserved biomass estimates and comparisons for grazing perissodactyls from the late Miocene Love Bone Bed, late Clarendonian of northern Florida*

Taxon	MNI	Body mass (kg)	Biomass (kg)
Equidae			
Neohipparion trampasense	86	142	12,212
Pseudhipparion skinneri	10	61	610
Hipparion cf. H. tehonense	3	151	453
Nannippus westoni	23	93	2,139
Cormohipparion ingenuum	62	139	8,618
Cormohipparion plicatile	83	214	17,762
Protohippus gidleyi	31	204	6,324
Calippus cerasinus	28	102	2,856
Calippus elachistus	9	49	441
Total Equidae	335		51,415
Rhinocerotidae			
Aphelops malacorhinus	22	889	19,558
Teleoceras proterum	42	635	26,670
Total Rhinocerotidae	64		46,228

Source: Data from MacFadden and Hulbert (1990).

about 6–7 Ma, however, faunal replacement declined, and extinctions within the grazing equid clades were greater than originations (Webb 1983). Thus, by the late Miocene, at about 5–6 Ma, horse diversity had dropped to eight genera (see Figure 8.9), as had diversity for other mammalian herbivores (Webb 1983) (see Figure 8.10). That resulted in a dramatic change in the character of the North American terrestrial mammalian community and is taken as the endpoint for the Clarendonian chronofauna. During the Pliocene, the equid genera *Equus*, *Nannippus*, and *Cormohipparion* persisted, but only *Equus* survived into the Pleistocene (Figure 8.11).

There may have been more than one cause for the decline of the equid fauna from its acme during the middle Miocene to the depauperate diversity seen in the Pliocene–Pleistocene. One possible factor that has been suggested was the advent of many groups of smaller-bodied herbivores (particularly rodents) that replaced the larger forms during the late Cenozoic (Webb 1969b, 1984). The principal factors, however, seem to have been related to further increases in global aridity and cooling, resulting in expansion of the steppe environment at the expense of the more productive savanna biome of the Clarendonian chronofauna.

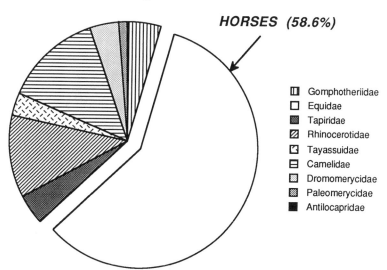

HORSES *(58.6%)*

- ⬚ Gomphotheriidae
- ☐ Equidae
- ▦ Tapiridae
- ▨ Rhinocerotidae
- ▧ Tayassuidae
- ⊟ Camelidae
- ▦ Dromomerycidae
- ▦ Paleomerycidae
- ■ Antilocapridae

Figure 13.6. *Pie diagram showing relative abundances of ungulates, including the large equid component of the fauna, from the late Miocene Love Bone Bed in northern Florida. Data from MacFadden and Hulbert (1990, Table 15.2).*

Patterns of community structure: selected topics related to fossil horses

Miocene climatic barriers, the ancient Neotropics, and faunal endemism

The Miocene equid faunas of North America started from relatively low diversity, then increased to their peak diversity during the Clarendonian chronofauna, and declined thereafter (see Figure 8.9). During that time there were many climatic changes throughout North America that resulted in ecological/biogeographic barriers and regional faunal endemism.

During the early Miocene (Arikareean–Hemingfordian) there seems to have been a climatic gradient that resulted in a relatively restricted tropical belt in southeastern North America termed the "Gulf Coast savanna corridor" (Webb 1974). That was the ancestral extension of the Neotropics into higher latitudes. That tropical belt and the climatic gradient between more northern climes apparently had a biogeographic filtering effect, so that among the ungulates, oreodonts and anchithere horses were common elsewhere, but relatively rare in the Gulf Coast region (MacFadden 1980).

As discussed in earlier chapters, the middle Miocene (late Hemingfordian and early Barstovian), between 18 and 15 Ma, was the time of the explosive adaptive radiation of hypsodont horses in North America. It also was a time of distinct endemism of equid faunas (Hulbert & MacFadden 1991). Although climatic barriers and distinctive biogeographic regions (e.g., the Gulf Coast savanna corridor) cannot be ruled out, the regional nature of equid faunas may also have involved taxonomic factors. Thus,

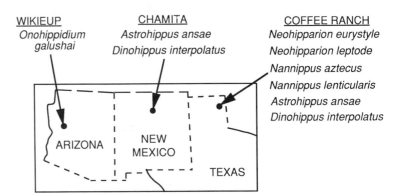

WIKIEUP
Onohippidium galushai

CHAMITA
Astrohippus ansae
Dinohippus interpolatus

COFFEE RANCH
Neohipparion eurystyle
Neohipparion leptode
Nannippus aztecus
Nannippus lenticularis
Astrohippus ansae
Dinohippus interpolatus

ARIZONA
NEW MEXICO
TEXAS

Figure 13.7. Late Hemphillian (5–6 Ma) equid faunas from the southwestern United States showing diversity differences at three well-sampled localities.

different clades may have originated in isolated regions (e.g., *Merychippus carrizoensis* in southern California, and several other merychippine species in the midcontinent and southeast) prior to opportunities for extensive intracontinental dispersal. That distinct regional endemism of equid faunas changed dramatically by about 14–15 Ma, during the middle to late Barstovian. Thus, at the beginning of the Clarendonian chronofauna, endemism was diminished, and the taxonomic compositions of equid faunas were generally more uniform over wider regions. That probably resulted from a more equable climate, with diminished barriers, more extensive savanna biomes, and the possibility of added interregional corridor dispersal. There still were local effects; for example, hipparions were generally more abundant than equines in late Miocene faunas in Florida. In contrast, equines were more abundant than hipparions in some of the rich faunas in the midcontinent, such as those from the classic sites in the Texas panhandle (this is also seen in the different relative abundances of the grazing rhinocerotids *Aphelops* and *Teleoceras* from these same localities) (Webb, MacFadden, & Baskin 1981). These differences may have resulted from local vegetation/biome pattern shifts within different regions in which the Clarendonian chronofauna existed. For example, if Shotwell (1961) was correct, the greater abundance of hipparions in the Gulf Coast region may have reflected a relatively higher forested aspect to the savanna/woodland mosaic and a generally more mesic climate.

MacFadden, Opdyke, & Johnson (1979) noted another interesting endemic pattern of equid faunas from roughly contemporaneous late Hemphillian sites in southwestern North America (Figure 13.7). The highest equid diversity is represented by six species in the Coffee Ranch local fauna (L.F.) in Texas. However, in the Chamita L.F. of New Mexico and the Wikieup L.F. of Arizona, equid diversity is significantly lower, with only two and one species, respectively. In many regions during the late Miocene (late Hemphillian), between about 7 and 5 Ma, ungulate faunas were devastated, principally because of the decline of the higher-productivity savanna biome and its replacement with extensive arid steppes. The highly endemic nature of equid faunas in the arid southwest undoubtedly was related to that dramatic climate change. However, because of our knowledge of the late Tertiary geological history of that region, another explanation is plausible. Rift basins and associated rugged topography were extensively developed at that time in areas such as the Santa Fe Graben in New Mexico and the ancestral Basin and Range regions of Arizona, Nevada, and southeastern California. In his study of wild *Equus caballus* from the Great Basin of Nevada, Berger (e.g., 1986) found that these topographic barriers (e.g., high-walled canyons and mountains) enhanced genetic isolation, resulting in stable resident populations, with few immigrants. Therefore, it is equally plausible that during the late Hemphillian, endemic equid faunas from New Mexico and Arizona evolved in biogeographic isolation from one another within separate fault-block-bounded structural/topographic basins. This seems further corroborated by the observation that certain other ungulate groups, such as camels and, perhaps most notably, antilocaprids, were highly endemic at late Hemphillian localities in the southwest.

Elsewhere in southern North America, late Hemphillian ungulate faunas were more cosmopolitan and exhibited greater diversity, including those in the Texas panhandle, the classic Bone Valley in Florida, and the El Ocote L.F., each of which contained five to six species of horses. Relative to elsewhere in North America during the latest Miocene, the Neotropical Gulf Coast savanna corridor was less affected by the onset of the drier and cooler climatic regime (Webb 1987), and those differences impacted southeastern equid faunas. In contrast to midcontinental North America, where many extinctions occurred, the Gulf Coast region apparently was a refugium for certain long-persisting, relictual taxa, including the horses *Pseudhipparion* and *Cormohipparion* (Webb & Hulbert 1986; Hulbert 1987) (see Figure 8.11). In summary, by the end of the Clarendonian chronofauna, at about 5–6 Ma, climatic barriers had disrupted the previously widespread and relatively uniform nature of terrestrial mammalian communities in North America.

Body size, again: cenograms and Hutchinsonian ratios

In Chapter 12 we discussed how important a knowledge of body size is for an understanding of many life-history strategies of mammals. In this section we shall investigate the distribution of body sizes among coexisting ungulates within selected Cenozoic land-mammal faunas and what those patterns indicate about the paleoecology of those terrestrial communities, particularly as they relate to extinct horses.

CENOGRAMS. The cenogram is a graphic means of analyzing the distribution of body sizes of all mammals within terrestrial communities. It was first used to analyze the pattern of small (<5 kg), medium (>5 to 250 kg), and large (>250 kg) mammals within extant communities (Valverde 1964). This method has also been used successfully with Eocene–Oligocene faunas in France (e.g., Legendre 1986, 1987). After comparisons with modern analogues, the slopes and "breakpoints" between small, medium, and large mammals in cenograms are useful predictors of the biomes in which ancient mammalian communities lived. A cenogram is simply a plot of the natural logarithm of body mass versus the body-mass rank within the fauna. Many characteristics of an ancient community can be interpreted from cenograms. For example, species-rich communities will have shallower slopes than species-poor communities, because the latter would have had fewer species within the ranked hierarchy. Furthermore, it follows that the species-rich communities characteristically inhabited more humid, forested biomes, whereas species-poor communities were adapted to relatively arid, open habitats. Another parameter, the breakpoint, usually is greater for the small, medium, and large mammals living in drier, more open environments (Legendre 1987; an exception to this is in the Subhumid category in Table 13.2).

Gingerich (1989) produced a cenogram for early Eocene Wasatchian mammals from the Bighorn Basin of Wyoming (Figure 13.8). That Wa_0 fauna contained two equid species: *Hyracotherium grangeri* and *H. sandrae*. Comparisons of several parameters of the Wa_0 cenogram allow an interpretation of the dominant paleoecological setting in which that ancient mammalian community lived: (1) The slope of -0.228 for the medium-sized mammals is similar to those for modern faunas in forests and woodland vegetation gradients and humid moisture gradients (Table 13.2). (2) The small–medium-mammal breakpoint (0.21) for the Wa_0 fauna is similar to those for modern-day communities living in humid and forested regions (Table 13.2). These results, based on both the slopes and breakpoints of the Wa_0 cenogram, are internally consistent and therefore constrain the early Eocene paleoenvironment and habitat in that region. A similar study of body-size distributions for middle Eocene faunas from the Bridger Basin of Wyoming (including the equid *Orohippus*) has indicated a forested habitat for those mammals (Gunnell 1990).

In summary, cenogram analysis provides a powerful indicator of the habitats in which ancient communities existed, and this

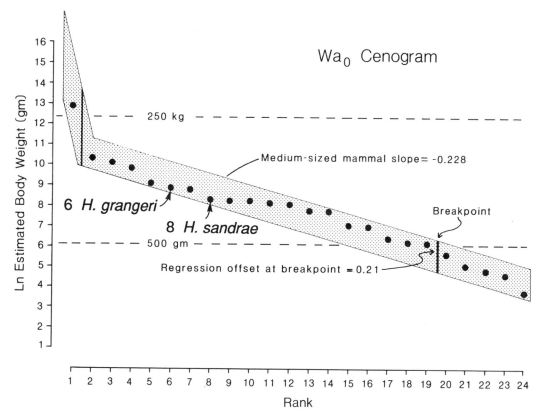

Figure 13.8. Cenogram of ranked distribution of body masses estimated for 24 mammalian species from the early Eocene Wasatchian (Wa₀) fauna from northern Wyoming. The two equid species, Hyracotherium grangeri *and* H. sandrae, *occupy, respectively, ranks 6 and 8 within this fauna. Modified from* Gingerich (1989); reproduced with permission of the Museum of Paleontology, University of Michigan.

method should prove even more useful as the body masses of other extinct mammalian faunas are estimated. As with all similar paleoecological studies, however, the results and interpretations derived from cenograms should be corroborated by the use of other independent data sources, such as functional morphology, paleobotany, and the sedimentary environments preserved with the ancient fauna.

HUTCHINSONIAN RATIOS. The great ecologist G. E. Hutchinson (1959) noted that sympatric, ecologically similar species differed in size by a factor of about 1.3 for linear structures related to feeding and by a factor of 2.0–2.2 for body mass. He believed that the underlying biological reasons for those ratios were explainable in terms of the minimum amount of morphological similarity sufficient to decrease competition for limited resources between sympatric species, so that coexistence would be pos-

Table 13.2. *Slope and offset parameters for cenograms for extant mammalian faunas in different environmental settings*

Gradient	Slope for medium mammals	Offset at breakpoint between small and medium mammals
Vegetation gradient		
Forest	−0.292	0.236
Woodland	−0.257	1.367
Savanna	−0.690	1.154
Moisture gradient		
Humid	−0.208	0.151
Subhumid	−0.313	1.788
Semiarid	−0.616	0.565
Arid	−1.058	1.000

Source: Adapted from Gingerich (1989, Table 31).

sible. Since the publication of that classic paper there have been numerous discussions of "Hutchinsonian ratios" in the literature (e.g., Maiorana 1978; Roth 1981; Eadie, Broekhoven, & Colgan 1987). Although few workers would deny that there are empirical data that support Hutchinsonian ratios, several have been critical of Hutchinson's explanations of their underlying biological origin, particularly when such ratios also are found in systems not involving competition, such as musical instruments and bicycles (Horn & May 1977; Maiorana 1978).

MacFadden and Hulbert (1990) calculated Hutchinsonian ratios for 21 sympatric ungulate mammals from the late Miocene Love Bone Bed in Florida, including nine sympatric species of fossil horses. In order to compare these ungulates, MacFadden and Hulbert (1990) measured several dental characters, which undoubtedly are related to feeding, from 480 specimens. In addition, body masses were estimated from regression equations derived from these and other preserved characters.

The patterns of body-size distributions for the 21 species of Love-site ungulates provide insight into several questions that have occupied other investigators. As seen in Figure 13.9, there are very few ungulates among these species that are either very small (10 kg for the artiodactyl *Pseudoceras*) or very large (3,440 kg for *Amebelodon* cf. *A. barbourensis*), and most of the species occur in the midrange of the distribution. Eadie et al. (1987) have suggested that the pattern of distribution of body sizes in mammalian faunas approximates a lognormal distribution. Using the body-size data from the Love site, MacFadden and Hulbert (1990) found a nonrandom distribution of ungulate body masses, with a slightly better fit to a lognormal curve than to other regres-

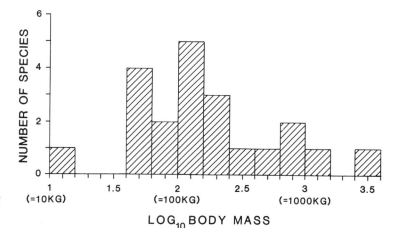

Figure 13.9. *Histogram of distribution of estimated body masses of the 21 ungulate species from the Love Bone Bed in northern Florida. From MacFadden and Hulbert (1990); reproduced with permission of Cambridge University Press.*

sion models, thus supporting the hypothesis of Eadie et al. (1987).

MacFadden and Hulbert (1990) also compared ratios of dental dimensions and estimated body masses for seven pairs of closely related and presumably ecologically similar species of Love-site ungulates (Table 13.3). Their results indicate mean values of 1.35 and 1.29 for linear dental dimensions and 2.33 for body mass. These are surprisingly close to the values predicted for Hutchinsonian ratios (1.3 and 2.0–2.2, respectively). However, as in the case of musical instruments and bicycles, as well as biological systems, these ratios may not always result from competition between sympatric, ecologically similar species. If character displacement can be interpreted from the fossil record, at present it will have to come from other kinds of data sets (as discussed later), although Hutchinsonian ratios, like those for the Love-site ungulates, still will remain of interest for future comparisons with other extinct mammalian faunas.

Niche partitioning, character displacement, and feeding: other possibilities

As we have seen, during the Miocene, at the height of the Clarendonian chronofauna, there were 10–15 contemporaneous, presumably sympatric species of equids in many local faunas in North America. With that great diversity, the question can rightly be asked, how did these phylogenetically and adaptively similar mammals coexist? Of that diversity, the two genera of browsing horses were sufficiently different from the others because presumably they were forest-adapted and fed on resources available within those types of biomes. However, even without

Table 13.3. *Ratios of upper and lower molar lengths and estimated body masses of ecologically and/or taxonomically "similar" species from the Love Site*

Taxa	UM1APL[a]	LM1APL[b]	Estimated body mass (kg)
Neohipparion trampasense / *Pseudhipparion skinneri*	$\frac{19.1}{13.7} = 1.39$	$\frac{18.7}{14.1} = 1.33$	$\frac{142}{61} = 2.38$
Cormohipparion plicatile / *Cormohipparion ingenuum*	$\frac{20.4}{18.2} = 1.12$	$\frac{20.9}{18.3} = 1.14$	$\frac{214}{139} = 1.54$
Protohippus gidleyi / *Calippus cerasinus*	$\frac{21.2}{16.7} = 1.27$	$\frac{20.7}{16.6} = 1.25$	$\frac{204}{102} = 2.00$
Calippus cerasinus / *Calippus elachistus*	$\frac{16.7}{13.0} = 1.28$	$\frac{16.6}{12.9} = 1.29$	$\frac{102}{49} = 2.08$
Procamelus grandis / *Hemiauchenia minor*	$\frac{26.8}{18.6} = 1.44$	$\frac{26.4}{17.0} = 1.55$	$\frac{498}{110} = 4.53$
Hemiauchenia minor / Antilocaprid	$\frac{18.6}{12.9} = 1.44$	$\frac{17.0}{—}$	$\frac{110}{55} = 2.00$
Peccary #1 / Peccary #2	$\frac{21.4}{11.4} = 1.49$	$\frac{14.9}{12.5} = 1.19$	$\frac{88}{50} = 1.76$
	$\bar{x} = 1.35$	$\bar{x} = 1.29$	$\bar{x} = 2.33$

[a]UM1APL, first upper molar, anteroposterior length.
[b]LM1APL, first lower molar, anteroposterior length.
Source: MacFadden and Hulbert (1990).

the browsers, we still have 8–13 species of hypsodont horses to consider, not to mention other grazers and mixed feeders, including the contemporaneous proboscideans, rhinoceroses, and artiodactyls.

In Chapter 12 it was suggested that during the Oligocene the contemporaneous horses *Mesohippus* and *Miohippus* partitioned their community so that the two genera inhabited biomes that were slightly different or very different. There has been some suggestion that a similar pattern occurred among Miocene horses. For example, on the basis of studies of Hemphillian horses from the Great Basin, Shotwell (1961) believed that the three-toed hipparions were better adapted to savanna/woodland mosaics (represented by the Arcto-Tertiary Geoflora). In contrast, he believed that the one-toed equines [= *Pliohippus* and *E. (Plesippus)* in Figure 7.2] were better adapted to the arid steppe, and they flourished (with a corresponding decline in hipparions) in response to the spread of the Madro-Tertiary Geoflora. Shotwell believed that functional differences in the monodactyl foot resulted in the equines being better adapted to the

more open country of the steppes relative to the tridactyl hipparions. On the other hand, there is evidence to suggest that
the tridactyl foot was better adapted for some settings, particularly muddy substrates (Renders 1984). Thus, although there
probably were ecological differences between hipparions and
equines, by the Hemphillian the tridactyl foot was functionally
monodactyl, and therefore there probably was little overall advantage in being truly monodactyl. Accordingly, other adaptive
complexes undoubtedly underwent functional changes to par-
tition the ecosystem, and the most striking of those were related
to feeding.

It is an empirical observation that various clades of coexisting
Miocene hypsodont equids evolved roughly similar crown
heights (MacFadden 1985a, 1988b; Hulbert 1987), and thus it
would be difficult to distinguish their specific feeding specializations on these characters alone. However, possible examples
of niche partitioning can be discerned on closer inspection of
the cheek-tooth enamel patterns and cropping mechanisms.

CHEEK-TOOTH ENAMEL. In addition to increased crown
height, another way to evolve more durable teeth is by increasing
the degree of enamel infolding (Van Valen 1960; Janis & Fortelius
1988) (see Figures 11.1 and 11.2). Therefore, comparisons of
different enamel patterns in coexisting equid species should (if
similar ontogenetic stages are compared) indicate niche partitioning of food resources. A possible example of this comes from
the late Miocene Love site in northern Florida. The equid species
in that fauna that had more complex enamel patterns, such as
Cormohipparion ingenuum (Figure 13.10A), would have been better adapted to feeding on more abrasive grasses. At the other
extreme, those species with less complex enamel patterns, but
similar in size and crown height, such as *Pseudhipparion skinneri*
(Figure 13.10C), would potentially have been adapted to feeding
on less abrasive grasses. Moderate enamel complexities, such as
those of *Nannippus westoni* (Figure 13.10B), might reflect more
generalized or moderately abrasive diets.

CROPPING MECHANISM. As discussed in Chapter 11, several
studies have shown that variations in the cropping mechanism,
including the morphology of the incisors, symphysis, and muzzle width, indicate niche partitioning and feeding specialization
in ungulate mammals (Owen-Smith 1985; Janis 1988; Janis &
Ehrhardt 1988). For example, among extant herbivore species
whose diets are known, those with relatively narrow muzzles
are more specialized than those with relatively broad muzzles.
Furthermore, a range of muzzle morphologies can be seen in
contemporaneous, presumably sympatric fossil horse species

Late Miocene Love Site

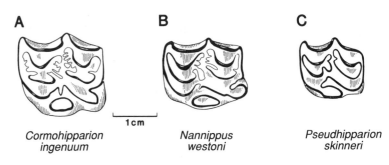

Figure 13.10. Comparisons of enamel complexities in coexisting fossil horses using the M2 of three selected species from the late Miocene Love Bone Bed in northern Florida. These taxa are all of roughly the same degree of hypsodonty and are not greatly different in size. Thus, as discussed in the text, the relative complexity of dental pattern may reflect different dietary specializations. Specimens are as follows: A, UF 32300; B, UF 50351; C, UF 50354.

A   B   C

Cormohipparion ingenuum   *Nannippus westoni*   *Pseudhipparion skinneri*

(see Figure 11.8) in many Miocene sites in North America, thus indicating a diversification of feeding adaptations.

In summary, differences in enamel complexity and in the cropping mechanism indicate feeding specializations and niche separation among hypsodont fossil horse species within contemporaneous faunas during the late Cenozoic. Although the time frames in which such character displacements occur generally are too short to reveal the transitional stages resulting from competitive interactions, the results of these evolutionary and ecological processes are preserved in the ancient morphologies of fossil horses.

Concluding comments on paleoecology of Cenozoic mammals

Chapters 12 and 13 have been devoted to discussions of mammalian paleoecology and what we can learn about both the population biology of extinct species and the community dynamics of extinct terrestrial faunas. The science of reconstructing patterns that occurred millions of years ago is dependent upon a firm philosophical and methodological framework. In this regard, the works of Deevey, Kurtén, Van Valen, and Voorhies are classics in population ecology, and those of Olson, Webb, and Van Valkenburgh are of similar importance in community ecology as applied to extinct mammalian faunas. Paleoecology has lagged behind alpha-level taxonomy for several reasons, including a lack of excellent quarry samples with remarkable morphological preservation (e.g., to determine the sexes of specimens) for population studies. Furthermore, to reconstruct the community ecology of ancient faunas, paleobiologists need to collect huge samples in order to recover both the common and rare taxa of "macromammals" and then employ screen-washing techniques to add the micromammals to the picture. All of these pursuits require immense and methodical efforts both in the field and in the laboratory. However, the payoff is correspondingly large, because it is only from the study of paleoecology that we can derive an understanding of the patterns of population, species, and community evolution through geological time.

14

Epilogue

Dominant themes

In addition to those topics that are obvious from the table of contents, as this book developed it became clear that there were several important and recurring themes that dominated many of the discussions.

A question of scale

There has recently been much interest in the scientific community in the phenomenon of scale and how it affects our perception of nature. In the earth sciences this has led to discussions of the applications of fractal geometry, namely, those processes and patterns that are scale-invariant, or in other words, similar at different scales (e.g., Plotnick 1986; Turcotte 1991). Here the relevant question is which phenomena interpreted from the fossil record of horses are scale-dependent, and which are scale-invariant?

SCALE-DEPENDENT PHENOMENA. The scale of observation can have a profound effect on interpretation of pattern and process, and this is particularly true for evolutionary studies. For example, Gingerich's observation (1983) that morphological rates of evolution are highly dependent on the time intervals being compared is an important discovery in evolutionary biology.

At the generic level, the evolution of fossil horses traditionally has been interpreted as both anagenetic and orthogenetic. However, more recent phylogenetic investigations of fossil horses on the species level indicate a highly complex pattern, with many instances of cladogenesis and anagenesis, or the balance between the two changing with different circumstances. In addition to interest in modes of speciation, within the past two decades there has been considerable discussion concerning the relative importance of phyletic gradualism versus punctuated equilibria as valid evolutionary processes. In general, the higher-level (superspecific) investigations of fossil horses have traditionally suggested gradual evolution, but that is almost certainly simplistic. For example, the evolution of body size described in Chapter 10 suggests that certain higher-level patterns are characterized by stasis at some times, followed by rapid diversification; a similar pattern has been noted for other taxa (Gould 1988c). At a lower level, evolving populations and species of fossil horses suggest a significantly more complex interpretation, one that may not be explainable by only a single mode. Thus, horse evolution is neither simply gradual nor always punctuated. Empirical data suggest that either mode can occur in different clades at different times, depending on many variables. This is also the case for taxonomic rates, so that at certain times clades evolved rapidly, whereas given other conditions they evolved more slowly. Furthermore, all organisms are evolutionary mosaics of humanly perceived characters. Even within a given lineage, some characters evolve very rapidly, such as crown height in hypsodont horses, whereas others, such as occlusal dimensions in the very same teeth, show normal, or horotelic, rates of evolution. Thus, to use the fossil record of the Equidae as a prime example of either rapid or gradual evolution is both simplistic and wrong, because factors such as the scale of observation, as well as the history of the particular lineage, or even the character being examined, must be considered.

SCALE-INVARIANT PHENOMENA. Given the immense complexity in nature and the recognition that there are so many examples of dependence on the scale of observation, another obvious question is which phenomena are similar at different scales? It seems that scientists have been principally interested in describing complex and unique systems and phenomena in nature, so that it has been only recently that questions of scale-invariant patterns and processes have received equal time (e.g., Vrba & Eldredge 1984).

With regard to evolution, it sometimes seems that paleontologists believe that higher taxa speciate and evolve. This mindset, however, seems proper, because studies of taxonomic rates indicate that long-term patterns (i.e., longevities, R, and extinction) within groups are similar on generic, familial, and ordinal levels. Furthermore, studies of species-level rates, such as the Miocene diversification of horses, although up to now rare for fossil groups, show that these same patterns extend to lower levels within the Linnean hierarchy, as they well should, because the species is the fundamental unit of biological change. In summary, scale affects our perception of nature, and it is interesting to know which phenomena are scale-dependent and which are scale-invariant.

Juggling the parameters: complexity and simplicity

COMPLEXITY OF NATURE. The natural world is exceedingly complex. Many topics in this book, such as the evolution of social systems throughout the Cenozoic, or the advent of hypsodonty during the Miocene, are discussed for their own sake. In reality, however, these are interrelated to a bewildering array of different phenomena, including climate and vegetation patterns, as well as coexisting plants and animals within ancient communities and ecosystems. To say that evolution is driven principally by coexisting species and natural selection or, on the other hand, by what is occurring with distantly interacting species or both random and nonrandom physical factors is senseless. Thus, in my mind the selectionist versus antiselectionist (or even gradualist versus punctuated) arguments are antitheses constructed to stimulate intellectual thought. In reality, all existing parameters and factors are involved and interrelated in natural systems. With the ever-increasing trend toward specialization in scientific inquiry, we need to remain aware that the old-time natural philosophers and natural historians undoubtedly had a more holistic, albeit less informed, view of the world.

SIMPLICITY IN SCIENCE. If the natural world is so complex, then how do we attempt descriptions of it? Statements such as "horses exhibit gradual evolution," "horses evolved high-crowned teeth as a response to eating grasses," or "they became extinct during the late Miocene because of better-adapted artiodactyls" have all been portrayed as scientific conclusions, or even truths in the secondary literature; yet such statements are grossly oversimplified. Simplification of scientific hypotheses and ideas so that they can more readily be communicated may seem justified to some, but ultimately that retards the progress of science. Rather than simplify, researchers, textbook writers, and designers of museum exhibits, to name a few, need to present up-to-date science in an intelligible form, but they should not sacrifice the most modern concepts for the sake of simplicity. This is indeed a difficult challenge, but if it is met we can accurately spread new ideas into the scientific and public domains as new advances are made.

Novelty and cyclicity in science

Although science advances through new discoveries and the addition of new data, many fundamental concepts have been known, even if in somewhat modified form, for a long time. For example, although the paradigm of evolution is almost universally associated with Darwin, the foundations of the theory extend back to Aristotle. Of relevance here, recognition of the advent of hypsodonty in Miocene horses stems from insightful work published in the nineteenth century (e.g., Kowalevsky 1873). However, today we have a much larger body of data to facilitate an understanding of this phenonemon. A corollary is that scientific thought often is cyclic. For example, general concepts such as the phylogenetic position of hyracoids close to perissodactyls, or the systematic utility of facial fossae, although popular now, were first considered by nineteenth-century paleontologists, albeit with fewer data and specimens.

Paleontologist and neontologist: We need them; they need us

Traditionally there has been a distinction between the paleontologist on the one hand and the neontologist on the other hand. This stems from the kinds of specimens they study, the research they do, or the results they present; yet the common goal of both is to understand nature. Some biologists have recently said that fossils cause problems because they do not always preserve all the needed characters, or, with the discovery of new taxa, they mess up the existing cladograms and classifications. Others might think that paleontology is the stepchild of biology because it must borrow from the present to interpret the past; for example, from an understanding of modern equid population biology or community ecology we can reconstruct past analogues. The lesson to be learned, however, is that without the unique window on the past provided by fossils, the study of present-day nature would be significantly less meaningful.

THE IMPORTANCE OF COLLECTIONS.   It surely is a tautology to say that without fossils there would be no paleontology. Along with this, it follows that without the *growth* of collections, paleontology would stagnate. In the early twentieth century, Matthew, as quoted in the second section of Chapter 4, felt fortunate that he had so many more good specimens of fossil horses to study than had Marsh, Cope, and Leidy during the late nineteenth century. It follows that many of the chapters in this book could not have been written, or else they would have been little improved over previous essays on these subjects, without the great paleontological efforts since Matthew's time that have added significantly to scientific collections. Finally, I hope that paleontologists during the twenty-first century will look back on the end of the twentieth century and consider themselves fortunate for the increased collections available for their research.

The renaissance continues

In addition to the popular interest in dinosaurs, fossil humans, and other extinct organisms, the current renaissance in paleontology and paleobiology stems from significant advances in three areas: (1) new technologies, (2) new theories and methodologies, and (3) new discoveries. For example, new technologies have resulted in increased precision in determining geological time and further refinement in our knowledge of dispersal patterns of fossil horses. New theories, such as cladistics, have allowed new interpretations of phylogenetic hypotheses, resulting in, for example, the splitting into natural, monophyletic clades the previously problematical merychippine horses. Among the many new fossil discoveries that have significantly advanced our understanding of the paleobiology of horses the following stand out: From the late Eocene *Lagerstätten* at Messel, the exquisitely preserved stomach contents of early palaeotheres have shed light on a century-old hypothesis about the diets of ancient equoids. In North America, the early Eocene Castillo Pocket sample of *Hyracotherium* from Huerfano, Colorado, resulted in an extraordinary view of the population biology of the earliest equid genus. There is every reason to believe that with further advances in the future, exciting interpretations of the natural history of fossil horses and other extinct life will emerge.

The future of paleontology

As trends continue, the gap between neobiology and paleobiology will decrease, although the latter will continue to provide the unique view of the long-term time dimension in natural history.

Increased precision in geological time

The development of relevant analytical instruments (e.g., sensitive mass spectrometers) over the past several decades has stimulated interest in more accurate calibration of geological time. Decades ago, radioisotopic age determinations using the $^{40}K-^{40}Ar$ decay series characteristically had analytical errors of 5–10%, whereas today the range is routinely 1–2%. Furthermore, with new techniques such as single-crystal laser fusion and new decay series such as $^{40}Ar/^{39}Ar$, errors have diminished to 0.1% or less for the Cenozoic (Swisher 1989). Up to now, the latest Pleistocene (i.e., the past 50,000 years) has been the only part of the geological time scale in which radioisotopic techniques would allow interpretation from the fossil record of biological phenomena such as dispersal. However, those past 50,000 years encompass only the smallest fraction of geological time and the history of paleobiological events. If current trends continue, and there is every reason to expect that they will, then radioisotopic errors of 0.1% or less will allow us to constrain events that occurred at 10 Ma to within about 10,000 years. That will be the precision we need to discern biological events back into the Cenozoic such as character displacement and intracontinental dispersal. Furthermore, it is conceivable that new geochemical systems (i.e., decay series) and techniques will be discovered that will allow us to date fossil vertebrate bones directly, rather than having to rely on interbedded volcanic ashes and sediments for, respectively, radioisotopic dating and paleomagnetic dating.

Biochemical and geochemical approaches to paleobiology

As mentioned earlier, within the past three decades there has been much emphasis on "absolute" dating techniques to provide a better understanding of the fossil record. This has been a logical goal because of our great concern for the positions of taxa in space and time. I now believe that a new wave in paleobiology will be the use of biochemical and geochemical techniques to ask questions that could not be answered earlier. There already are hints of this from studies involving DNA extraction from Miocene plants and the biochemistry of bone collagen in Pleistocene proboscideans, but these are merely the tip of the proverbial iceberg.

BONE CHEMISTRY. If current trends continue, the familiar saying "you are what you eat" will be a new motto for mammalian paleontologists. Within the past two decades, archaeologists have conducted interesting studies of the diets of ancient human cultures through the use of stable isotopes and trace elements. When the problem of diagenesis in fossil bones is resolved, or as similar studies are carried out on denser dental materials (enamel and dentine; e.g., Thorp & van der Merwe 1987), then these techniques will allow interpretations of the diets of extinct mammals independent of the traditional studies of functional morphology. Regarding fossil horses, a topic of great interest in the literature has been the browsing–grazing transition inferred from the acquisition of hypsodonty during the Miocene. This could be tested using carbon isotopes. Other isotopic studies could analyze feeding differences and possibly migration patterns within coexisting herbivore communities, including the time of maximum equid diversity during the Clarendonian chronofauna. In so doing, we can expect corroboration of hypotheses and interpretations based on traditional data, or even those previously outside the domain of scientific inquiry.

SYSTEMATICS. Many biochemical techniques have recently been developed that allow new hypotheses about the phylogenetic interrelationships of various taxa with extant representatives. This is exemplified by the study of mtDNA from the recently extinct quagga and its close relatives within *Equus*. However, there has been a rift between biochemists working with soft tissues and paleontologists working with characters preserved in bones and teeth. Hypotheses derived from these different sets of data have not always been in agreement, most notably regarding biochemical divergence times and paleontological evidence for certain higher-level mammalian taxa. Yet each side believes that it has valid evidence of the timing and sequence of evolutionary diversification for the groups that it studies. Even this disparity, however, is changing. For example, the phylogenetic divergence time, estimated to have been at about 4 Ma based on mtDNA of extant species of *Equus* (George & Ryder 1986) (see Figure 5.24), gives surprisingly good agreement with paleontological evidence constraining the early adaptive radiation of this genus during the Pliocene (Lindsay et al. 1980) (see Figure 6.9). Furthermore, if biochemically useful data can be extracted not only from soft anatomical structures and preserved hides (as in the quagga) but also from fossilized specimens, then new data can be brought to bear on the enigmatic interrelationships of extinct and living species of *Equus*. In summary, recent research has demonstrated extraordinary potential for closing the gap between the biochemist and paleontologist. As DNA or other biochemically useful data are found in older fossils (e.g., as in the recently discovered 18-myr-old *Magnolia* leaves, and perhaps someday in 18-myr-old fossil horses), an entirely new field of "paleobiochemical systematics" will surely develop.

A hypothetical scenario

Given the foregoing possibilities, it is conceivable that an abstract presented at a scientific meeting at some time in the future will read something like the following:

THE BROWSING–GRAZING TRANSITION IN MIOCENE FOSSIL HORSES FROM NORTH AMERICA. Laboratory analyses of newly collected specimens from critical sites, as well as specimens existing in museum collections, have allowed an interpretation of the phylogeny, biogeographic patterns, and paleoecology of the transition from low-crowned parahippines to hypsodont merychippines. Biochemical data from fossil bones and teeth have corroborated previous hypotheses based on osteological characters that *Parahippus leonensis* is the sister group of the equine radiation. The oldest bone samples of this out-group, represented by *P. leonensis floridanus*, yield ages of 18.012 Ma. Although limited intracontinental dispersal occurred by 17.245 Ma, thereafter, local, *in situ* evolution occurred until about 15.876 Ma. Analysis of selected temporal successions of paleo-populations indicates that morphological and ecological niche specialization occurred within 0.001 to 0.01 myr. Studies of bone and teeth using both stable isotopes and trace elements from these same successions indicate that the dietary shift from browsing to grazing occurred early during cladogenesis, despite an evolutionary lag of several hundred thousand years in hyp-sodonty. Early populations had more similarities in isotopic sig-natures, indicating similar diets, followed by rapid shifts toward dietary specializations soon after cladogenesis of sympatric sub-species. In conjunction with previous studies of population dy-namics and functional morphology, new isotopic data indicate that some equid species became generalized feeders, exploiting local resources, whereas others, particularly those of large body size, underwent seasonal migrations to exploit more specialized, patchy resources. After 15.876 Ma, several 100,000-year cycles of climate change, as well as a long-term global temperature decrease, resulted in more widespread and uniform habitats, and consequently fewer dispersal barriers both within North America and across Beringia during lowered stands of sea level.

This scenario may not be as far off as it might seem.

Concluding comments: So, where do we go from here?

One of the principal goals of this book is to present a synthesis of our knowledge on a variety of subjects dealing with fossil horses. In so doing, I believe that rather than wrapping up this subject, this book reveals exciting potential and need for further studies, of which the following are just a few examples: Despite the flurry of activity in recent years, there are several systematic projects that need to be tackled. These include analyses of the anchitheres and *Desmatippus* in North America and a thorough revision of all Old World hipparions (the latter will be a mon-umental task). Although the recent work on the paraphyletic nature of *Hyracotherium* is a great step forward, perissodac-tyl and equid origins remain enigmatic. As additional well-calibrated sequences containing fossil horses are studied, com-parisons can be made with existing studies of taxonomic and morphological rates of evolution. These should be done at both the microevolutionary and macroevolutionary levels. It would be particularly interesting to see how extinct Old World horses compare in terms of evolutionary rates with those of the North American radiation. With regard to population biology, addi-tional studies need to be done on well-preserved quarry samples to better understand autecological parameters such as sexual dimorphism, socionomic ratio, behavioral ecology, and other life-history strategies in extinct equid clades other than the few that have already been studied. With regard to synecology, further studies of niche specialization and ungulate feeding guilds, as well as body-size distributions, will yield a better understanding of the evolution of terrestrial communities during the Cenozoic.

The days when paleontologists could be considered akin to stamp collectors are long gone. Thus, as distinguished earlier, today the study of fossils is more properly called paleobiology, or the pursuit of understanding varied aspects of the biology of extinct organisms. In the broad context of modern scientific inquiry, paleobiology is unique in providing a long-term time dimension to study natural history. Because of their great abundance worldwide during the Cenozoic, as well as their ties with the familiar and beloved modern *Equus*, horses have enjoyed a position of high visibility in the scientific literature. Our state of knowledge will continue to advance as new discoveries are made and new theories, methodologies, and technologies are developed. There are many interesting and exciting questions that still need to be answered, and fossil horses undoubtedly will remain in the limelight as a popular example that can be used to better understand the natural history of ancient life.

Appendix I: Abbreviations and conventions used in the text

AMNH: American Museum of Natural History

c: lower canine

C: upper canine

C3: three-carbon compounds produced during the pentose phosphate photosynthetic pathway; characteristic of many woody plants

C4: four-carbon compounds produced during the dicarboxylic photosynthetic pathway; characteristic of plants such as temperate grasses

d: darwin, a rate of evolutionary change in morphology

$$\delta^{13}C = \left(\frac{(^{13}C/^{12}C) \text{ sample}}{(^{13}C/^{12}C) \text{ standard}} - 1 \right) \times 1{,}000 \text{ parts per mil } (^0\!/_{00})$$

$$\delta^{18}O = \left(\frac{(^{18}O/^{16}O) \text{ sample}}{(^{18}O/^{16}O) \text{ standard}} - 1 \right) \times 1{,}000 \text{ parts per mil } (^0\!/_{00})$$

Δt: amount of elapsed time (usually in myr)

DPOF: dorsal (also called lacrimal or nasomaxillary) preorbital facial fossa

FAD: first-appearance datum plane (= LSD, lowest stratigraphic datum)

F:AM: Frick American Mammals, now part of Department of Vertebrate Paleontology, AMNH

i: lower incisor, e.g., i1 is first lower incisor

I: upper incisor, e.g., I3 is third upper incisor

INT: intermediate tubercle on proximal humerus

L.F.: local fauna

m: lower molar, e.g., m1 is first lower molar; also meters

M: upper molar, e.g., M3 is third upper molar

ma: macarthur, a rate of taxonomic extinction (μma, micro-macarthur $= 10^{-6}$ ma)

Ma: megannum, a radioisotopic age determination or point in time, in millions of years ago on the geological time scale

MC: metacarpal, e.g., MC III is the third metacarpal (in the forefoot)

MP: metapodial, e.g., MP IV is the fourth metapodial (usually used in discussions of forelimbs and hindlimbs together)

MPTS: magnetic-polarity (or geomagnetic-polarity) time scale

MT: metatarsal, e.g., MT II is the second metatarsal (in the hindfoot)

mtDNA: mitochondrial DNA

myr: millions of years, usually in reference to a duration of geological time

NALMA: North American Land Mammal Age

OTU: operational taxonomic unit

p: lower premolar, e.g., p1 is first lower premolar

P: upper premolar, e.g., P3 is third upper premolar

PAUP: Phylogenetic Analysis Using Parsimony (see Swofford 1985)

R: intrinsic rate of taxonomic increase

s.l.: *sensu lato*, in the broad sense

s.s.: *sensu stricto*, in the strict sense

UF: University of Florida, Florida Museum of Natural History, Vertebrate Paleontology Collection

UF-M: University of Florida, Florida Museum of Natural History, Mammalogy Collection

USNM: National Museum of Natural History, Vertebrate Paleontology Collection

USNM-M: National Museum of Natural History, Mammalogy Collection

V: coefficient of variation

YPM: Yale Peabody Museum

†: denotes an extinct taxon

Appendix II: Classification of the Equidae

Class Mammalia Linnaeus 1758
  Order Perissodactyla Owen 1848
    Family Equidae Gray 1821
      †Subfamily "Hyracotheriinae" Cope 1881
        †*Hyracotherium* Owen 1840; early Eocene (58–51 Ma), NA, Eu
        †*Gobihippus* Dashzeveg 1979; Eocene, As[a]
        †*Xenicohippus* Bown & Kihm 1981; early Eocene (56–51 Ma) NA[a]
        †*Orohippus* Marsh 1872; early–middle Eocene (52–48 Ma), NA
        †*Haplohippus* McGrew 1953; early Oligocene (38.5–33 Ma), NA
        †*Epihippus* Marsh 1877; late Eocene (48–38 Ma), NA
      †Subfamily "Anchitheriinae" Osborn 1910
        †*Mesohippus* Marsh 1875; early–late Oligocene (38.5–29 Ma), NA
        †*Miohippus* Marsh 1874; early Oligocene–early Miocene (38.5–19 Ma), NA
        †*Kalobatippus* Osborn 1915; late Oligocene–middle Miocene (29–13 Ma), NA[b]
        †*Anchitherium* Meyer 1844; early Miocene, Eur, Af
        †*Hypohippus* Leidy 1858; middle–late Miocene (18–9 Ma), NA
        †*Sinohippus* Zhai 1962; Miocene, As
        †*Megahippus* McGrew 1937; middle–late Miocene (16–9 Ma), NA
        †*Archaeohippus* Gidley 1906; early–middle Miocene (23–13 Ma), NA
        †*Desmatippus* Scott 1893; late Oligocene–early Miocene (29–18 Ma), NA[b]
        †*Parahippus* Leidy 1858; early–middle Miocene (23–15 Ma), NA

Note: NA, North America; SA, South America; Eu, Europe; As, Asia; Eur, Eurasia (when distribution is known to include both Europe and Asia); Af, Africa. Biostratigraphic ranges (in Ma) are given for North American occurrences and follow the calibrations of Woodburne (1987).

[a]May not be generically distinct, but included here pending further study.

[b]Despite synonymy in Simpson (1945), considered a distinct genus here.

[c]The interrelationships of *Cormohipparion*, *Stylohipparion*, *Proboscidipparion*, and *Cremohipparion* are complex, and elsewhere in this book they usually are referred to as "Old World hipparions," e.g., in Figures 5.14 (phylogeny) and 5.15 (cladogram).

335

Subfamily Equinae Gray 1821
  †Tribe Hipparionini Quinn 1955
    †*Merychippus* Leidy 1857; middle Miocene (18–10 Ma), NA
    †*Pseudhipparion* Ameghino 1904; middle Miocene–earliest Pliocene (13–4.5 Ma), NA
    †*Hipparion* de Cristol 1832; middle–late Miocene (13.5–6 Ma), NA; late Miocene, Eur, Af
    †*Neohipparion* Gidley 1903; late Miocene–earliest Pliocene (11–4.5 Ma), NA
    †*Nannippus* Matthew 1926; middle Miocene–late Pliocene (12–2 Ma), NA
    †*Cormohipparion* Skinner & MacFadden 1977; middle Miocene–late Pliocene (16–2 Ma), NA; middle–late Miocene, Eur, ?Af
    †*Cremohipparion* Qiu et al. 1987; Miocene, Eur[c]
    †*Proboscidipparion*[b,c] Sefve 1927; Pliocene, As
    †*Stylohipparion* van Hoepen 1932; ?Miocene–?Pleistocene, Af[a,c]
  Tribe Equini Gray 1821
    †*Protohippus* Leidy 1858; middle–late Miocene (16–6 Ma), NA
    †*Calippus* Matthew & Stirton 1930; middle–late Miocene (15–6 Ma), NA
    †*Pliohippus* Marsh 1874; middle–late Miocene (15–6 Ma), NA
    †*Astrohippus* Stirton 1940; late Miocene–earliest Pliocene (6–4.5 Ma), NA
    †*Dinohippus* Quinn 1955; late Miocene–earliest Pliocene (10–4.5 Ma), NA
    †*Hippidion* Owen 1869; late Miocene–?middle Pleistocene (8–1 Ma; discontinuous), NA; ?Pliocene, early–late Pleistocene, SA
    †*Onohippidium* Moreno 1891; late Miocene (6–5 Ma), NA; ?Pliocene, early–late Pleistocene, SA
    †*Parahipparion* C. Ameghino 1904; Pleistocene, SA[a]
    Equus Linnaeus 1758; earliest Pliocene–late Pleistocene, NA (4.5–0 Ma); early–late Pleistocene, SA; late Pliocene–Recent, Eur, Af

References

Abelson, P. H. (1988). Isotopes in earth science. *Science* 242:1357.

Abrahamson, W. G., & Hartnett, D. C. (1990). Pine flatwoods and dry prairies. In *Ecosystems of Florida*, ed. R. L. Meyers & J. J. Ewel, pp. 103–49. Orlando: University of Central Florida Press.

Abusch-Siewert, S. (1983). Gebissmorphologische Untersuchungen an eurasiatischen Anchitherien (Equidae, Mammalia) unter besonder Berücksichtigung der Fundstelle Sandelzhausen. *Courir Forschungsinstitut Senkenberg* 62:1–361.

Akersten, W. A., Lowenstam, H. A., & Walker, A. (1984). "Pigmentation" of soricine teeth: composition, ultrastructure, and function. In *Am. Soc. Mammal. 64th Annual Meeting, Abstracts*, Humboldt State University, Arcata, California, no. 153, p. 40.

Albritton, C. C. (1980). *The Abyss of Time*. San Francisco: Freeman, Cooper & Co.

Allen, J. A. (1878). The geographical distribution of mammals. *U.S. Geological Surv.* 4:313–76.

Allison, P. A. (1988). Konservat-Lagerstätten: cause and classification. *Paleobiology* 14:331–44.

Ambrose, S. E. (1985). *Pegasus Bridge: June 6, 1944*. New York: Simon & Schuster.

Ambrose, S. H., & DeNiro, M. J. (1989). Climate and habitat reconstruction using stable carbon and nitrogen isotope ratios of collagen in prehistoric herbivore teeth from Kenya. *Quat. Res.* 31:407–22.

Anderson, E. (1984). Who's who in the Pleistocene: a mammalian bestiary. In *Quaternary Extinctions: A Prehistoric Revolution*, ed. P. S. Martin & R. G. Klein, pp. 40–89. Tucson: University of Arizona Press.

Arambourg, C., & Piveteau, J. (1929). Les vertébrés du Pontian de Salonique. *Ann. Paléont.* 18:59–138, 12 plates.

Archibald, J. D., Gingerich, P. D., Lindsay, E. H., Clemens, W. A., Krause, D. W., & Rose, K. D. (1987). First North American

land mammal ages of the Cenozoic era. In *Cenozoic Mammals of North America: Geochronology and Biostratigraphy*, ed. M. O. Woodburne, pp. 24–76. Berkeley: University of California Press.

Axelrod, D. I., & Bailey, H. P. (1969). Paleotemperature analysis of Tertiary floras. *Palaeogeogr., Palaeoclimatol., Palaeoecol.* 6:163–95.

Ayala, F. J. (1988). Can "progress" be defined as a biological concept? In *Evolutionary Progress*, ed. M. H. Nitecki, pp. 75–96. Chicago: University of Chicago Press.

Azzaroli, A. (1982). On Villafranchian Palaearctic *Equus* and their allies. *Palaeontograph. Italica*, N. Ser. 72:74–97.

   (1985). *An Early History of Horsemanship*. Leiden: E. J. Brill.

   (1988). On the equid genera *Dinohippus* Quinn 1955 and *Pliohippus* Marsh 1874. *Boll. Soc. Paleont. Italiana* 27:61–72.

   (1990). The genus *Equus* in Europe. In *European Neogene Mammal Chronology*, ed. E. H. Lindsay, V. Fahlbusch, & P. Mein, pp. 339–56. New York: Plenum Press.

Bader, R. S. (1956). A quantitative study of the Equidae of the Thomas Farm Miocene. *Bull. Mus. Comp. Zool.* 115:47–78.

Badgley, C., & Tauxe, L. (1990). Paleomagnetic stratigraphy and time in sediments: studies in alluvial Siwalik rocks of Pakistan. *J. Geol.* 98:457–77.

Badgley, C., Tauxe, L., & Bookstein, F. L. (1986). Estimating the error of age interpolation in sedimentary rocks. *Nature* 319:139–41.

Barry, J. C., Lindsay, E. H., & Jacobs, L. L. (1982). A biostratigraphic zonation of the middle and upper Siwaliks of the Potwar Plateau of northern Pakistan. *Palaeogeogr., Palaeoclimatol., Palaeoecol.* 37:95–130.

Baskett, J. (1980). *The Horse in Art*. Boston: New York Graphic Society.

Bennett, D. K. (1980). Stripes do not a zebra make. Part I: A cladistic analysis of *Equus*. *Syst. Zool.* 29:272–88.

Benton, M. J. (1990). *Vertebrate Palaeontology.* London: Unwin Hyman.

Berger, J. (1977). Organizational systems and dominance in feral horses in the Grand Canyon. *Behav. Ecol. Sociobiol.* 2:131–46.

——— (1981). The role of risks in mammalian combat: zebra and onager fights. *Z. Tierpsychol.* 56:297–304.

——— (1983a). Predation, sex ratios, and male competition in equids (Mammalia, Perissodactyla). *J. Zool.* 201:205–16.

——— (1983b). Induced abortion and social factors in wild horses. *Nature* 303:59–61.

——— (1983c). Ecology and catastrophic mortality in wild horses: implications for interpreting fossil assemblages. *Science* 220:1403–4.

——— (1986). *Wild Horses of the Great Basin: Social Competition and Population Size.* Chicago: University of Chicago Press.

——— (1987). Reproductive fates of dispersers in a harem-dwelling ungulate: the wild horse. In *Mammalian Dispersal Patterns: The Effects of Social Structure on Population Genetics,* ed. B. D. Chepko-Sade & Z. T. Halpin, pp. 41–54. Chicago: University of Chicago Press.

Berggren, W. A., Kent, D. V., Flynn, J. J., & Van Couvering, J. A. (1985). Cenozoic geochronology. *Bull. Geol. Soc. Am.* 96:1407–18.

Berggren, W. A., & Van Couvering, J. A. (1974). The late Neogene: biostratigraphy, geochronology and paleoclimatology of the last 15 million years in marine and continental sequences. *Palaeogeogr., Palaeoclimatol., Palaeoecol.* 16:1–216.

Bernor, R. L., & Tobien, H. (1989). Two small species of *Cremohipparion* (Equidae, Mamm.) from Samos, Greece. *Mitt. Bayer. Staatsslg. Paläont. hist. Geol.* 29:207–26.

Berry, W. B. N. (1987). *Growth of a Prehistoric Time Scale: Based on Organic Evolution.* Palo Alto: Blackwell Scientific Publications.

Bonaparte, J. F. (1986). A new and unusual late Cretaceous mammal from Patagonia. *J. Vert. Paleontol.* 6:264–70.

Boné, E. L., & Singer, R. (1965). *Hipparion* from Langebaanweg, Cape Province and a revision of the genus in Africa. *Ann. South African Mus.* 48: 273–397.

Bowler, P. J. (1976). *Fossils and Progress: Paleontology and the Idea of Progressive Evolution in the Nineteenth Century.* New York: Science History Publications.

——— (1989). Holding your head up high: degeneration and orthogenesis in theories of human evolution. In *History, Humanity, and Evolution,* ed. J. R. Moore, pp. 329–53. Cambridge University Press.

Bown, T. M., & Kihm, A. J. (1981). *Xenicohippus,* an unusual new hyracothere (Mammalia, Perissodactyla) from lower Eocene rocks of Wyoming, Colorado, and New Mexico. *J. Paleontol.* 55:257–70.

Bown, T. M., Kraus, M. J., Wing, S. L., Fleagle, J. G., Tiffney, B. H., Simons, E. L., & Vondra, C. F. (1982). The Fayum primate forest revisited. *J. Human Evol.* 11:603–32.

Brown, B. (1926). Is this the earliest-known fossil collected by man? *Nat. Hist.* 26:535.

Brown, W. L., Jr. (1987). Punctuated equilibria excused: the original examples fail to support it. *Biol. J. Linnean Soc.* 31:383–404.

Bruce, R. V. (1987). *The Launching of Modern American Science 1846–1876.* Ithaca: Cornell University Press.

Buffetaut, E. (1987). *A Short History of Vertebrate Palaeontology.* London: Croom Helm.

Burmeister, H. (1875). *Los Caballos Fósiles de la Pampa Argentina.* Buenos Aires: La Tribuna.

Butler, P. B. (1952). The milk-molars of Perissodactyla, with remarks on molar occlusion. *Proc. Zool. Soc. London* 121:777–817.

Buzas, M. A., & Culver, S. J. (1984). Species duration and evolution: benthic Foraminifera on the Atlantic continental margin of North America. *Science* 225:829–30.

Cain, A. J. (1954). *Animal Species and Their Evolution.* London: Hutchinson's University Library.

Cain, S. A. (1944). *Foundations of Plant Geography.* New York: Harper.

Calder, W. A., III (1984). *Size, Function, and Life History.* Cambridge, Mass: Harvard University Press.

Camp, C. L., & Smith, N. (1942). Phylogeny and functions of the digital ligaments of the horse. *Univ. California Mem.* 13:69–124.

Carroll, R. L. (1988). *Vertebrate Paleontology and Evolution.* New York: Freeman.

Case, J. A. (1988). Paleogene floras from Seymour Island, Antarctic Peninsula. In *Geology and Paleontology of Seymour Island, Antarctic Peninsula,* ed. R. M. Feldmann & M. O. Woodburne, pp. 523–30. Geol. Soc. Amer. Mem. 169.

Cerling, T. E., Quade, J., Wang, Y., & Bowman, J. R. (1989). Carbon isotopes in soils and palaeosols as ecology and palaeoecology indicators. *Nature* 341:138–9.

Chow, M., & Qi, T. (1978). Paleocene mammalian faunas from Nomogen Formation of Inner Mongolia. *Vert. Palasiatica* 16:77–85.

Chubb, S. H. (1934). Frontal protuberances in horses: an explanation of the so-called "horned" horse. *Am. Mus. Novitates* 740:1–9.

Cifelli, R. L. (1981). Patterns of evolution among the Artiodactyla and Perissodactyla (Mammalia). *Evolution* 35:433–40.

Clark, J., Beerbower, J. R., & Kietzke, K. K. (1967). Oligocene sedimentation, stratigraphy, paleoecology and paleoclimatology in the Big Badlands of South Dakota. *Fieldiana: Geol. Mem.* 5:1–158.

Clutton-Brock, J. (1981). *Domesticated Animals from Early Times*. Austin: University of Texas Press. London: British Museum (Natural History).

Clutton-Brock, T. H., Guinness, F. E., & Albon, S. D. (1982). *Red Deer: Behavior and Ecology of Two Sexes*. Chicago: University of Chicago Press.

Colbert, E. H. (1935). Distributional and phylogenetic studies on Indian fossil mammals. II. The correlation of the Siwaliks of India as inferred by the migrations of *Hipparion* and *Equus*. *Am. Mus. Novitates* 797:1–15.

(1968). *Men and Dinosaurs: The Search in Field and Laboratory*. New York: Dutton.

(1980). *Evolution of the Vertebrates: A History of the Backboned Animals through Time*. New York: Wiley.

Conroy, G. C., & Vannier, M. W. (1984). Noninvasive three-dimensional computer imaging of matrix-filled fossil skulls by high-resolution computed tomography. *Science* 226:456–8.

Cope, E. D. (1884). *The Vertebrata of the Tertiary Formations of the West, Book 1*. Washington, D.C.: U.S. Geological Survey.

(1889). A review of the North American species of *Hippotherium*. *Proc. Am. Philos. Soc.* 26:429–58.

Cracraft, J. (1981). Patterns and process in paleobiology: the role of cladistic analysis in systematic paleontology. *Paleobiology* 7:456–68.

Cunningham, C., & Berger, J. (1986). Wild horses of the Granite Range. *Nat. Hist.* 95:32–8.

Cuvier, G. (1836). *Recherches sur les Ossemens Fossiles*. Paris: Edmond D'Ocagne.

Dalquest, W. W. (1978). Phylogeny of American horses of Blancan and Pleistocene age. *Acta Zool. Fennica* 15:191–9.

(1988). *Astrohippus* [sic] and the origin of Blancan and Pleistocene horses. *Occas. Pap. Mus. Texas Tech. Univ.* 116:1–23.

Dalrymple, G. B. (1979). Critical tables for conversion of K-Ar ages from old to new constants. *Geology* 7:558–60.

Dalrymple, G. B., & Lanphere, M. A. (1969). *Potassium-Argon Dating: Principles, Techniques and Applications to Geochronology*. San Francisco: Freeman.

Damuth, J. D. (1981). Population density and body size in mammals. *Nature* 290:699–700.

Damuth, J. D., & MacFadden, B. J. (1990). Introduction: body size and its estimation. In *Body Size in Mammalian Paleobiology: Estimation and Biological Implications*, ed. J. D. Damuth & B. J. MacFadden, pp. 1–10. Cambridge University Press.

Danforth, K. C. (ed.) (1982). *Journey into China*. Washington, D.C.: National Geographic Society.

Darlington, P. J. (1957). *Zoogeography: The Geographical Distribution of Animals*. New York: Wiley.

(1965). *Biogeography of the Southern End of the World*. Cambridge, Mass.: Harvard University Press.

Darwin, C. (1852). *Journal of Researches into the Natural History and Geology of the Countries Visited During the Voyage of the H.M.S. Beagle Round the World*, new ed. London: J. Murray.

(1859). *On the Origin of Species* (a facsimile of 1st edition, with introduction by Ernst Mayr, 1964). Cambridge, Mass.: Harvard University Press.

Dashzeveg, D. (1979). On an archaic representative of the equoids (Mammalia, Perissodactyla) from the Eocene of central Asia. *Trans. Joint Soviet-Mongolian Paleontol. Exped.* 8:10–22.

de Beer, G. (1954). Archaeopteryx [sic] and evolution. *Adv. Sci.* 42:1–11.

De Blase, A. F., & Martin, R. E. (1981). *A Manual of Mammalogy with Keys to the Families of the World*. Dubuque, Iowa: Wm. C. Brown Co.

Deevey, E. S., Jr. (1947). Life tables for natural populations of animals. *Quart. Rev. Biol.* 22:283–314.

Deino, A., Tauxe, L., Monaghan, M., & Drake, R. (1990). ^{40}Ar/^{39}Ar age calibration of the litho- and paleomagnetic stratigraphies of the Ngorora Formation, Kenya. *J. Geol.* 98:567–87.

DeNiro, M. J. (1987). Stable isotopy and archaeology. *Am. Sci.* 75:182–91.

Dennell, R. (1986). Needles and spear-throwers. *Nat. Hist.* 95(10):70–8.

De Porta, J. (1960). Los equidos fósiles de la Sabana de Bogotá. *Bol. Geol., Univ. Industrial Santander* (Colombia) 4:51–78.

Desmond, A. (1982). *Archetypes and Ancestors: Palaeontology in Victorian London 1850–1875*. Chicago: University of Chicago Press.

Devillers, C., Mahe, J., Ambrose, D., Bauchot,

R., & Chatelain, E. (1984). Allometric studies on the skull of living and fossil Equidae (Mammalia: Perissodactyla). *J. Vert. Paleontol.* 4:471–80.

Doherty, T. (1975). *The Anatomical Works of George Stubbs.* London: Secker & Warburg.

Donoghue, M. J., Doyle, J. A., Gauthier, J., Kluge, A. G., & Rowe, T. (1989). The importance of fossils in phylogeny reconstruction. *Ann. Rev. Ecol. Syst.* 20:431–60.

Downs, T. (1961). A study of variation and evolution in Miocene *Merychippus. Los Angeles Co. Mus. Contrib. Sci.* 45:1–75.

Dunbar, C. O., & Waage, K. M. (1969). *Historical Geology,* 3rd ed. New York: Wiley.

Durham, J. W. (1970). The fossil record and the origin of the Deuterostomata. *Proc. North Am. Paleontol. Conv.* H:1104–32.

Eadie, J. M., Broekhoven, L., & Colgan, P. (1987). Size ratios and artifacts: Hutchinson's rule revisited. *Am. Nat.* 129:1–17.

Edinger, T. (1948). Evolution of the horse brain. *Geol. Soc. Am. Mem.* 25:1–177.

Eicher, D. L. (1976). *Geologic Time,* 2nd ed. Englewood Cliffs, N.J.: Prentice-Hall.

Eisenberg, J. F. (1981). *The Mammalian Radiations: An Analysis of Trends in Evolution, Adaptations, and Behavior.* Chicago: University of Chicago Press.

(1990). The behavioral/ecological significance of body size in the Mammalia. In *Body Size in Mammalian Paleobiology: Estimation and Biological Implications,* ed. J. D. Damuth & B. J. MacFadden, pp. 25–37. Cambridge University Press.

Eisenmann, V. (1975). Nouvelles interprétations des restes D'Équidés (Mammalia, Perissodactyla) de Nihowan (Pléistocène Inférieur de La Chine du Nord): *Equus teilhardi* nov. sp. *Géobios* 8:125–34.

(1977). Les Hipparions africains: Valeur et signification de quelques caractères des jugales inférieures. *Bull. Mus. Natl. Hist. Nat. (France)* 438: 69–86.

(1980). Les chevaux (*Equus* sensu lato) fossiles et actuels: crânes et dents jugales supérieures. *Cahiers de Paléont., Cent. Natl. Recherche Sci.* 1–186, 22 plates.

(1981). Étude des dents jugales inférieures des *Equus* (Mammalia, Perissodactyla) actuels et fossiles. *Palaeovertebrata* 10:127–226, 4 plates.

(1986). Comparative osteology of modern and fossil horses, half-asses, and asses. In *Equids in the Ancient World,* ed. R. H. Meadow & H.-P. Uerpman, pp. 67–116. Wiesbaden: Dr. Ludwig Reichart Verlag.

Eisenmann, V., Sondaar, P. Y., Alberdi, M.-T.,

& DeGuili, C. (1987). Is horse phylogeny becoming a playfield in the game of theoretical evolution? *J. Vert. Paleontol.* 7:224–9.

Elderfield, H. (1986). Strontium isotope stratigraphy. *Palaeogeogr., Palaeoclimatol., Palaeoecol.* 57: 71–90.

Eldredge, N., & Gould, S. J. (1972). Punctuated equilibria: An alternative to phyletic gradualism. In *Models in Paleobiology,* ed. T. J. M. Schopf, pp. 82–115. San Francisco: Freeman, Cooper & Co.

Elias, M. K. (1942). Tertiary prairie grasses and other herbs from the High Plains. *Geol. Soc. Am. Spec. Pap.* 41:1–176.

Emry, R. J., Russell, L. S., & Bjork, P. J. (1987). The Chadronian, Orellan, and Whitneyan North American land mammal ages. In *Cenozoic Mammals of North America: Geochronology and Biostratigraphy,* ed. M. O. Woodburne, pp. 118–52. Berkeley: University of California Press.

Evander, R. L. (1989). Phylogeny of the family Equidae. In *The Evolution of Perissodactyls,* ed. D. R. Prothero & R. M. Schoch, pp. 109–27. Oxford: Clarendon Press.

Evernden, J. F., Savage, D. E., Curtis, G. H., & James, J. T. (1964). Potassium-argon dates and the Cenozoic mammalian chronology of North America. *Am. J. Sci.* 262:145–98.

Faure, G. (1986). *Principles of Isotope Geology,* 2nd ed. New York: Wiley.

Feist, J. D., & McCullough, D. R. (1975). Reproduction in feral horses. *J. Reprod. Fert., Suppl.* 23:13–18.

(1976). Behavior patterns and communication in feral horses. *Z. Tierpsychol.* 41:337–71.

Fischer, M. S. (1989). Hyracoids, the sister-group of perissodactyls. In *The Evolution of Perissodactyls,* ed. D. R. Prothero & R. M. Schoch, pp. 37–56. Oxford: Clarendon Press.

Flower, W. H. (1892). *The Horse: A Study in Natural History.* New York: D. Appleton & Co.

Flynn, J. J. (1986). Faunal provinces and the Simpson coefficient. In *Vertebrates, Phylogeny, and Philosophy,* ed. K. M. Flanagan & J. A. Lillegraven, pp. 317–38. Contrib. Geol., Univ. Wyoming, Spec. Pap. 3.

Flynn, J. J., MacFadden, B. J., & McKenna, M. C. (1984). Land-mammal ages, faunal heterochrony, and temporal resolution in Cenozoic terrestrial sequences. *J. Geol.* 92:687–705.

Ford, S. M. (1980). Callitrichids as phyletic dwarfs, and the place of the Callitrichidae in Platyrrhini. *Primates* 21:31–43.

Forsten, A. M. (1970). Variation in and between three populations of Mesohippus bairdii [*sic*]

Leidy from the Big Badlands, South Dakota. *Acta Zool. Fennica* 126:1–16.

——— (1975). The fossil horses of the Texas Gulf Coastal Plain: a revision. *Pearce-Sellards Ser., Texas Mem. Mus.* 22:1–86.

——— (1982a). The taxonomic status of the Miocene horse genus *Sinohippus. Palaeontology* 25:673–9.

——— (1982b). The status of the genus *Cormohipparion* Skinner and MacFadden (Mammalia, Equidae). *J. Paleontol.* 56:1332–5.

——— (1986). Chinese fossil horses of the genus *Equus. Acta Zool. Fennica* 181:1–40.

Forster-Cooper, C. (1932). The genus *Hyracotherium*. A revision and description of new specimens found in England. *Phil. Trans. Roy. Soc. London, Ser. B* 221:431–448.

Fortelius, M. (1985). Ungulate cheek teeth: developmental, functional, and evolutionary interrelations. *Acta Zool. Fennica* 180:1–76.

Franzen, J. L. (1976). Die Fossilfundstelle Messel: ihre Bedeutung für die palaontologische Wissenschaft. *Naturwissenschaften* 63:418–25.

——— (1977). *Urpferdchen und Krokodile Messel vor 50 Millionen Jahren.* Frankfurt: Senckenbergischen Naturforschenden Gesellschaft.

——— (1984). Die Stammesgeschichte der Pferde in ihrer wissenschaftshistorischen Entwicklung. *Natur Mus.* 114:149–62.

——— (1985). Exceptional preservation of Eocene vertebrates in the lake deposit of Grube Messel (West Germany). *Phil. Trans. Roy. Soc. London, Ser. B* 311:181–6.

French, E. G. (1951). *Good-bye to Boot and Saddle: Or the Tragic Passing of British Cavalry.* London: Hutchinson & Co.

Frick, C. (1937). Horned ruminants of North America. *Bull Am. Mus. Nat. Hist.* 69:1–669.

Froehlich, J. (1989). A small hyracothere from the San Jose Formation (Lower Eocene) of New Mexico. *J. Vert. Paleontol.* 9:21A.

Gaffney, E. S. (1979). An introduction to the logic of phylogeny reconstruction. In *Phylogenetic Analysis and Paleontology*, ed. J. Cracraft & N. Eldredge, pp. 79–111. New York: Columbia University Press.

Galusha, T. (1975). Childs Frick and the Frick Collection of fossil mammals. *Curator* 18:5–15.

Gamkrelidze, T. V., & Ivanov, V. V. (1990). The early history of Indo-European languages. *Sci. Am.* 262(3):110–116.

Gaudry, A. (1867). *Animaux Fossiles et Géologie de l'Attique*, 4th ed. Paris, 475 p., atlas, 75 plates [not seen].

——— (1896). *Essai de Paléontologie Philosophique.* Paris:

Masson (1980 facsimilie reprint by Arno Press, New York).

Gazin, C. L. (1936). A study of the fossil horse remains from the Upper Pliocene of Idaho. *Proc. U.S. Natl. Mus.* 83:281–319.

Geist, V. (1974). On the relationship of social evolution and ecology in ungulates. *Am. Zool.* 14: 205–20.

Geneser, F. (1986). *Textbook of Histology.* Philadelphia: Lea & Febiger.

George, T. N. (1956). Biospecies, chronospecies, and morphospecies. *Syst. Assoc. Pub.* 2:123–37.

George, M., & Ryder, O. A. (1986). Mitochondrial DNA evolution in the genus *Equus. Mol. Biol. Evol.* 3:535–46.

Getty, R. (1975). *Sisson and Grossman's The Anatomy of Domesticated Animals.* Philadelphia: W. B. Saunders.

Gidley, J. W. (1907). Revision of the Miocene and Pliocene Equidae of North America. *Bull. Am. Mus. Nat. Hist.* 23:865–934.

Gillette, D. D. (1991). Seismosaurus halli, gen. et sp. nov., a new sauropod dinosaur from the Morrison Formation (Upper Jurassic/Lower Cretaceous) of New Mexico, USA. *J. Vert. Paleontol.* 11:417–33.

Gillette, D. D., & Colbert, E. H. (1976). Catalog of type specimens of fossil vertebrates, Academy of Natural Sciences, Philadelphia. Part II: Terrestrial mammals. *Proc. Acad. Nat. Sci.* 128:25–8.

Gingerich, P. D. (1976). Paleontology and phylogeny: patterns of evolution at the species level in early Tertiary mammals. *Am. J. Sci.* 276:1–28.

——— (1977). Patterns of evolution in the mammalian fossil record. In *Patterns of Evolution, as Illustrated by the Fossil Record*, ed. A. Hallam, pp. 469–500. Amsterdam: Elsevier.

——— (1981). Variation, sexual dimorphism, and social structure in the early Eocene horse *Hyracotherium* (Mammalia, Perissodactyla). *Paleobiology* 7:443–55.

——— (1982). Time resolution in mammalian evolution: sampling, lineages, and faunal turnover. *Proc. 3rd N. Am. Paleontol. Conv.* I:205–10.

——— (1983). Rates of evolution: effects of time and temporal scaling. *Science* 222:159–61.

——— (1984a). Smooth curve of evolutionary rate: a psychological and mathematical artifact (response to comment by S. J. Gould). *Science* 226:995.

——— (1984b). Punctuated equilibria – Where is the evidence? *Syst. Zool.* 33:335–8.

——— (1987a). Simpson as a model. *Palaios* 2:111.

(1987b). Evolution and the fossil record: patterns, rates, and processes. *Can. J. Zool.* 65:1053–60.

(1989). New earliest Wasatchian mammalian fauna from the Eocene of northwestern Wyoming: composition and diversity in a rarely sampled high-floodplain assemblage. *Univ. Michigan Pap. Paleontol.* 28:1–97.

Glaessner, M. F. (1984). *The Dawn of Animal Life: A Biohistorical Study.* Cambridge University Press.

Glen, W. (1982). *The Road to Jaramillo: Critical Years of the Revolution in Earth Science.* Stanford: Stanford University Press.

Glimcher, M. J., Cohen-Solal, L., Kossiva, D., & de Ricqles, A. (1990). Biochemical analyses of fossil enamel and dentin. *Paleobiology* 16:219–32.

Goldschmidt, R. (1940). *The Material Basis of Evolution.* New Haven: Yale University Press.

Golenberg, E. M., Giannasi, D. E., Clegg, M. T., Smiley, C. J., Durbin, M., Henderson, D., & Zurawski, G. (1990). Chloroplast DNA sequence from a Miocene *Magnolia* species. *Nature* 344: 656–8.

Gould, S. J. (1983). *Hen's Teeth and Horse's Toes.* New York: Norton.

(1984). Smooth curve of evolutionary rate: a psychological and mathematical artifact. *Science* 226:994–5.

(1987). Life's little joke: the evolutionary histories of horses and humans share a dubious distinction. *Nat. Hist.* 96:16–25.

(1988a). The case of the creeping fox terrier clone. *Nat. Hist.* 97(1):16–24.

(1988b). On replacing the idea of progress with an operational notion of directionality. In *Evolutionary Progress,* ed. M. H. Nitecki, pp. 319–38. Chicago: University of Chicago Press.

(1988c). Trends as changes in variance: a new slant on progress and directionality in evolution. *J. Paleontol.* 62:319–29.

(1989a). An essay on a pig roast. *Nat. Hist.* 98(1):14–25.

(1989b). *Wonderful Life: The Burgess Shale and the Nature of History.* New York: Norton.

(1991). Fall in the house of Ussher: How foolish was the archbishop's precise date for Creation? *Nat. Hist.* 100(11):12–21.

Gould, S. J., & Eldredge, N. (1977). Punctuated equilibria: the tempo and mode of evolution reconsidered. *Paleobiology* 3:115–51.

Grand, T. I. (1990). The functional anatomy of body mass. In *Body Size in Mammalian Paleobiology: Estimation and Biological Implications,* ed. J. D. Damuth & B. J. MacFadden, pp. 39–47. Cambridge University Press.

Gray, J.E. (1821). On the natural arrangement of vertebrose animals. *London Med. Reposit. Rev.* 15:296–310 [not seen].

Gray, E. T. (1978). Horse genealogy: the Oregon connection. *Geology* 6:587–91.

Grayson, D. K. (1989). The chronology of North American late Pleistocene extinctions. *J. Archaeol Sci.* 16:153–66.

Gregory, J. T. (1942). Pliocene vertebrates from Big Spring Canyon, South Dakota. *Univ. California Publ., Dept. Geol. Sci.* 26:307–446.

(1971). Speculations on the significance of fossil vertebrates for the antiquity of the Great Plains of North America. *Abh. hess. L.-Amt. Bodenforsch* 60:64–72.

(1979). North American vertebrate paleontology, 1776–1976. In *Two Hundred Years of Geology in America: Proceedings of the New Hampshire Bicentennial Conference on the History of Geology,* ed. C. J. Schneer, pp. 305–35. Hanover, N.H.: University Press of New England.

Gregory, W. K. (1910). The orders of mammals. *Bull. Am. Mus. Nat. Hist.* 27:1–524.

(1920). Studies in comparative myology and osteology, no. V: On the anatomy of the preorbital fossae of Equidae and other ungulates. *Bull. Am. Mus. Nat. Hist.* 42:265–84.

Gromova, V. (1952). Le genre *Hipparion* (translated from Russian by St. Aubin). *Bur. Res. Min. Geol.* 12:1–288.

Groves, C. P., & Willoughby, D. P. (1981). Studies on the taxonomy and phylogeny of the genus *Equus.* 1: Subgeneric classification of the recent species. *Mammalia* 45:321–55.

Gunnell, G. F. (1990). Cenogram analysis of the Bridger B mammalian fauna (middle Eocene). *J. Vert. Paleontol.* 10:26A.

Haacke, W. (1893). *Gestaltung und Vererbung.* Leipzig: Weigel [not seen].

Haldane, J. B. S. (1949). Suggestions as to quantitative measurement of rates of evolution. *Evolution* 3:51–6.

(1954). The measurement of natural selection. *Proc. IX Int. Congr. Genet.* 1:480–7.

Harris, J. M., & Jefferson, G. T. (eds.) (1985). *Rancho La Brea: Treasures of the Tar Pits.* Los Angeles Co. Nat. Hist. Mus. Sci. Ser. 31.

Haubitz, B., Prokop, M., Döhring, W., Ostrom, J. H., & Wellnhofer, P. (1988). Computed tomography of *Archaeopteryx.* *Paleobiology* 14:206–13.

Hauff, B., & Hauff, R. B. (1981). *Das Holzmadenbuch.* Holzmaden: published by authors.

Hay, O. P. (1902). Bibliography and catalogue of the fossil Vertebrata of North America. *Bull. U.S. Geol. Surv.* 179:1–868.

Heilprin, A. (1887). *The Geographical and Geological*

Distribution of Animals. New York: D. Appleton & Co.

Hermanson, J. W., & Hurley, K. J. (1990). Architectural and histochemical analysis of the biceps brachii muscle of the horse. *Acta Anat.* 137:146–56.

Hermanson, J. W., & MacFadden, B. J. (1992). Evolutionary and functional morphology of the shoulder joint and stay apparatus in fossil and extant horses (Equidae). *J. Vert. Paleontol.*

Hickey, L. J. (1980). Paleocene stratigraphy and flora of the Clark's Fork Basin. In *Early Cenozoic Paleontology and Stratigraphy of the Bighorn Basin, Wyoming,* ed. P. D. Gingerich, pp. 33–49. Univ. Michigan Pap. Paleontol. 24.

Hickey, L. J., West, R. M., Dawson, M. R., & Choi, D. K. (1983). Arctic terrestrial biota: paleomagnetic evidence of age disparity with mid-northern latitudes during the late Cretaceous and early Tertiary. *Science* 221:1153–6.

Higuchi, R., Bowman, B., Freiberger, M., Ryder, O. A., & Wilson, A. C. (1984). DNA sequences from the quagga, an extinct member of the horse family. *Nature* 312:282–4.

Hildebrand, M. (1987). The mechanics of horse legs. *Am. Sci.* 75:594–601.

Hirsch, K. F., Stadtman, K. L., Miller, W. E., & Madsen, J. (1989). Upper Jurassic dinosaur egg from Utah. *Science* 243:1711–13.

Hodell, D. A., Mueller, P. A., & Garrido, J. R. (1991). Variations in strontium isotopic composition of seawater during the Neogene. *Geology* 19:24–7.

Hoffstetter, R. (1952). Les Mammifères Pléistocènes de La République de L'Équateur. *Soc. Géol. France. Mém.* 66:1–391, 8 plates.

Holledge, S. (1988). *Xi'an: Ancient Capital of China.* Lincolnwood, Ill.: Passport Books.

Hooker, J. J. (1984). A primitive ceratomorph (Perissodactyl, Mammalia) in the English early Eocene. *Bull. British Mus. (Nat. Hist.), Geol.* 33:101–14.

(1989). Character polarities in early perissodactyls and their significance for *Hyracotherium* and infraordinal relationships. In *The Evolution of Perissodactyls,* ed. D. R. Prothero & R. M. Schoch, pp. 79–101. Oxford: Clarendon Press.

Horn, H. S., & May, R. M. (1977). Limits to similarity among coexisting competitors. *Nature* 270:660–1.

Hsü, K. J. (1986). *The Great Dying: Cosmic Catastrophe, Dinosaurs, and the Theory of Evolution.* New York: Ballantine Books.

Hudson, C., Smith, M. T., DePratter, C. B., & Kelley, E. (1989). The Tristán de Luna expedition, 1559–1561. In *First Encounters: Spanish Explorations in the Caribbean and the United States, 1492–1570,* ed. J. T. Milanich & S. Milbrath, pp. 119–34. Gainesville: University of Florida Press.

Hughes, A. L. (1988). The quagga case: molecular evolution of an extinct species. *Trends Ecol. Evol.* 3:95–6.

Hulbert, R. C., Jr. (1982). Population dynamics of the three-toed horse *Neohipparion* from the late Miocene of Florida. *Paleobiology* 8:159–67.

(1984). Paleoecology and population dynamics of the early Miocene (Hemingfordian) horse *Parahippus leonensis* from the Thomas Farm site, Florida. *J. Vert. Paleontol.* 4:547–58.

(1987). A new *Cormohipparion* (Mammalia, Equidae) from the Pliocene (latest Hemphillian and Blancan) of Florida. *J. Vert. Paleontol.* 7:451–68.

(1988a). *Calippus* and *Protohippus* (Mammalia, Perissodactyla, Equidae) from the Miocene (Barstovian–early Hemphillian) of the Gulf Coastal Plain. *Bull. Florida State Mus., Biol. Sci.* 32:221–340.

(1988b). *Cormohipparion* and *Hipparion* (Mammalia, Perissodactyla, Equidae) from the late Neogene of Florida. *Bull. Florida State Mus., Biol. Sci.* 33:229–338.

(1989). Phylogenetic interrelationships and evolution of North American late Neogene Equidae. In *The Evolution of Perissodactyls,* ed. D. R. Prothero & R. M. Schoch, pp. 176–96. Oxford: Clarendon Press.

Hulbert, R. C., Jr., & MacFadden, B. J. (1991). Morphological transformation and cladogenesis at the base of the adaptive radiation of Miocene hypsodont horses. *Am. Mus. Novitates* 3000:1–61.

Hull, D. L. (1987). Genealogical actors in ecological roles. *Biol. & Philos.* 2:168–84.

(1988). Progress in ideas of progress. In *Evolutionary Progress,* ed. M. K. Nitecki, pp. 27–48. Chicago: University of Chicago Press.

Hull, D. L., Tessner, P. D., & Diamond, A. M. (1978). Planck's principle. *Science* 202:717–23.

Hunt, R. M., Jr. (1990). Taphonomy and sedimentology of Arikareean (lower Miocene) fluvial, eolian, and lacustrine paleoenvironments, Nebraska and Wyoming: a paleobiota entombed in fine-grained volcaniclastic rocks. In *Volcanism and Fossil Biotas,* ed. M. G. Lockley & A. Rice, pp. 69–111. Geol. Soc. Am. Spec. Pap. 244.

Hussain, S. T. (1975). Evolutionary and functional anatomy of the pelvic limb in fossil and Recent Equidae (Perissodactyla, Mammalia). *Anat., Histol., Embryol.* 4:179–222.

Hutchinson, G. E. (1959). Homage to Santa Rosalia, or why are there so many kinds of animals? *Am. Nat.* 93:145–59.

Huxley, J. S. (1940). *The New Systematics.* Oxford University Press.

Jablonski, D. (1987). How pervasive is Cope's rule? A test using Late Cretaceous mollusks. *Geol Soc. Am., Abstr. Prog.* 19(7):713–14.

Janis, C. M. (1976). The evolutionary strategy of the Equidae and the origins of rumen and cecal digestion. *Evolution* 30:757–74.

(1982). Evolution of horns in ungulates: ecology and paleoecology. *Biol. Rev.* 57:261–318.

(1984). The use of fossil ungulate communities as indicators of climate and environment. In *Fossils and Climate,* ed. P. Brenchley, pp. 85–104. New York: Wiley.

(1986). Evolution of horns and related structures in hoofed mammals. *Discovery* 19:8–17.

(1988). An estimation of tooth volume and hypsodonty indices in ungulate mammals, and the correlation of these factors with dietary preference. In *Teeth Revisited: Proceedings VIIth International Symposium Dental Morphology, Paris 1986,* ed. D. E. Russell, J.-P. Santoro, & D. Sigogneau-Russell, pp. 367–87. Mém. Mus. natn. Hist. nat., Paris, Sér. C, 53.

(1989). A climatic explanation for patterns of evolutionary diversity in ungulate mammals. *Palaeontology* 32:463–81.

(1990a). Correlation of reproductive and digestive strategies in the evolution of cranial appendages. In *Horns, Pronghorns, and Antlers: Evolution, Morphology, Physiology, and Social Significance,* ed. G. A. Bubenik & A. B. Bubenik, pp. 114–33. Berlin: Springer-Verlag.

(1990b). The correlation between diet and dental wear in herbivorous mammals, and its relationship to the determination of diets in extinct species. In *Evolutionary Paleobiology of Behavior and Coevolution,* ed. A. J. Boucot, pp. 241–59. Amsterdam: Elsevier.

Janis, C. M., & Ehrhardt, D. (1988). Correlation of relative muzzle width and relative incisor width with dietary preference in ungulates. *Zool. J. Linnean Soc.* 92:267–84.

Janis, C. M., & Fortelius, M. (1988). On the means whereby mammals achieve increased functional durability of their dentitions, with special reference to limiting factors. *Biol. Rev.* 63:197–230.

Jarman, P. J. (1974). The social organisation of antelope in relation to their ecology. *Behaviour* 48:213–67.

Jepsen, G. L. (1952). [Essay review] Horses; by George Gaylord Simpson. *Am. J. Sci.* 250:834–6.

Jepsen, G. L., Mayr, E., & Simpson, G. G. (eds.) (1949). *Genetics, Paleontology, and Evolution.* Princeton: Princeton University Press.

Jepsen, G. L., & Woodburne, M. O. (1969). Paleocene hyracothere from Polecat Bench Formation, Wyoming. *Science* 164:543–7.

Jerison, H. (1973). *Evolution of the Brain and Intelligence.* New York: Academic Press.

Joeckel, R. M. (1990). The auditory region of Leptauchenine oreodonts (Merycoidodontidae, Leptaucheniinae): a unique artiodactyl ear. *J. Vert. Paleontol.* 10:29A.

Johnson, N. M., Opdyke, N. D., & Lindsay, E. H. (1975). Magnetic polarity stratigraphy of Plio-Pleistocene terrestrial deposits and vertebrate faunas, San Pedro Valley, Arizona. *Geol. Soc. Am. Bull.* 86:5–12.

Kay, L. E. (1989). Selling pure science in wartime: the biochemical genetics of G. W. Beadle. *J. Hist. Biol.* 22:73–101.

Kennedy, W. J. (1977). Ammonite evolution. In *Patterns of Evolution,* ed. A. Hallam, pp. 251–304. Amsterdam: Elsevier.

Kent, D. V., McKenna, M. C., Opdyke, N. D., Flynn, J. J., & MacFadden, B. J. (1984). Technical comments: Arctic biostratigraphic heterochroneity. *Science* 224:173–4.

Kiltie, R. A. (1984). Seasonality, gestation time, and large mammal extinctions. In *Quaternary Extinctions: A Prehistoric Revolution,* ed. P. S. Martin & R. G. Klein, pp. 299–314. Tucson: University of Arizona Press.

Kitts, D. B. (1956). American *Hyracotherium* (Perissodactyla, Equidae). *Bull. Am. Mus. Nat. Hist.* 110:1–60.

(1957). A revision of the genus *Orohippus* (Perissodactyla, Equidae). *Am. Mus. Novitates* 1864:1–40.

(1977). *The Structure of Geology.* Dallas: Southern Methodist University Press.

Kleiber, M. (1961). *The Fire of Life.* New York: Wiley.

Klingel, H. (1975). Social organization and reproduction in equids. *J. Reprod. Fert., Suppl.* 23:7–11.

(1977). Observations on social organization and behaviour of African and Asiatic wild asses (*Equus africanus* and *E. hemionus*). *Z. Tierpsychol.* 44:323–31.

Knopf, A. (1949). Time in earth history. In *Genetics, Paleontology, and Evolution,* ed. G. L. Jepsen, G. G. Simpson, & E. Mayr, pp. 1–9. Princeton: Princeton University Press.

Koenigswald, W. von, & Schaarschmidt, F. (1983). Ein Urpferd aus Messel, das Weinbeeren frass. *Natur Mus.* 115:55–60.

Koepnick, R. B., Denison, R. E., & Dahl, D. A. (1988). The Cenozoic seawater $^{87}Sr/^{86}Sr$ curve: data review and implications for correlation of marine strata. *Paleoceanography* 3:743–56.

Kowalevsky, V. (1873). Sur *L'Anchitherium aurelianense* Cuv. et sur L'Historie paléontologique des chevaux. *Mém. L'Acad. Impériale Sci. St. Pétersbourg, 7th Sér.* 20(5):1–73, 3 plates.

Kragh, H. (1987). *An Introduction to the Historiography of Science.* Cambridge University Press.

Kraus, F., & Miyamoto, M. M. (1991). Rapid cladogenesis among the pecoran ruminants: evidence from mitochondrial DNA sequences. *Syst. Zool.* 40:117–30.

Krishtalka, L., Stuckey, R. K., & Redline, A. D. (1988). Geological remote sensing of Palaeogene rocks in the Wind River Basin, Wyoming, USA. In *Proc. Internat. Geosci. Remote Sensing Symp.* 2:1069–72.

Krishtalka, L., West, R. M., Black, C. C., Dawson, M. R., Flynn, J. J., Turnbull, W. D., Stuckey, R. K., McKenna, M. C., Bown, T. M., Golz, D. J., & Lillegraven, J. A. (1987). Eocene (Wasatchian through Duchesnean) biochronology of North America. In *Cenozoic Mammals of North America: Geochronology and Biostratigraphy*, ed. M. O. Woodburne, pp. 77–117. Berkeley: University of California Press.

Kurtén, B. (1953). On the variation and population dynamics of fossil and Recent mammal populations. *Acta Zool. Fennica* 76:1–122.

(1955). Contribution to the history of a mutation during 1,000,000 years. *Evolution* 9:107–18.

(1957). Mammal migrations, Cenozoic stratigraphy, and the age of Peking man and the australopithecines. *J. Paleontol.* 31:215–27.

(1960). Rates of evolution in fossil mammals. *Cold Spring Harbor Symp. Quantit. Biol.* 24:205–15.

(1966). Pleistocene bears of North America. 1: Genus *Tremarctos*, spectacled bears. *Acta Zool. Fennica* 115:1–120.

(1969). Sexual dimorphism in fossil mammals. In *Sexual Dimorphism in Fossil Metazoa and Taxonomic Implications*, ed. G. E. G. Westermann, pp. 226–33. Stuttgart: E. Schweizerbart'sche Verlagsbuchhandlung.

Lance, J. F. (1950). Paleontología y estratráfia del Plioceno de Yepómera, Estado de Chihuahua: 1ª parte: Equidos, excepto *Neohipparion. Univ. Nac. Autónoma México, Inst. Geol. Bol. Núm.* 54:1–81.

Lande, R. (1977). On comparing coefficients of variation. *Syst. Zool.* 26:214–17.

Lanham, U. (1973). *The Bone Hunters.* New York: Columbia University Press.

Legendre, S. (1986). Analysis of mammalian communities from the late Eocene and Oligocene of southern France. *Palaeovertebrata* 16:191–212.

(1987). Les communautés de mammifères du Paléogène (Eocène supérieur et Oligocène) d'Europe occidentale: structures, milieux, et évolution. Thèse du Doctorat d'Etat, Montpellier, France, Université de Montpellier [not seen].

Leidy, J. (1847). On the fossil horse of America. *Proc. Acad. Nat. Sci.* 3:262–6.

(1851). [no title; description of *Palaeotherium proutii*]. *Proc. Acad. Nat. Sci.* 5:122, 170–1.

Leroi-Gourhan, A. (1965). *Préhistorie de L'Art Occidental.* Paris: Éditions D'Art Lucien Mazenod.

(1967). *Treasures of Prehistoric Art.* New York: Harvey N. Abrams.

(1968). The evolution of Paleolithic art. *Sci. Am.* 218(2):58–70.

(1982). *The Dawn of European Art: An Introduction to Palaeolithic Cave Painting.* Cambridge University Press.

Leuthold, W., & Leuthold, B. M. (1975). Patterns of social grouping in ungulates in Tsavo National Park, Kenya. *J. Zool.* 175:405–20.

Lewin, R. (1980). Evolutionary theory under fire. *Science* 210:883–7.

(1982). Biology is not postage stamp collecting. *Science* 216:718–20.

(1986). Punctuated equilibrium is now old hat. *Science* 231:672–3.

Lincoln, R. J., Boxshall, G. A., & Clark, P. F. (1982). *A Dictionary of Ecology, Evolution and Systematics.* Cambridge University Press.

Lindsay, E. H., Johnson, N. M., & Opdyke, N. D. (1975). Preliminary correlation of North American land mammal ages and geomagnetic chronology. In *Studies in Cenozoic Paleontology and Stratigraphy*, ed. G. R. Smith & N. E. Friedland, pp. 111–19. Univ. Michigan Pap. Paleontol. 12.

Lindsay, E. H., Johnson, N. M., Opdyke, N. D., & Butler, R. F. (1987). Mammalian chronology and the magnetic polarity time scale. In *Cenozoic Mammals of North America: Geochronology and Biostratigraphy*, ed. M. O. Woodburne, pp. 267–84. Berkeley: University of California Press.

Lindsay, E. H., Opdyke, N. D., & Johnson, N. D. (1980). Pliocene dispersal of the horse *Equus* and late Cenozoic mammalian dispersal events. *Nature* 287:135–8.

Linnaeus, C. (1758). *Systema Naturae per Regna tria Naturae, secundum Classes, Ordines, Genera, Species, cum Characteribus, Differentiis, Synonymis, Locis,* 10th ed. Stockholm [not seen].

Lull, R. S. (1917). *Organic Evolution.* New York: Macmillan.

(1921). *The Theory of Evolution: With Special Reference to the Evidence Upon Which it is Founded.* New York: Macmillan.

(1931). The evolution of the horse family. *Yale Peabody Mus. Nat. Hist., Spec. Guide* 1:1–31.

Lull, R. S., & Wright, N. E. (1942). Hadrosaurian dinosaurs of North America. *Geol. Soc. Am. Spec. Pap.* 40:1–242, 31 plates.

Lundelius, E. L., Jr., Downs, T., Lindsay, E. H., Semken, H. A., Zakrzewski, R. J., Churcher, C. S., Harington, C. R., Schultz, G. E., & Webb, S. D. (1987). The North American Quaternary sequence. In *Cenozoic Mammals of North America: Geochronology and Biostratigraphy,* ed. M. O. Woodburne, pp. 211–35. Berkeley: University of California Press.

Lydekker, R. (1896). *A Geographical History of Mammals.* Cambridge University Press.

Maas, M., Krause, D. W., & Strait, S. G. (1988). The decline and extinction of the Plesiadapiformes (Mammalia: ?Primates) in North America: Displacement or replacement? *Paleobiology* 14:410–31.

McDonald, H. G. (1978). A supernumerary tooth in the ground sloth, *Megalonyx* (Edentata Mammalia). *Florida Sci.* 41:12–14.

Macdonald, L. (1987). *1914.* New York: Atheneum Press.

McDougall, I., & Harrison, T. M. (1988). *Geochronology and Thermochronology by the $^{40}Ar/^{39}Ar$ Method.* Oxford: Clarendon Press.

MacFadden, B. J. (1976). Cladistic analysis of primitive equids, with notes on other perissodactyls. *Syst. Zool.* 25:1–14.

(1977). "Eohippus" to *Equus:* fossil horses in the Yale Peabody Museum. *Discovery* 12:69–76.

(1980). An early Miocene land mammal (Oreodonta) from a marine limestone in northern Florida. *J. Paleontol.* 54:93–101.

(1981). Comments on Pregill's appraisal of historical biogeography of Caribbean vertebrates: vicariance, dispersal, or both? *Syst. Zool.* 30:370–2.

(1984a). Systematics and phylogeny of *Hipparion, Neohipparion, Nannippus,* and *Cormohipparion* (Mammalia, Equidae) from the Miocene and Pliocene of the New World. *Bull. Am. Mus. Nat. Hist.* 179:1–196.

(1984b). *Astrohippus* and *Dinohippus* from the Yepómera Local Fauna (Hemphillian, Mexico) and implications for the phylogeny of one-toed horses. *J. Vert. Paleontol.* 4:273–83.

(1985a). Patterns of phylogeny and rates of evolution in fossil horses: hipparions from the Miocene and Pliocene of North America. *Paleobiology* 11:245–57.

(1985b). Drifting continents, mammals, and time scales: current developments in South America. *J. Vert. Paleontol.* 5:169–74.

(1986). Late Miocene monodactyl horses (Mammalia, Equidae) from the Bone Valley Formation of central Florida. *J. Paleontol.* 60:466–75.

(1987a). Fossil horses from "Eohippus" (*Hyracotherium*) to *Equus:* scaling, Cope's law, and the evolution of body size. *Paleobiology* 12:355–69.

(1987b). Systematics, phylogeny and evolution of fossil horses: a rational alternative to Eisenmann et al. *J. Vert. Paleontol.* 7:230–5.

(1988a). Horses, the fossil record, and evolution – a current perspective. In *Evolutionary Biology,* Vol. 22, ed. M. K. Hecht, B. Wallace, & G. T. Prance, pp. 131–58. New York: Plenum.

(1988b). Fossil horses from "Eohippus" (*Hyracotherium*) to *Equus.* 2: Rates of dental evolution revisited. *Biol. J. Linnean Soc.* 35:37–48.

(1989). Dental character variation in paleopopulations and morphospecies of fossil horses and extant analogues. In *The Evolution of Perissodactyls,* ed. D. R. Prothero & R. M. Schoch, pp. 128–41. Oxford: Clarendon Press.

(1990). Chronology of Cenozoic primate localities in South America. *J. Human Evol.* 19:7–21.

In press. Magnetic polarity stratigraphy of vertebrate-bearing sediments: methods, assumptions, and selected applications. *Geol. Soc. Amer. Spec. Pap.*

MacFadden, B. J., & Azzaroli, A. (1987). Cranium of *Equus insulatus* (Mammalia, Equidae) from the middle Pleistocene of Tarija, Bolivia. *J. Vert. Paleontol.* 7:230–5.

MacFadden, B. J., Bryant, J. D., & Mueller, P. A. (1991). Sr-isotopic, paleomagnetic, and biostratigraphic calibration of horse evolution: evidence from the Miocene of Florida. *Geology* 19:242–5.

MacFadden, B. J., Campbell, K. E., Jr., Cifelli, R. L., Siles, O., Johnson, N. M., Naeser, C. W., & Zeitler, P. K. (1985). Magnetic polarity stratigraphy and mammalian fauna of the Deseadan (late Oligocene–early Miocene) Salla Beds of northern Bolivia. *J. Geol.* 93:223–50.

MacFadden, B. J., & Hulbert, R. C., Jr. (1988). Explosive speciation at the base of the adaptive radiation of Miocene grazing horses. *Nature* 336:466–8.

(1990). Body size estimates and size distribution of ungulate mammals from the late Miocene Love Bone Bed of Florida. In *Body Size in Mammalian Paleobiology: Estimation and Biological Implications*, ed. J. Damuth & B. J. MacFadden, pp. 337–63. Cambridge University Press.

MacFadden, B. J., Opdyke, N. D., & Johnson, N. M. (1979). Magnetic polarity stratigraphy of the Mio-Pliocene mammal-bearing Big Sandy Formation of western Arizona. *Earth Planet. Sci. Letters* 44:349–64.

MacFadden, B. J., & Skinner, M. F. (1979). Diversification and biogeography of the one-toed horses *Onohippidium* and *Hippidion*. *Yale Peabody Mus. Nat. Hist. Postilla* 175:1–9.

(1981). Earliest Holarctic hipparion, *Cormohipparion goorisi* n. sp. (Mammalia, Equidae), from the Barstovian (medial Miocene) Texas Gulf Coastal Plain. *J. Paleontol.* 55:619–27.

(1982). Hipparion horses and modern phylogenetic interpretation: comments on Forsten's view of *Cormohipparion*. *J. Paleontol.* 56:1336–42.

MacFadden, B. J., Swisher, C. C., III, Opdyke, N. D., & Woodburne, M. O. (1990). Paleomagnetism, geochronology, and possible tectonic rotation of the middle Miocene Barstow Formation, Mojave Desert, southern California. *Bull. Geol. Soc. Am.* 102:478–93.

MacFadden, B. J., & Woodburne, M. O. (1982). Systematics of the Neogene Siwalik hipparions (Mammalia, Equidae) based on cranial and dental morphology. *J. Vert. Paleontol.* 2:185–218.

MacGinitie, H. D. (1962). The Kilgore Flora: a late Miocene flora from northern Nebraska. *Univ. California Pub. Geol. Sci.* 35:67–158.

(1969). The Eocene Green River flora of northwestern Colorado and northeastern Utah. *Univ. California Pub. Geol. Sci.* 83:1–203.

McGrew, P. O. (1938). The Burge Fauna, a lower Pliocene mammalian assemblage from Nebraska. *Univ. California Pub. Bull. Dept. Geol. Sci.* 24:309–28.

(1944). An early Pleistocene (Blancan) fauna from Nebraska. *Field. Mus. Nat. Hist., Geol. Ser.* 9:33–66.

(1953). A new and primitive early Oligocene horse from Trans-Pecos Texas. *Fieldiana, Geol.* 10:167–71.

McKenna, M. C. (1960). Fossil Mammalia from the early Wasatchian Four Mile Fauna, Eocene of northwest Colorado. *Univ. California Pub. Geol. Sci.* 37:1–130.

(1972). Was Europe connected directly to North America prior to the middle Eocene?

In *Evolutionary Biology, Vol. 6.*, ed. T. Dobzhansky, M. Hecht, & W. C. Steere, pp. 179–88. New York: Appleton-Century-Crofts.

(1973). Sweepstakes, filters, corridors, Noah's arks, and beached Viking funeral ships in palaeogeography. In *Implications of Continental Drift to the Earth Sciences, Vol. 1*, ed. D. H. Tarling & S. K. Runcorn, pp. 295–308. London: Academic Press.

(1975a). Fossil mammals and early Eocene North Atlantic land continuity. *Ann. Missouri Bot. Garden* 62:335–53.

(1975b). Toward a phylogenetic classification of the Mammalia. In *Phylogeny of the Primates*, ed. W. P. Luckett & F. S. Szalay, pp. 21–46. New York: Plenum Press.

(1983). Cenozoic paleogeography of North Atlantic land bridges. In *Structure and Development of the Greenland-Scotland Ridge*, ed. M. H. P. Bott, S. Savox, M. Talwani, & J. Thiede, pp. 351–99. New York: Plenum Press.

(1987). Molecular and morphological analysis of high-level mammalian interrelationships. In *Molecules and Morphology in Evolution: Conflict or Compromise?* ed. C. Patterson, pp. 55–93. Cambridge University Press.

McKenna, M. C., Chow, M., Ting, S., & Zhexi, L. (1989). *Radinskya yupingae*, a perissodactyl-like mammal from the late Paleocene of China. In *The Evolution of Perissodactyls*, ed. D. R. Prothero & R. M. Schoch, pp. 24–36. Oxford: Clarendon Press.

McKinney, M. L. (1988). *Heterochrony in Evolution: A Multidisciplinary Approach*. New York: Plenum Press.

McNab, B. K. (1963). Bioenergetics and determination of home range size. *Am. Nat.* 97:133–40.

McNaughton, S. J., Tarrants, J. L., McNaughton, M. M., & Davis, R. H. (1985). Silica as a defense against herbivory and a growth promoter in African grasses. *Ecology* 66:528–35.

Maiorana, V. C. (1978). An explanation of ecological and developmental constraints. *Nature* 273:375–7.

Malmgren, B. A., Berggren, W. A., & Lohmann, G. P. (1984). Species formation through punctuated gradualism in planktonic foraminifera. *Science* 225:317–19.

Marcus, L. F. (1964). Measurement of natural selection in natural populations. *Nature* 202:1033–4.

Marsh, O. C. (1876). Notice of new Tertiary mammals. *Am. J. Sci.* 12:401–4.

(1879). Polydactyle horses, recent and extinct. *Am. J. Sci.* 17:499–505.

(1884). Dinocerata: a monograph of an extinct order of gigantic mammals. *U.S. Geol. Surv. Monogr.* 10:1–237.

(1892). Recent polydactyle horses. *Am. J. Sci.* 43:339–55.

(1895). Thomas Henry Huxley. *Am. J. Sci.* 50:177–83.

Marshall, C. R. (1990). Confidence intervals on stratigraphic ranges. *Paleobiology* 16:1–10.

Martin, E. T. (1981). Megalonyx, mammoth, and Mother Earth (Thomas Jefferson). In *Language of the Earth*, ed. F. H. T. Rhodes & R. O. Stone, pp. 284–8. New York: Pergamon Press.

Martin, L. D. (1975). Biostratigraphic relationships of the early Miocene Gering Fauna. *Proc. Nebraska Acad. Sci.* 85:41.

Martin, P. S., & Klein, R. G. (eds.) (1984). *Quaternary Extinctions: A Prehistoric Revolution.* Tucson: University of Arizona Press.

Martin, P. S., & Wright, H. E., Jr. (eds.) (1967). *Pleistocene Extinctions: The Search for a Cause.* New Haven: Yale University Press.

Matthew, W. D. (1903). The evolution of the horse. *Am. Mus. Nat. Hist., Suppl. Am. Mus. J. Guide Leaflet* 9(3):1–30.

(1924). Third contribution to the Snake Creek Fauna. *Bull. Am. Mus. Nat. Hist.* 50:59–210.

(1926). The evolution of the horse: a record and its interpretation. *Quart. Rev. Biol.* 1:139–85.

(1930). Pattern of evolution. *Sci. Am.* 143:192–6.

(1939). *Climate and Evolution,* 2nd ed. New York Academy of Sciences, Special Publication 1.

Matthew, W. D., & Stirton, R. A. (1930). Equidae from the Pliocene of Texas. *Univ. California Pub. Bull. Dept. Geol. Sci.* 19:349–96.

Mayr, E. (1969). *Principles of Systematic Zoology.* New York: McGraw-Hill.

(1970). *Populations, Species, and Evolution.* Cambridge, Mass.: Belknap Press, Harvard University.

Merriam, J. C., & Stock, C. (1932). *The Felidae of Rancho La Brea.* Washington, D.C.: Carnegie Institution Publication 422.

Miller, H. S. (1981). Fossil and free enterprisers. In *Language of the Earth*, ed. F. H. T. Rhodes & R. O. Stone, pp. 104–8. New York: Pergamon Press.

Mitchell, P. C. (1905). On the intestinal tract of mammals. *Trans. Zool. Soc. London* 17:437–536.

Miyamoto, M. M., & Goodman, M. (1986). Biomolecular systematics of eutherian mammals: phylogenetic patterns and classification. *Syst. Zool.* 35:230–40.

Miyamoto, M. M., Slightom, J. L., & Goodman, M. (1987). Phylogenetic relations of humans and African apes from DNA sequences in the $\psi\nu$-globin region. *Science* 238:369–73.

Moir, R. J. (1968). Ruminant digestion and evolution. In *Handbook of Physiology, Vol. 5, Chapter 126, sect. 6 (Alimentary Canal),* pp. 2673–94, Washington, D.C.: Am. Physiol. Soc.

Mones, A. (1982). An equivocal nomenclature: What means hypsodonty? *Palaeontol. Zeitschrift* 56:107–11.

Moody, P. A. (1970). *Introduction to Evolution,* 3rd ed. New York: Harper & Row.

Moore, R. C. (ed.) (1953–86). *The Treatise on Invertebrate Paleontology.* Lawrence: University of Kansas and Geological Society of America.

Morris, W. J. (1968). A new early Tertiary perissodactyl, *Hyracotherium seekinsi,* from Baja California. *Los Angeles Co. Mus. Contrib. Sci.* 151:1–11.

Murray, A. (1866). *The Geographical Distribution of Mammals.* London: Day & Son.

Muybridge, E. (1899). *Animals in Motion: An Electro-photographic Investigation of Consecutive Phases of Animal Progressive Movements.* London: Chapman & Hall.

Naeser, C. W. (1979). Fission track dating and geological annealing of fission tracks. In *Lectures in Isotope Geology*, ed. E. Jaeger & J. L. Hunziker, pp. 154–69. Berlin: Springer-Verlag.

Neilson, W. A. (1952). *Webster's New International Dictionary,* 2nd ed., unabridged. Springfield, Mass.: G. & C. Merriam Co.

Nelson, G., & Platnick, N. (1981). *Systematics and Biogeography: Cladistics and Vicariance.* New York: Columbia University Press.

Nelson, G., & Rosen, D. E. (eds.) (1981). *Vicariance Biogeography: A Critique.* New York: Columbia University Press.

Neville, C., Opdyke, N. D., Lindsay, E. H., & Johnson, N. M. (1979). Magnetic stratigraphy of Pliocene deposits of the Glenns Ferry Formation, Idaho, and its implications for North American mammalian biostratigraphy. *Am. J. Sci.* 279:503–26.

Nicholson, T. D., Schaeffer, B., Galusha, T., McKenna, M. C., Skinner, M. F., Taylor, B. E., & Tedford, R. H. (1975). The fossil mammal collections of the American Museum of Natural History. *Curator* 18:18–38.

Nitecki, M. H. (ed.) (1979). *Mazon Creek Fossils.* New York: Academic Press.

(1988a). Discerning the criteria for concepts of progress. In *Evolutionary Progress*, ed. M. H.

Nitecki, pp. 3–24. Chicago: University of Chicago Press.

(ed.) (1988b). *Evolutionary Progress.* Chicago: University of Chicago Press.

Novacek, M. J. (1982). Information for molecular studies from anatomical and fossil evidence on higher eutherian phylogeny. In *Macromolecular Sequences in Systematic and Evolutionary Biology,* ed. M. Goodman, pp. 3–41. New York: Plenum Press.

(1985). Evidence for echolocation in the oldest known bats. *Nature* 315:140–1.

(1986). The skull of leptictid insectivores and the higher-level classification of Eutherian mammals. *Bull. Am. Mus. Nat. Hist.* 183:1–112.

(1988). Fossils, missing data, and phylogenies. *J. Vert. Paleontol.* 8:23A.

Novacek, M. J., Ferrusquía-Villafranca, I., Flynn, J. J., Wyss, A. R., & Norell, M. (1991). Wasatchian (early Eocene) mammals and other vertebrates from Baja California, Mexico: the Lomas las Tetas de Cabra Fauna. *Bull. Am. Mus. Nat. Hist.* 208:1–88.

Novacek, M. J., & Norell, M. A. (1982). Fossils, phylogeny, and taxonomic rates of evolution. *Syst. Zool.* 31:366–75.

Novacek, M. J., Wyss, A., & McKenna, M. C. (1988). The major groups of eutherian mammals. In *The Phylogeny and Classification of the Tetrapods, Volume 2: Mammals,* ed. M. J. Benton, pp. 31–71. Systematics Association Special Volume No. 35B. Oxford: Clarendon Press.

Olsen, S. L. (1989). Solutré: A theoretical approach to the reconstruction of Upper Palaeolithic hunting strategies. *J. Human Evol.* 18:295–327.

Olson, E. C. (1952). The evolution of a Permian vertebrate chronofauna. *Evolution* 6:181–96.

(1966). Community evolution and the origin of mammals. *Ecology* 47:291–302.

Olson, S. L. (1985). The fossil record of birds. *Avian Biol.* 8:79–252.

Opdyke, N. D. (1990). Magnetic stratigraphy of Cenozoic terrestrial sediments and mammalian dispersal. *J. Geol.,* 98:621–37.

Opdyke, N. D., Jones, D. S., MacFadden, B. J., Smith, D. L., Mueller, P. A., & Shuster, R. D. (1987). Florida as an exotic terrane: paleomagnetic and geochronologic investigation of lower Paleozoic rocks from the subsurface of Florida. *Geology* 15:900–3.

Osborn, H. F. (1902). Dolichocephaly and brachycephaly in the lower mammals. *Bull. Am. Mus. Nat. Hist.* 16:77–89.

(1908). *From the Greeks to Darwin: An Outline of the Development of the Evolution Idea.* New York: Macmillan.

(1910). *The Age of Mammals: In Europe, Asia, and North America.* New York: Macmillan.

(1912a). Integument of the iguanodont dinosaur *Trachodon. Am. Mus. Nat. Hist., Mem., N. Ser.,* 1:33–54.

(1912b). Craniometry of the Equidae. *Am. Mus. Nat. Hist., Mem.* 3:55–100.

(1915). [no title; description of *Kalobatippus* gen. nov.]. In *Hitherto Unpublished Plates of Tertiary Mammalia and Permian Vertebrata,* ed. E. D. Cope & W. D. Matthew, plate CVIII. Am. Mus. Nat. Hist., Monograph Ser. 2.

(1917). *The Origin and Evolution of Life: On the Theory of Action, Reaction and Interaction of Energy.* New York: Charles Scribner's Sons.

(1918). Equidae of the Oligocene, Miocene, and Pliocene of North America. Iconographic type revision. *Mem. Am. Mus. Nat. Hist., N. Ser.* 2:1–326.

(1922). *Hesperopithecus,* the first anthropoid primate found in America. *Am. Mus. Novitates* 37:1–5.

(1930). *Fifty-two Years of Research, Observation and Publication, 1877–1929.* New York: Charles Scribner's Sons.

(1934). Aristogenesis, the creative principle in the origin of species. *Am. Nat.* 68:193–235.

Osborn, H. F., & Matthew, W. D. (1909). Cenozoic mammal horizons of western North America. *U.S. Geol. Surv. Bull.* 361:1–138.

Owen, R. (1840a). Description of the fossil remains of a mammal, a bird, and a serpent, from the London Clay. *Proc. Geol. Soc. London* 3:162–6.

(1840b). *The Zoology of the Voyage of the H.M.S. Beagle. Part I. Mammalia.* London: Smith Elder & Co.

(1848). Description of teeth and portions of jaws of two extinct anthracotheroid quadrapeds (*Hyopotamus vectianus* and *H. bovinus*) discovered by the Marchioness of Hastings in the Eocene deposits on the N.W. coast of the Isle of Wight, with an attempt to develop Cuvier's idea of the classification of pachyderms by the number of their toes. *Quart. J. Geol. Soc. London* 4:104–41.

(1868). *On the Anatomy of Vertebrates. Vol. 3. Mammals.* London: Longmans, Green & Co. [not seen].

Owen-Smith, N. (1985). Niche separation among African ungulates. In *Species and Speciation,* ed. E. S. Vrba, pp. 167–71. Pretoria: Transvaal Museum Monograph No. 4.

(1988). *Megaherbivores: The Influence of Very Large Body Size on Ecology.* Cambridge University Press.

Patterson, B., & Pascual, R. (1972). The fossil mammal fauna of South America. In *Evolution, Mammals, and Southern Continents,* ed. A. Keast, F. C. Erk, & B. Glass, pp. 247–309. Albany: State University of New York Press.

Patterson, C. (1978). Verifiability in systematics. *Syst. Zool.* 27:218–22.

(1987). *Evolution.* London: British Museum (Natural History); Ithaca, NY: Cornell University Press.

Penguilly, D. (1984). Developmental versus functional explanations for patterns of variability and correlation in the dentitions of foxes. *J. Mammal.* 65:34–43.

Peters, R. H. (1983). *The Ecological Implications of Body Size.* Cambridge University Press.

Pirlot, P.-L. (1952). Les canines chez *Hipparion* et L'apparition D'un caractère sexuel secondaire des mammifères. *Bull. Mus. (Nat. Hist.), 2nd Sér.* 24:419–22.

Plotnick, R. E. (1986). A fractal model for the distribution of stratigraphic hiatuses. *J. Geol.* 94:885–90.

Pratt, A. E. (1990). Taphonomy of the large vertebrate fauna from the Thomas Farm locality (Miocene, Hemingfordian), Gilchrist County, Florida. *Bull. Florida Mus. Nat. Hist., Biol. Sci.* 35:35–130.

Preston, D. J. (1986). *Dinosaurs in the Attic: An Excursion into the American Museum of Natural History.* New York: Ballantine Books.

Prothero, D. R. (1988). Mammals and magnetostratigraphy. *J. Geol. Education* 36:227–36.

(1989). Stepwise extinctions and climatic decline during the later Eocene and Oligocene. In *Mass Extinctions, Processes and Evidence,* ed. S. K. Donovan, pp. 217–34. New York: Columbia University Press.

Prothero, D. R., Manning, E. M., & Fischer, M. (1988). The phylogeny of ungulates. In *The Phylogeny and Classification of the Tetrapods. Volume 2: Mammals,* ed. M. E. Benton, pp. 201–34. Oxford: Clarendon Press.

Prothero, D. R., & Schoch, R. M. (1989a). Origin and evolution of the Perissodactyla: summary and synthesis. In *The Evolution of Perissodactyls,* ed. D. R. Prothero & R. M. Schoch, pp. 504–29. Oxford: Clarendon Press.

(1989b). Classification of the Perissodactyla. In *The Evolution of Perissodactyls,* ed. D. R. Prothero & R. M. Schoch, pp. 530–7. Oxford: Clarendon Press.

Prothero, D. R., & Shubin, N. (1989). The evolution of Oligocene horses. In *The Evolution of Perissodactyls,* ed. D. R. Prothero & R. M. Schoch, pp. 142–75. Oxford: Clarendon Press.

Prothero, J., & Jürgens, K. D. (1987). Scaling of maximal lifespan in mammals: a review. In *Evolution of Longevity in Animals,* ed. A. D. Woodhead & K. H. Thompson, pp. 49–74. New York: Plenum Press.

Prout, H. A. (1847). Description of a fossil maxillary bone of a Palaeotherium [*sic*] from near White River. *Am. J. Sci.* 3:248–50.

Provine, W. (1988). Progress of evolution and the meaning of life. In *Evolutionary Progress,* ed. M. H. Nitecki, pp. 49–74. Chicago: University of Chicago Press.

Qiu, Z., Huang, W., & Guo, Z. (1987). The Chinese hipparionine fossils. *Palaeont. Sinica, Ser. C* 25:1–250, 47 plates.

Quade, J., Cerling, T. E., & Bowman, J. R. (1989a). Systematic variations in the carbon and oxygen isotopic composition of pedogenic carbonate along elevation transects in the southern Great Basin, United States. *Bull. Geol. Soc. Am.* 101:464–75.

(1989b). Development of Asian monsoon revealed by marked ecological shift during the latest Miocene in northern Pakistan. *Nature* 342:163–5.

Quinn, J. H. (1955). Miocene Equidae of the Texas Gulf Coastal Plain. *Bur. Econ. Geol. Univ. Texas. Pub.* 5516:1–102.

Racle, F. A. (1979). *Introduction to Evolution.* Englewood Cliffs, N.J.: Prentice-Hall.

Radinsky, L. R. (1966). The adaptive radiation of the phenacodontid condylarths and the origin of the Perissodactyla. *Evolution* 20:408–17.

(1969). The early evolution of the Perissodactyla. *Evolution* 23:308–26.

(1976). Oldest horse brains: more advanced than previously realized. *Science* 194:626–7.

(1978a). Do albumin clocks run on time? *Science* 200:1182–3.

(1978b). Evolution of brain size in carnivores and ungulates. *Am. Nat.* 112:815–31.

(1983). Allometry and reorganization in horse skull proportions. *Science* 221:1189–91.

(1984). Ontogeny and phylogeny in horse skull evolution. *Evolution* 38:1–15.

Rainger, R. (1981). The continuation of the morphological tradition: American paleontology, 1880–1910. *J. Hist. Biol.* 14:129–58.

(1986). Just before Simpson: William Diller Matthew's understanding of evolution. *Proc. Am. Philos. Soc.* 130:453–74.

(1989). What's the use: William King Gregory and the functional morphology of fossil vertebrates. *J. Hist. Biol.* 22:103–39.

Raphael, M. (1945). *Prehistoric Cave Paintings.* Bollingen Series IV. Washington, D.C.: Pantheon Books.

Rau, R. E. (1986). The quagga and its kin. *Sagittarius* 1(2):8–10.

Raup, D. M. (1975). Taxonomic survivorship curves and Van Valen's law. *Paleobiology* 1:82–96.

(1978). Cohort analysis of generic survivorship. *Paleobiology* 4:1–15.

Raup, D. M., & Boyajian, G. E. (1988). Patterns of extinction in the fossil record. *Paleobiology* 14:109–25.

Reeve, E. C. R., & Murray, P. D. F. (1942). Evolution in the horse's skull. *Nature* 150:402–3.

Renders, E. (1984). The gait of *Hipparion* sp. from fossil footprints in Laetoli, Tanzania. *Nature* 308:179–81.

Renders, E., & Sondaar, P. Y. (1987). Hipparion [*sic*]. In *Laetoli: A Pliocene Site in Northern Tanzania,* ed. M. D. Leakey & J. M. Harris, pp. 471–481. Oxford: Clarendon Press.

Rensberger, J. M., Forsten, A., & Fortelius, M. (1984). Functional evolution of the cheek tooth pattern and chewing direction in Tertiary horses. *Paleobiology* 10:439–52.

Rensch, B. (1959). *Evolution Above the Species Level.* New York: Columbia University Press.

Repenning, C. H. (1967). Subfamilies and genera of the Soricidae. *U.S. Geol. Surv. Prof. Pap.* 565:1–74.

Retallack, G. J. (1981). Fossil soils: indicators of ancient terrestrial environments. In *Paleobotany, Paleoecology and Evolution, Vol. 1,* ed. K. J. Niklas, pp. 55–102. New York: Praeger.

(1982). Paleopedological perspectives on the development of grasslands during the Tertiary. *Proc. 3rd North Am. Paleontol. Conv.* 2:417–21.

(1983a). Paleopedological approach to the interpretation of terrestrial sedimentary rocks: the mid-Tertiary fossil soils of Badlands National Park, South Dakota. *Geol. Soc. Am. Bull.* 94:823–40.

(1983b). Late Eocene and Oligocene paleosols from Badlands National Park, South Dakota. *Geol. Soc. Am. Spec. Pap.* 193:1–82.

Rich, P. V., Rich, T. H., Wagstaff, B. E., McEwen Mason, J., Douthitt, C. B., Gregory, R. T., & Felton, E. A. (1988). Evidence for low temperatures and biologic diversity in Cretaceous high latitudes of Australia. *Science* 242:1403–6.

Richey, K. A. (1948). Lower Pliocene horses from Black Hawk Ranch, Mount Diablo, California. *Univ. California Bull. Dept. Geol. Sci.* 28:1–44.

Rickards, R. B. (1977). Patterns of evolution in graptolites. In *Patterns of Evolution,* ed. A. Hallam, pp. 333–58. Amsterdam: Elsevier.

Rieppel, O. C. (1988). *Fundamentals of Comparative Biology.* Basel: Birkhäuser Verlag.

Robb, R. C. (1935). A study of mutations in evolution, parts I and II. *J. Genetics* 31:39–52.

Rogers, D. J., Fleming, H. S., & Estabrook, G. (1967). Use of computers in studies of taxonomy and evolution. In *Evolutionary Biology, Vol. 1,* ed. T. Dobzhansky, M. K. Hecht, & W. C. Steere, pp. 169–96. New York: Appleton-Century-Crofts.

Romer, A. S. (1966). *Vertebrate Paleontology.* Chicago: University of Chicago Press.

Root, R. B. (1967). The niche exploitation pattern of the blue-gray gnatcatcher. *Ecol. Monogr.* 37:317–50.

Rose, K. D. (1990). Postcranial skeletal remains and adaptations in early Eocene mammals from the Willwood Formation, Bighorn Basin, Wyoming. In *Dawn of the Age of Mammals in the Northern Part of the Rocky Mountain Interior, North America,* ed. T. M. Bown & K. D. Rose, pp. 107–33. Geol. Soc. Am. Spec. Pap. 243.

Rosen, D. E. (1976). A vicariance model of Caribbean biogeography. *Syst. Zool.* 24:431–6.

Roth, V. L. (1981). Constancy in size ratios of sympatric species. *Am. Nat.* 118:394–404.

Rudwick, M. J. S. (1972). *The Meaning of Fossils: Episodes in the History of Palaeontology.* London: Macdonald.

Sack, W. O., & Habel, R. E. (1977). *Rooney's Guide to the Dissection of the Horse.* Ithaca, N.Y.: Veterinary Textbooks.

Savage, D. E. (1951). Late Cenozoic vertebrates of the San Francisco Bay region. *Univ. California Pub. Dept. Geol. Sci.* 28:215–314.

Savage, J. M. (1977). *Evolution,* 3rd ed. New York: Holt, Rinehart & Winston.

Savage, R. J. G., & Long, M. R. (1986). *Mammal Evolution: An Illustrated Guide.* New York: Facts on File Publications.

Schiebout, J. A. (1974). Vertebrate paleontology and paleoecology of the Paleocene Black Peaks Formation, Big Bend National Park, Texas. *Texas Mem. Mus. Bull.* 24:1–88.

Schmidt-Nielsen, K. (1984). *Scaling: Why Is Animal Size so Important?* Cambridge University Press.

Schmitz-Münker, M., & Franzen, J. L. (1988). Die Rolle von Bakterien in Verdauungstrakt Mitteleozäner Vertebraten aus der Grube Messel bei Darmstadt und ihr Beitrag zur Fossildiagenese und Evolution. *Cour. Forsch.-Inst. Senckenberg* 107:129–46.

Schoch, R. M. (1986). *Phylogeny Reconstruction in Paleontology*. New York: Van Nostrand Reinhold.

——— (1989). A brief historical review of perissodactyl classification. In *The Evolution of Perissodactyls*, ed. D. R. Prothero & R. M. Schoch, pp. 13–23. Oxford: Clarendon Press.

Schopf, T. J. M., Raup, D. M., Gould, S. J., & Simberloff, D. S. (1975). Genomic versus morphological rates of evolution: influence of morphological complexity. *Paleobiology* 1:63–70.

Schuchert, C., & LeVene, C. M. (1940). *O. C. Marsh: Pioneer in Paleontology*. New Haven: Yale University Press.

Sclater, P. L. (1858). On the general geographic distribution of the members of the class Aves. *J. Linnean Soc. Zool.* 2:130–45.

——— (1874). The geographical distribution of mammals. *Manchester Sci. Lect., Ser. 5–6*:202–19.

Sclater, W. L., & Sclater, P. L. (1899). *The Geography of Mammals*. London: Kegan Paul, Trench, Trühner.

Scott, E. (1990). New record of *Equus conversidens* Owen, 1869 (Mammalia; Perissodactyla; Equidae) from Rancho La Brea. *J. Vert. Paleontol.* 10:41A.

Scott, W. B. (1913). *A History of Land Mammals in the Western Hemisphere*. New York: Macmillan.

——— (1917). *The Theory of Evolution*. New York: Macmillan.

——— (1941). The mammalian fauna of the White River Oligocene. Part V. Perissodactyla. *Trans. Am. Philos. Soc.* 28:747–980, plates 79–100.

Sefve, I. (1912). Die fossilen Pferde Südamerikas. *Kungl. Svenska Vetenskapsakademiens Handlingar.* 48:1–185, 3 plates.

——— (1927). Die Hipparionen Nord-Chinas. *Paleontol. Sinica, Ser. C* 4:1–91.

Seilacher, A. (1970). Begriff und Bedeutung der Fossil-Lagerstätten. *N. Jb. Geol. Palaont. Mh.* 1970:34–9.

Seilacher, A., Reif, W.-E., & Westphal, F. (1985). Sedimentological, ecological and temporal patterns of fossil Lagerstätten. *Phil. Trans. Roy. Soc. London, Ser. B*, 311:5–23.

Sellers, C. C. (1980). *Mr. Peale's Museum: Charles Wilson Peale and the First Popular Museum of Natural Science and Art*. New York: Norton.

Sepkoski, J. J., Jr. (1982). A compendium of fossil marine families. *Milwaukee Public Mus. Contrib. Biol. Geol.* 51:1–125.

Shackleton, N. J., & Kennett, J. P. (1975). Paleotemperature history of the Cenozoic and the initiation of Antarctic glaciation: oxygen and carbon isotope analyses in DSDP sites 277, 279, and 281. In *Initial Reports of the Deep Sea Drilling Project, Vol. 29*, ed. J. P. Kennett, R. E. Houtz, et al., pp. 743–55. Washington, D.C.: U.S. Government Printing Office.

Shackleton, N. J., & Opdyke, N. D. (1973). Oxygen isotope and palaeomagnetic stratigraphy of Equatorial Pacific Core V28–238: oxygen isotope temperatures and ice volumes on a 10^5 year and 10^6 year scale. *Quat. Res.* 3:39–55.

——— (1977). Oxygen isotope and paleomagnetic evidence for early Northern Hemisphere glaciation. *Nature* 270:216–19.

Sheets, M., Smith, B., Longnecker, A., & Kneib, A. P. (1984). *The Horse in Folk Art*. La Jolla, Calif.: Mingel International Museum of World Folk Art.

Sheets-Pyenson, S. (1987). Cathedrals of science: the development of colonial natural history museums during the late nineteenth century. *Hist. Sci.* 25:279–300.

Shoshoni, J., Lowenstein, J. M., Walz., D. A., & Goodman, M. (1985). Proboscidean origins of mastodon and woolly mammoth demonstrated immunologically. *Paleobiology* 11:429–37.

Shotwell, J. A. (1961). Late Tertiary biogeography of horses in the northern Great Basin. *J. Paleontol.* 35:203–17.

Simpson, G. G. (1933). Glossary and correlation charts of North American Tertiary mammal-bearing formations. *Bull. Am. Mus. Nat. Hist.* 67:79–121.

——— (1936). Data on the relationships of local and continental mammalian faunas. *J. Paleontol.* 5:410–14.

——— (1940a). Resurrection of the dawn-horse. *Nat. Hist.* 46:194–9.

——— (1940b). Mammals and land bridges. *J. Wash. Acad. Sci.* 30:137–63.

——— (1942). The beginnings of vertebrate paleontology in North America. *Proc. Am. Philos. Soc.* 86:130–88.

——— (1944). *Tempo and Mode in Evolution*. New York: Columbia University Press.

——— (1945). The principles of classification and a classification of mammals. *Bull. Am. Mus. Nat. Hist.* 85:1–350.

——— (1949). Rates of evolution in animals. In *Genetics, Paleontology, and Evolution*, ed. G. L. Jepsen, G. G. Simpson, & E. Mayr, pp. 205–28. Princeton: Princeton University Press.

——— (1951). *Horses: The Story of the Horse Family in*

the Modern World and through Sixty Million Years of History. Oxford University Press.

(1952a). Notes on British hyracotheres. *Zool. J. Linnean Soc. London* 42:195–206.

(1952b). Probabilities of dispersal in geological time. *Bull. Am. Mus. Nat. Hist.* 99:163–76.

(1953a). *The Major Features of Evolution.* New York: Columbia University Press.

(1953b). *Evolution and Geography.* Eugene: Oregon State System of Higher Education.

(1965). *The Geography of Evolution: Collected Essays.* Philadelphia: Chilton.

(1980). *Why and How: Some Problems and Methods in Historical Biology.* Oxford: Pergamon Press.

(1981). Historical science. In *Language of the Earth,* ed. F. H. T. Rhodes & R. O. Stone, pp. 132–9. New York: Pergamon Press.

(1984). *Discoverers of the Lost World.* New Haven: Yale University Press.

Simpson, G. G., Roe, A., & Lewontin, R. C. (1960). *Quantitative Zoology.* New York: Harcourt, Brace & Co.

Skinner, M. F. (1972). Order Perissodactyla, pp. 117–30 in Skinner, M. F., & Hibbard, C. W., Early Pleistocene pre-glacial and glacial rocks and faunas of north-central Nebraska. *Bull. Am. Mus. Nat. Hist.* 148:1–148.

Skinner, M. F., & MacFadden, B. J. (1977). *Cormohipparion* n. gen. (Mammalia, Equidae) from the North American Miocene (Barstovian–Clarendonian). *J. Paleontol.* 51:912–26.

Skinner, M. F., Skinner, S. M., & Gooris, R. J. (1977). Stratigraphy and biostratigraphy of late Cenozoic deposits in central Sioux County, western Nebraska. *Bull. Am. Mus. Nat. Hist.* 158:263–370.

Skinner, M. F., & Taylor, B. E. (1967). A revision of the geology and paleontology of the Bijou Hills, South Dakota. *Am. Mus. Novitates* 2300:1–53.

Smith, A. B., & Patterson, C. (1988). The influence of taxonomic method on the perception of patterns of evolution. In *Evolutionary Biology, Vol. 23,* ed. M. K. Hecht & B. Wallace, pp. 127–216. New York: Plenum Publishing.

Sokal, R. R., & Braumann, C. A. (1980). Significance tests for coefficients of variation and variability profiles. *Syst. Zool.* 29:50–66.

Solounias, N., Teaford, M., & Walker, A. (1988). Interpreting the diet of extinct ruminants: the case of the nonbrowsing giraffid. *Paleobiology* 14:287–300.

Sondaar, P. Y. (1968). The osteology of the manus of fossil and Recent Equidae. *Ver-*

hand. *Koninkijke Nederlandse Akad. Weten., afd. Naturkunde* 25:1–76.

Spinage, C. A. (1972). African ungulate life tables. *Ecology* 53:645–52.

Stanley, S. M. (1974a). Effects of competition on rates of evolution, with special reference to bivalve mollusks and mammals. *Syst. Zool.* 22:486–506.

(1974b). Relative growth of the titanothere horn: a new approach to an old problem. *Evolution* 28:447–57.

(1979). *Macroevolution: Pattern and Process.* San Francisco: Freeman.

(1981). *The New Evolutionary Timetable: Fossils, Genes, and the Origin of Species.* New York: Basic Books.

(1985). Rates of evolution. *Paleobiology* 11:13–26.

Stanton, S. L. (1987). *Anatomy of a Division: The 1st Cav in Viet Nam.* Novato, Calif.: Presidio Press.

Stevens, C. E. (1988). *Comparative Physiology of Vertebrate Digestion.* Cambridge University Press.

Stirton, R. A. (1940). Phylogeny of North American Equidae. *Univ. California Pub. Bull. Dept. Geol. Sci.* 25:165–98.

(1947). Observations on evolutionary rates in hypsodonty. *Evolution* 1:32–41.

Stock, C. (1925). *Cenozoic Gravigrade Edentates of Western North America with Special Reference to the Pleistocene Megalonychinae and Mylodontidae of Rancho La Brea.* Washington, D.C.: Carnegie Institution Publication 331.

Strauss, D., & Sadler, P. M. (1989). Classical confidence intervals and Bayesian probability estimates for ends of local taxon ranges. *Mathemat. Geol.* 21:411–27.

Stuckey, R. K., Lang, H. R., Krishtalka, L., & Redline, A. D. (1987). Analysis of Eocene depositional environments: preliminary TM and TIMS results, Wind River Basin, Wyoming. *Proc. Internat. Geosci. Remote Sensing Symp.* 1163–8.

Sutcliffe, A. J. (1985). *On the Track of Ice Age Mammals.* Cambridge, Mass.: Harvard University Press.

Swisher, C. C., III (1989). Single-crystal ^{40}Ar/^{39}Ar dating and its application to the calibration of North American land mammal "ages." *J. Vert. Paleontol.* 9:40A.

Swisher, C. C., III, & Prothero, D. R. (1990). Single-crystal ^{40}Ar/^{39}Ar dating of the Eocene–Oligocene transition in North America. *Science* 249:760–2.

Swofford, D. (1985). *PAUP, Phylogenetic Analysis*

Using Parsimony. Champaign, Ill.: published by author.

Tarling, D. H. (1983). *Palaeomagnetism: Principles and Applications in Geology, Geophysics, and Archaeology.* London: Chapman & Hall.

Tauxe, L., & Clark, D. R. (1987). New paleomagnetic results from the Eureka Sound Group: implications for the age of early Tertiary Arctic biota. *Bull. Geol. Soc. Am.* 99:739–47.

Teaford, M. F., & Walker, A. C. (1984). Quantitative differences in dental microwear between primate species with different diets and a comment on the presumed diet of *Sivapithecus. Am. J. Phys. Anthrop.* 64:191–200.

Tedford, R. H. (1970). Principles and practices of mammalian geochronology in North America. *Proc. 3rd N. Am. Paleontol. Conv.* F:666–703.

Tedford, R. H., Skinner, M. F., Fields, R. W., Rensberger, J. M., Whistler, D. P., Galusha, T., Taylor, B. E., Macdonald, J. R., & Webb, S. D. (1987). Faunal succession and biochronology of the Arikareean through Hemphillian interval (late Oligocene through earliest Pliocene epochs) in North America. In *Cenozoic Mammals of North America: Geochronology and Biostratigraphy*, ed. M. O. Woodburne, pp. 153–210. Berkeley: University of California Press.

Thewissen, J. G. M. (1990). Evolution of Paleocene and Eocene Phenacodontidae (Mammalia, Condylarthra). *Univ. Michigan Pap. Paleontol.* 29:1–107.

Thomason, J. J. (1985). The relationship of trabecular architecture to inferred loading patterns in the third metacarpals of the extinct equids *Merychippus* and *Mesohippus. Paleobiology* 11:232–335.

(1986). The functional morphology of the manus in the tridactyl equids *Merychippus* and *Mesohippus*: paleontological inferences from neontological models. *J. Vert. Paleontol.* 6:143–61.

Thorp, J. L., & van der Merwe, N. J. (1987). Carbon isotope analysis of fossil bone apatite. *S. Afr. J. Sci.* 83:712–15.

Tomida, Y. (1981). "Dragonian" fossils from the San Juan Basin and status of the "Dragonian" land mammal "age." In *Advances in San Juan Basin Paleontology*, ed S. G. Lucas, J. K. Rigby, Jr., & B. S. Kues, pp. 242–63. Albuquerque: University of New Mexico Press.

Tomida, Y., & Butler, R. F. (1980). Dragonian mammals and Paleocene magnetic polarity stratigraphy, North Horn Formation, central Utah. *Am. J. Sci.* 280:787–811.

Troxell, E. L. (1915). The vertebrate fossils of Rock Creek, Texas. *Am. J. Sci.* 39:613–38.

Trumler, E. (1959). Das "Rossigkeitsgesicht" und ähnliches Ausdrucksverhalten bei Einhufern. *Z. Tierpsychol.* 16:478–88.

Turcotte, D. L. (1991). Fractals in geology: What are they and what are they good for? *GSA Today* 1:1–4.

Ucko, P. J., & Rosenfeld, A. (1967). *Palaeolithic Cave Art.* London: Weidenfeld & Nicolson.

Udvardy, M. D. F. (1969). *Dynamic Zoogeography with Special Reference to Land Animals.* New York: Van Nostrand Reinhold.

(1981). The riddle of dispersal: dispersal theories and how they affect vicariance biogeography. In *Vicariance Biogeography: A Critique*, ed. G. Nelson & D. E. Rosen, pp. 6–38. New York: Columbia University Press.

Valentini, M. B. (1704). *Museum Museorum.* Frankfurt am Main [not seen].

Valverde, J. A. (1964). Remarques sur la structure et l'évolution des communautés de vertébrés terrestres. I. Structure d'une communauté. II. Rapports entre prédateurs et proies. *La Terre et la Vie* 111:121–54.

Van der Merwe, N. J. (1982). Carbon isotopes, photosynthesis, and archaeology. *Am. Sci.* 70:596–606.

Van Soest, P. J. (1982). *Nutritional Ecology of the Ruminant: Ruminant Metabolism, Nutritional Strategies, the Cellulolytic Fermentation and the Chemistry of Forages and Plant Fibers.* Ithaca: Cornell University Press.

Van Valen, L. (1960). A functional index of hypsodonty. *Evolution* 14:531–2.

(1963). Selection in natural populations: *Merychippus primus*, a fossil horse. *Nature* 197:1181–3.

(1964). Age in two fossil horse populations. *Acta Zool.* 45:93–106.

(1965). Selection in natural populations. III. Measurement and estimation. *Evolution* 19:514–28.

(1973). A new evolutionary law. *Evol. Theory* 1:1–30.

(1978). Why not to be a cladist. *Evol. Theory* 3:285–99.

Van Valkenburgh, B. (1982). Evolutionary dynamics of terrestrial, large, predator guilds. *Proc. 3rd North Am. Paleontol. Conv.* 2:557–62.

(1985). Locomotory diversity within past and present guilds of large predatory mammals. *Paleobiology* 11:406–28.

(1988). Trophic diversity in past and present guilds of large predatory mammals. *Paleobiology* 14:155–73.

(1989). Carnivore dental adaptations and diet: a study of trophic diversity within guilds. In *Carnivore Behavior, Ecology, and Evolution*, ed. J. L. Gittleman, pp. 410–36. Ithaca: Cornell University Press.

Viohl, G. (1985). Geology of the Solnhofen lithographic limestone and the habitat of *Archaeopteryx*. In *The Beginnings of Birds: Proceedings of the International Archaeopteryx Conference Eichstätt*, ed. M. K. Hecht, J. H. Ostrom, G. Viohl, & P. Wellnhofer, pp. 31–44. Eichstätt: Jura-Museums.

Voorhies, M. R. (1969). Taphonomy and population dynamics of an early Pliocene vertebrate fauna, Knox County, Nebraska. *Contrib. Geol., Univ. Wyoming, Spec. Pap.* 1:1–69.

(1981). Dwarfing the St. Helens eruption: ancient ashfall creates a Pompeii of prehistoric animals. *Natl. Geogr.* 159:66–75.

Voorhies, M. R., & Thomasson, J. R. (1979). Fossil grass antothecia within Miocene rhinoceros skeletons: diet of an extinct species. *Science* 206:331–3.

Vrba, E. S., & Eldredge, N. (1984). Individuals, hierarchies, and processes: towards a more complete evolutionary theory. *Paleobiology* 10:146–71.

Walker, E. P. (1975). *Mammals of the World*, 3rd ed. Baltimore: Johns Hopkins University Press.

Wallace, A. R. (1876). *The Geographical Distribution of Animals, Vol. 1*. New York: Harper & Brothers.

Webb, S. D. (1969a). The Burge and Minnechaduza Clarendonian mammalian faunas of north-central Nebraska. *Univ. California Pub. Geol. Sci.* 78:1–191.

(1969b). Extinction–origination equilibria in late Cenozoic land mammals of North America. *Evolution* 23:688–702.

(1974). Chronology of Florida Pleistocene Mammals. In *Pleistocene Mammals of Florida*, ed. S. D. Webb, pp. 5–31. Gainesville: University Presses of Florida.

(1977). A history of savanna vertebrates in the New World. Part I: North America. *Ann. Rev. Ecol. Syst.* 8:355–80.

(1978). A history of savanna vertebrates in the New World. Part II. South America and the Great Interchange. *Ann. Rev. Ecol. Syst.* 9:393–426.

(1983). The rise and fall of the late Miocene ungulate fauna in North America. In *Coevolution*, ed. M. H. Nitecki, pp. 267–306. Chicago: University of Chicago Press.

(1984). Ten million years of mammal extinctions in North America. In *Quaternary Extinctions: A Prehistoric Revolution*, ed. P. S. Martin & R. G. Klein, pp. 189–210. Tucson: University of Arizona Press.

(1987). Community patterns in extinct terrestrial vertebrates. In *Organization of Communities Past and Present*, ed. J. H. R. Gee & P. S. Giller, pp. 439–66. Oxford: Blackwell.

Webb, S. D., & Hulbert, R. C., Jr. (1986). Systematics and evolution of *Pseudhipparion* (Mammalia, Equidae) from the late Neogene of the Gulf Coastal Plain and the Great Plains. In *Vertebrates, Phylogeny, and Philosophy*, ed. K. M. Flanagan & J. A. Lillegraven, pp. 237–72. Contrib. Geol., Univ. Wyoming, Spec. Pap. 3.

Webb, S. D., MacFadden, B. J., & Baskin, J. A. (1981). Geology and paleontology of the Love Bone Bed from the late Miocene of Florida. *Am. J. Sci.* 281:513–44.

Webb, S. D., Morgan, G. S., Hulbert, R. C., Jr., Jones, D. S., MacFadden, B. J., & Mueller, P. A. (1989). Geochronology of a rich early Pleistocene vertebrate fauna, Leisey Shell Pit, Tampa Bay, Florida. *Quat. Res.* 32:96–110.

West, R. M. (1976). The North American Phenacodontidae (Mammalia, Condylarthra). *Contrib. Biol. Geol., Milwaukee Pub. Mus.* 6:1–78.

West, R. M., Dawson, M. R., & Hutchison, J. H. (1977). Fossils from the Paleogene Eureka Sound Formation, N.W.T., Canada: occurence, climatic and paleogeographic implications. *Milwaukee Pub. Mus., Spec. Pub. Biol. Geol.* 2:77–93.

Western, D. (1979). Size, life history and ecology of mammals. *E. Afr. Wild. J.* 13:265–86.

White, T. E. (1959). The endocrine glands and evolution, no. 3: os cementum, hypsodonty, and diet. *Contrib. Mus. Paleontol. Univ. Michigan* 13:211–65.

Whitehead, A. N. (1957). *The Concept of Nature*. Ann Arbor: University of Michigan Press.

Whittington, H. B. (1985). *The Burgess Shale*. New Haven: Yale University Press.

Wible, J. R. (1986). Transformations in the extracranial course of the internal carotid artery in mammalian phylogeny. *J. Vert. Paleontol.* 6:313–25.

(1987). The eutherian stapedial artery: character analysis and implications for superordinal relationships. *Zool. J. Linnean Soc.* 91:107–35.

Wiley, E. O. (1981). *Phylogenetics: The Theory and Practice of Phylogenetic Systematics*. New York: Wiley.

Wilson, J. T. (1968). Static or mobile earth: the current scientific revolution. *Proc. Am. Philos. Soc.* 112:309–20.

Wilson, R. W. (1960). Early Miocene rodents and insectivores from northeastern Colorado. *Univ. Kansas Paleontol. Contrib., Vertebrata* 7:1–92.

Winans, M. C. (1985). Revision of North American fossil species of the genus *Equus* (Mammalia: Perissodactyla: Equidae). Unpublished PhD dissertation, University of Texas, Austin.

(1989). A quantitative study of the North American fossil species of the genus *Equus*. In *The Evolution of Perissodactyls*, ed. D. R. Prothero & R. M. Schoch, pp. 262–97. Oxford: Clarendon Press.

Wing, S. L. (1980). Fossil floras and plant-bearing beds of the central Bighorn Basin. In *Early Cenozoic Paleontology and Stratigraphy of the Bighorn Basin, Wyoming*, ed. P. D. Gingerich, pp. 119–25. Univ. Michigan Pap. Paleontol. 24.

Wolfe, J. A. (1971). Tertiary climatic fluctuations and methods of analysis. *Palaeogeogr., Palaeoclimatol., Palaeoecol.* 9:27–57.

(1978). A paleobotanical interpretation of Tertiary climates in the Northern Hemisphere. *Am. Sci.* 66:694–703.

Wood, D. (1989). *The Deconstruction of Time*. Atlantic Highlands, N.J.: Humanities Press International.

Wood, H. E., II (1937). Perissodactyl suborders. *J. Mammal.* 18:106.

Wood, H. E., II, Chaney, R. W., Clark, J., Colbert, E. H., Jepsen, G. L., Reeside, J. B., Jr., & Stock, C. (1941). Nomenclature and correlation of the North American continental Tertiary. *Bull. Geol. Soc. Am.* 52:1–48.

Woodburne, M. O. (1987). *Cenozoic Mammals of North America: Geochronology and Biostratigraphy*. Berkeley: University of California Press.

Woodburne, M. O., & Bernor, R. L. (1980). On superspecific groups of some Old World hipparionine horses. *J. Paleontol.* 54:1319–48.

Woodburne, M. O., MacFadden, B. J., & Skinner, M. F. (1981). The North American "*Hipparion*" datum, and implications for the Neogene of the Old World. *Géobios* 14:1–32.

Yablokov, A. (1974). *Variability of Mammals*. New Delhi: Amerind Publishing.

Zeuner, F. E. (1963). *A History of Domesticated Animals*. London: Hutchinson & Co.

Zhai, R.-J. (1962). On the generic character of "Hypohippus zitteli." [*sic*] *Vert. Palasiatica* 6:48–56.

Taxonomic index

f indicates figure; n indicates footnote

Subject index

f indicates figure; n indicates footnote

abbreviations used in text, 333–4
adaptive radiation, 181–3, 307
  Artiodactyla, 184
  Perissodactyla, 184–7
adaptive zone, 178
allometry, 196–200
  canines, 296–7
  evolutionary, 197, 198
  ontogenetic, 197, 198
  positive, 196, 197f, 197–8
  static, 197
alpha-level taxonomy, 170
American Museum of Natural History, 56–8
anagenesis, 171–3, 172f
Aristotle, 27–8, 79, 326
Ashfall Fossil Beds, 71–3, 72f, 276, 309
assemblages, fossil
  attritional, 64, 275, 276f, 277, 288
  catastrophic, 64, 70–1, 167, 276, 276f, 277, 278, 286, 288
astragalus, 86, 87f
atavism, 224
  horses, 224, 225f
attritional assemblages, 64, 275, 276f, 277, 288

barriers, *see* biogeography
behavioral displays
  horses, 265–8
behavioral ecology, 264–72, *also see* social organization
Bergmann's rule, 213
Bering land bridge, 155f, 158, 160–1f
Beringia, 154
biochemical divergence times, 19, 329
biochemical systematics and techniques, 19, 114–7, 328, 329
biogeographic regions, 144, 145f, 151
biogeography, 144–5
  barriers, 146, 153f, 313
  centers of origin, 150–2
  climate change, 147–8
  continental drift, 146, 152
  corridors, 146, 153f

  filters, 146–7, 153f
  land bridges, 146
  organic interactions, 148
  plate tectonics, 146, 152–4
  sea-level fluctuation, 146
  sweepstakes, 146–7, 153f
  vicariance, 151–2
biomass estimate, 310
biostratigraphic ranges, *see* ranges
body size, 216–23, 258–9, 269–70, 279
  estimating, 218–20, 219f
  evolution, 221–2
  life-history parameters, 269–70
bradytely, 200
brain
  evolutionary trends, 214, 215f
  fossil endocasts, 214, 215f
  orthogenesis, 43f
Brongniart, A., 123
Brown, B., 49
browsers, 239, 241, 283, 308f
Brú, J. B., 50
Buffon, G., 14, 122, 144
Burge quarry, 276, 288

C3, 18, 244f
C4, 18, 244f
California
  Merychippus from, 167
carbon isotopes, 271, *also see* isotopes
carrying capacity, 178
Castillo Pocket, 65–6, 66f, 278, 327
catastrophic assemblages, 64, 167, 276, 276f, 277, 286, 288
cave paintings, 1–3, 2f
cecum, 88f, 237
cenograms, 314–5
  Bighorn Basin, 314, 315f
  Bridger Basin, 314
  interpreting ancient communities, 314
Cenozoic, 133, 135, 136, 140
  mammalian chronofaunas, 301
  primary-consumer diversity, 186f